Human–Computer Interaction Series

Editors-in-Chief

Desney Tan
Microsoft Research, Redmond, WA, USA

Jean Vanderdonckt
Louvain School of Management, Université catholique de Louvain,
Louvain-La-Neuve, Belgium

The Human–Computer Interaction Series, launched in 2004, publishes books that advance the science and technology of developing systems which are effective and satisfying for people in a wide variety of contexts. Titles focus on theoretical perspectives (such as formal approaches drawn from a variety of behavioural sciences), practical approaches (such as techniques for effectively integrating user needs in system development), and social issues (such as the determinants of utility, usability and acceptability).

HCI is a multidisciplinary field and focuses on the human aspects in the development of computer technology. As technology becomes increasingly more pervasive the need to take a human-centred approach in the design and development of computer-based systems becomes ever more important.

Titles published within the Human–Computer Interaction Series are included in Thomson Reuters' Book Citation Index, The DBLP Computer Science Bibliography and The HCI Bibliography.

More information about this series at http://www.springer.com/series/6033

Jean-François Uhl · Joaquim Jorge ·
Daniel Simões Lopes · Pedro F. Campos
Editors

Digital Anatomy

Applications of Virtual, Mixed
and Augmented Reality

 Springer

Editors
Jean-François Uhl
Paris Descartes University
Paris, France

Daniel Simões Lopes
INESC-ID Lisboa
Instituto Superior Técnico
Universidade de Lisboa
Lisbon, Portugal

Joaquim Jorge
INESC-ID Lisboa
Instituto Superior Técnico
Universidade de Lisboa
Lisbon, Portugal

Pedro F. Campos
Department of Informatics
University of Madeira
Funchal, Portugal

ISSN 1571-5035   ISSN 2524-4477 (electronic)
Human–Computer Interaction Series
ISBN 978-3-030-61907-7   ISBN 978-3-030-61905-3 (eBook)
https://doi.org/10.1007/978-3-030-61905-3

This Springer imprint is published by the registered company Springer Nature Switzerland AG
The registered company address is: Gewerbestrasse 11, 6330 Cham, Switzerland

Foreword by Mark Billinghurst

As a discipline, anatomy can trace its roots back to at least ancient Egypt and papyri that described the heart and other internal organs. In the thousands of years since knowledge of anatomy improved dramatically, in most of that history the main way to understand the body was through dissection and looking beneath the skin. At the close of the nineteenth century, the development of the X-ray machine enabled images to be captured from within the body, and this was followed by many other imaging technologies. At the same time at the invention of the CT scanner, the first interactive computer graphics was also demonstrated. Since then advances in computer graphics and medical imaging have gone hand in hand. **Digital anatomy** has emerged as an essential subfield of **anatomy** that processes the human body in a computer-accessible format.

The many different revolutions brought about by computer graphics, visualization, and interactive techniques have meant that digital anatomy is a rapidly evolving discipline. One of the main goals is to provide a greater understanding of the spatial structure of the body and its internal organs. Virtual Reality (VR) and Augmented Reality (AR) are particularly valuable for this. Using VR, people can immerse themselves in a graphical representation of the body, while with AR virtual anatomy can be superimposed back into the real body. Both provide an intuitive way to view and interact with digital anatomy, which could bring about a revolution in many health-related fields. This book provides a valuable overview of applications of Virtual, Mixed, and Augmented Reality in Digital Anatomy.

Just as X-ray machines and CT scanners changed twentieth-century healthcare, medical image analysis is revolutionizing twenty-first-century medicine, ushering in powerful new tools designed to assist the clinical diagnosis and to better model, simulate, and guide the patient's therapy more efficiently. Doctors can perform image-guided surgery through small incisions in the body and use AR and VR tools to help with pre-operative planning. In academic settings, digital technologies are changing anatomy education by improving retention and learning outcomes with new learning tools such as virtual dissection. This book provides comprehensive coverage of many medical applications relying on digital anatomy ranging from educational to pathology to surgical uses.

An outstanding example of the value of digital anatomy was the Visible Human Project (VHP) unveiled by the United States National Library of Medicine at the end of the twentieth century. The VHP made high-resolution digital images of anatomical parts of a cryosectioned body freely available on the Internet, accompanied by a collection of volumetric medical images acquired before slicing. These images are used by computer scientists worldwide to create representations of most organs with exceptional precision. Many other projects, including the Visible Korean and Chinese visible human male and female projects, achieved greater fidelity of images and improved anatomical details. Digital atlases based on the VHP have made it possible to navigate the human body in three dimensions, and have proven to be immensely valuable as an educational tool.

These original virtual human reconstructions were very labor-intensive and only depicted a single person's particular structures. There remains much arduous work to be done to make the data derived from visible human projects meet the application-oriented needs of many fields. Researchers are making significant progress toward developing new datasets, segmenting and creating computer-assisted medicine platforms. In the second decade of the twenty-first century, a new approach is used: we no longer seek to visualize individual representations of anatomy but rather model and display statistical representations within an entire population. Modern approaches use powerful algorithmic and mathematical tools applied to massive image databases to identify and register anatomical singularities to visualize statistical representations of shapes and appearances within a population.

A fascinating aspect of digital anatomy is its strong multidisciplinary flavor, and willingness of researchers in the field to experiment with new technology. Augmented and Virtual Reality play a vital role in growing this emerging field. For example, AR interfaces overlay three-dimensional images in the real world and can make intricate structures and delicate relationships easier to discern between trained and untrained eyes. Similarly, VR can be used to enable groups of medical students to be immersed in digital anatomy and learn together about the human body.

Future advances in digital patient research will rely heavily on algorithmic, statistical, and mathematical techniques in image processing, advances in digital modeling of the human body's anatomy and physiology, and methods and algorithms for customizing structural body models from measurements. These are likely to benefit from new medical imaging technologies and foreseeable improvements in hardware performance, both in computing speed and in the capacity to store and transmit information. This book provides a comprehensive coverage of many of these advances, ranging from educational to pathology to surgical applications.

Just like X-ray machines and CT scanners revolutionized medicine hundreds of years ago, AR and VR will have an impact on medicine for the next hundred years and beyond. Readers will find glimpses of this future in pages of this book. I hope that readers will find inspiration in this valuable collection of articles and it will stimulate their endeavors to advance research in this critical field.

Auckland, New Zealand   Mark Billinghurst
September 2020

Foreword by Nicholas Ayache

As both the patient data and medical practice become more and more digital, it is also the case for the anatomy discipline that undergoes a computational revolution.

This book presents all the aspects of this computational revolution; for instance, how to create dissections from 3D models which are useful for anatomical research and teaching, how to tailor those models to patient-specific anatomies from medical images, how to compute statistics based on digital anatomical models, how to introduce novel human–computer interfaces to perform digital dissection tasks, how Extended Reality opens new avenues for dissecting digital anatomical representations …

Not only this book presents methodological concepts and methods, but it also showcases practical tools and algorithms that are useful for physicians, anatomists, and computer scientists interested in digital anatomy: from students to researchers, from teachers to industry practitioners from various backgrounds including not only medicine and biology but also paleontology, history, arts, computer science, and applied mathematics.

Finally, this book will contribute to advance research in e-medicine as the Extended Reality applications, tools, methods, and algorithms presented in this book are relevant for computer-aided diagnosis, prognosis, and therapy that, in turn, heavily rely on a faithful digital representation of a patient's anatomy or, in other words, a patient's *digital twin*. Such advancements presented in this book will be paramount for the physicians and surgeons to improve their medical practice's quality and precision. Therefore, in the end, these advancements will contribute to the benefit of all the real patients in the world.

Sophia Antipolis, France   Nicholas Ayache
September 2020

Acknowledgments

The editors would like to wholeheartedly thank the anonymous reviewers' effort, who provided solid insights and worked to improve the scientific quality to this book. We are also grateful to all the book's authors for sharing their valuable technical and scientific standpoints regarding Digital Anatomy. We hope the book will help spearhead their research.

Jean-François Uhl thanks the UNESCO Chair in Teaching and Research in Digital Anatomy Paris Descartes for its support.

Joaquim Jorge is grateful to the Portuguese Foundation for Science and Technology and the New Zealand Ministry of Business, Innovation, and Employment, which partially funded this work through Grants UIDB/50021/2020 and SFRH/BSAB/150387/2019, and CATALYST 19-VUW-015-ILF, respectively.

Daniel Simões Lopes acknowledges INESC-ID Lisboa, the University of Lisbon, and the Portuguese Foundation for Science and Technology.

Pedro F. Campos acknowledges ITI/LARSys and the Portuguese Foundation for Science and Technology.

Contents

Chapter 1
Introduction to Digital Anatomy

Joaquim Jorge

Abstract **Anatomy** studies the morphology and structure of organisms. The word originates from the **Greek** *ana-*, up; and *tome-*, meaning cutting. As its name implies, anatomy relies heavily on dissection and studies the human body parts' arrangement and interaction. Heir from a rich Greco-Roman tradition and background, Vesalius (1543) is arguably the father of modern anatomy (Fig. 1.1). Since its early origins, there has been a clear connection between the study of anatomy, graphics depictions, and illustrative visualizations. True to its origins, computer-based three-dimensional modeling of the human body, also known as Digital Anatomy, has strong visualization roots. Digital Anatomy has benefited from the computer and communications technological revolution. It lies at the intersection of converging different disciplines, ranging from Medical Imaging, Medical Visualization, 3D printing, and Computer Graphics to Artificial Intelligence and Robotics. This book offers a perspective on current developments and a road map into the future for this exciting pillar of modern medicine.

Five hundred years after the birth of Andreas Vesalius, Digital Anatomy has evolved in significant part due to the ubiquitous availability of Medical Images (Ayache 2015). These images are prevalent in modern clinical and hospital practice. They serve to direct the diagnosis, prepare, and direct therapy. The proliferation in number, quality, and resolution of medical images has had a clear impact on medical practice and research. In effect, three recent studies have yielded a more extensive view of human anatomy. The Visible Human Project (Waldby 2003), the Visible Korean Human (Park et al. 2006), and the Chinese Visible Human (Zhang et al. 2006) performed the serial cryotomy of entire cadavers, creating cross-section images that were methodically applied to catalog gross human anatomy. Advanced visualization and segmentation techniques can convert these cross-section images into three-dimensional vectorial models of all anatomical structures. Compiling these 3D images has created an essential archive for medical education and science. Initially

J. Jorge (✉)
INESC-ID / Universidade de Lisboa, Lisboa, Portugal
e-mail: jorgej@tecnico.ulisboa.pt

© The Author(s), under exclusive license to Springer Nature Switzerland AG 2021
J.-F. Uhl et al. (eds.), *Digital Anatomy*, Human–Computer Interaction Series,
https://doi.org/10.1007/978-3-030-61905-3_1

focused on educational and expository goals, these efforts have expanded to new applications. These include virtual surgery, virtual endoscopy, surgical planning, and simulation of medical procedures. The emergence of virtual dissection software has popularized Digital Anatomy both as a subject of study and as a simulation, diagnosis, and research tool.

Medical images contain much information. An organ's anatomical image, or even the entire body, can have several hundred million voxels stored in megabytes or even gigabytes of memory. The nature, number, and resolution of medical images continue to grow with the constant advancement of image acquisition technology. In addition to X-rays, clinicians and hospitals commonly use five primary imaging modalities: CT scan (Computerized X-ray Tomography), MRI (Magnetic Resonance Imaging), echography (ultrasound imaging), and PET (Positron Emission Tomography) or Gamma Scan scintigraphy (nuclear medicine). Except for scintigraphy and X-rays, the images generated by these modalities are volumetric: they present details at each point of the organism in small volume elements called voxels (volume elements) by analogy to pixels (picture elements). Indeed, information quantities increase rapidly when different images in different modalities are acquired from the same patient to exploit their complementarity. Alternatively, when we use pictures spaced in time to follow an evolution: these become 4D images with three spatial dimensions and one temporal component. These dynamic images help capture both the function and movement of organs and the disease's progression over time.

Big Data: towards Digital Patient Models.

Furthermore, as if all these data were not enough, large databases of images are gradually becoming accessible on the Web. Anonymized data often enrich these images, with the patient's history and their pathology and can be viewed remotely to confirm a diagnosis or support statistical studies. To cope with the explosion of medical image data, computer science and information technologies have become fundamental to exploit this overabundance of complex, rich data to extract clinically relevant information. This information explosion partially explains and motivates the advent of new three-dimensional extraction and modeling methods available today. Fueled by advances in medical imaging, research has devised new tools for 3D reconstruction of human body structures. These have been applied to learning and teaching anatomy, which has leaped forward thanks to Virtual Reality (VR), Augmented Reality (AR), Mixed Reality (MR), and stereoscopic immersive 3D visualizations.

Digital patient models (Ayache 2016) provide a methodological framework that unifies the analysis, synthesis, and simulation of images. These models collect different computational models of the human body. These computational models combine numerical data and algorithms to simulate tissue and organ anatomy and physiology in the human body, using mathematical, biological, physical, and chemical models of living things at different spatial and temporal scales.

There is a robust statistical basis for modern Digital Anatomy. Indeed, most contemporary computational models are generic: they are controlled by standard parameters to describe and simulate the average shape and function of organs in

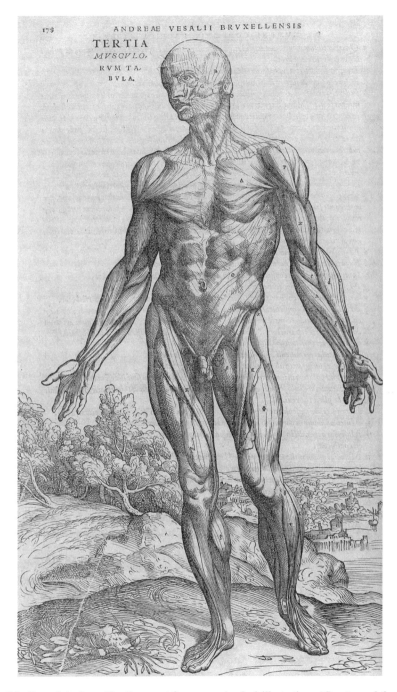

Fig. 1.1 One of Andreas Vesalius most famous anatomical illustrations (*Courtesy of the U.S. National Library of Medicine*) published in (Vesalius 1543)

a population. By comparing these models with specific medical images and data acquired for a single patient, we can adapt the generic model parameters to reproduce that individual's organs' shape and function more precisely (Pennec 2008). One of the critical challenges facing Digital Anatomy and computational medicine is to progress from generic computational models to personalized computational representations.

Further research will allow these personalized computational models to be projected in patients' medical images to facilitate their interpretation and assist diagnosis. The customized model also makes it pòssible to help the prognosis by simulating a pathology, then assist therapy by planning and simulating a surgical intervention, or even controlling it using intraoperative images, via image-guided surgery (Peters 2000). This exciting application of digital anatomy is revolutionizing therapeutic practices. Indeed, image-guided procedures are emerging in many surgical specialties. These range from rectal, digestive, breast, orthopedics, implantology, neurosurgery, to ophthalmology. These prefigure tomorrow's computational medicine, at the service of the doctor and the patient. Of particular relevance is a minimally invasive surgery, whether using classic laparoscopy or robotic-assisted operations, where image-guided techniques are bound to make significant inroads soon.

The operating room of the future will thus take advantage of the techniques empowered by Digital Anatomy, towards the advent of Computational Medicine. Clinicians will use preoperative images to build a personalized digital model of the patient. These models will make it possible to plan and simulate interventions using VR software. These will make it possible, for example, to simulate laparoscopic surgery on virtual organs with visual feedback and multi-sensory feedback, including force, smell, and other senses. It will be possible to train surgeons to perform complex procedures using the same principles and apparatus. These might range from hepatectomies or the most delicate eye and brain surgeries. Figure 1.2 illustrates ongoing research on using AR in the operating room.

Towards Computational Anatomy

Our book focuses on the three-dimensional reconstruction techniques used in human anatomy and discusses the leading 3D and visualization methods and assesses various human–computer interfaces' effectiveness.

This book can be used as a practical manual for students and trainers in anatomy, wishing to develop these new tools. In this perspective, the UNESCO Chair in Digital Anatomy was created at the Paris Descartes University in 2015 (Chair and in Digital Anatomy 2015). It aims to federate the teaching of anatomy around university partners from all over the world, wishing to use these new 3D modeling techniques of the human body.

This book results from an interdisciplinary collaboration between anatomists, clinicians, physicians, statisticians, and computer scientists, addressing topics as diverse as big data, anatomy, visualization, virtual reality, augmented reality, to computational anatomy. It is divided into four sections to provide both historical perspectives and address critical issues in Digital Anatomy.

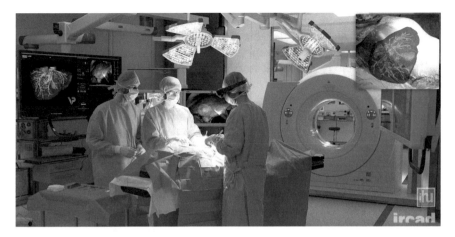

Fig. 1.2 Using augmented Reality in the Operating Room for Hepatic Surgery. Image courtesy of **IRCAD-IHU-Visible Patient** project

1.1 From Dissection to Digital Anatomy

The introductory section provides context for the book as a whole. The first chapter, **From Anatomical to Digital Dissection: a Historical Perspective since Antiquity towards the Early 21st Century**, provides a historical framework from the origins of anatomy to the twentieth and twenty-first century discussing techniques, approaches, methods, and innovations brought about by the marriage of Computer Science and traditional anatomy into the data-oriented disciplines of today. One critical remark is that Anatomy has been a collaborative endeavor since its inception. The second chapter, **A Tool for Collaborative Anatomy,** discusses recent research on Anatomy Studio. Anatomy Studio is a Mixed Reality (MR) tool for virtual dissection that combines computerized tablets with styli and see-through MR head-mounted displays. Its goal is to assist anatomists in describing and segmenting organs, and other morphological structures, easing manual tracing from cryosection images. These two contributions, taken together, explore the full gamut of anatomical research spanning over 500 years of work from Andreas Vesalius's generation to present-day multidisciplinary endeavors.

1.2 Imaging and Reconstruction

Thanks to ever-evolving image acquisition technologies, the abundance, quality, and resolution of medical images have directly affected medical practice and science. This section delves into the many emerging approaches to converting these images to vectorial data (curves, surfaces, and volumes) for educational and clinical processes. The Visible Human, Visible Korean, and Chinese Visible Man and Woman projects have provided a wealth of publicly available digitized images. In this part of the book, the first chapter, **3D modeling from anatomical and histological slices: Methodology and results of Computer-Assisted Anatomical Dissection** by Uhl and Chahim, describes Computer-Assisted Anatomical Dissection, a method to reconstruct anatomical structures from histological sections, mostly for educational purposes, using commercial software. The second chapter, **Volume Rendering Technique from DICOM data (MDCT) applied to the study of Virtual Anatomy** by Merino, Ovelar, and Cedola, focuses volumetric images originating from either MRI, CT, or ultrasound in Digital Imaging and Communication in Medicine (DICOM) format and using operator-assisted volume reconstruction techniques to produce graphical renderings of anatomical structures. The third chapter, **The virtual dissection table: a 3D atlas of the human body using vectorial modeling from anatomical slices** by Uhl, Mogorrón, and Ovelar, approaches manually segmenting 3D structures using anatomical sections from the Visible Korean Human project. They propose using commercially-available software and explain how their approach differs from that of the Korean team. The fourth chapter, **Segmentation and 3D printing of anatomical models from CT angiograms** by Prat et al., proposes a protocol to obtain 3D anatomical models from Computed Tomography Angiograms (CTA) data for 3D printing. Authors explain the features of freely available software and introduce them to those taking their first steps in the matter, including a brief discussion on the benefits and drawbacks of 3D anatomy to education and surgery. In a similar vein, the fourth chapter, **3D Reconstruction of CT images using Free Software Tools,** by Paulo, Lopes, and Jorge, discusses using software to generate 3D models from digital image data. The authors describe a 3D reconstruction pipeline to create 3D models from CT images and volume renderings for medical visualization purposes. They show through examples how to segment 3D anatomical structures with high-contrast detail, namely the skull, mandible, trachea, and colon, relying solely on free, open-source tools. Finally, Sect. 2.6 **Statistical analysis of organ's shapes and deformations: the Riemannian and the affine settings in computational anatomy** by Xavier Pennec, addresses the fundamental issue in Computational Anatomy of analyzing and modeling biological variability of organ shapes at the population level. Indeed modeling shape changes either due to variations within a population or due to the evolution of a degenerative disease such as Alzheimer's can provide significant inroads into the more general issue of moving from generic parameterizable anatomical models to personalized digital patient descriptions.

1.3   Virtual and Augmented Reality and Applications

On a par with progress in medical imaging, there are significant advances in VR and AR hardware. Much research focuses on educational and visualization applications. However, VR and AR can bring substantial advantages to whole other fields. This section focuses on the current and foreseeable impacts that accurate interactive visualizations can have on Digital Anatomy. The first chapter, **High fidelity 3D**

anatomical visualization of the fiber bundles of the muscle's facial expression as in situ, by Li et al. is an excellent example of how VR can provide insight into how the muscles individually and collectively contribute to the shaping and stiffening of facial soft tissues. The main novelty brought by this chapter is the reconstruction of entire muscular structures in 3D space as in situ, thanks to new techniques that build upon, e.g., volumetric musculoaponeurotic 3D digitized data of the human masseter. The current framework enables 3D structure and relationships to be explored and quantified, rather than interpreted from collections of images. The team digitized 22 muscles in total. As mentioned by the authors, the data revealed important geometrical information regarding asymmetry between the homologous muscles on both sides of the face.

Another application area of VR and AR techniques is medical training. The second chapter of this part presents **a VR simulation aimed at radiation therapy education**. Bannister et al. motivate medical students who have minimal access to practical training opportunities due to the high demand for radiation therapy equipment—cancer is the cause of over 16% of deaths globally, and radiation therapy continues to be the most common form of treatment. The chapter describes LINACVR, a VR prototype for radiation therapy simulation. LINACVR offers an approach to immersive training that is a first in several ways. The collaborative VR tool models a radiation therapy environment obviating the need for actual LINAC equipment. The authors solidly demonstrate that multi-user functionality increases realism and accuracy and improves medical students' training effectiveness.

In line with these two initial chapters, the third one of this part is titled **Multi-Touch Surfaces and Patient-Specific Data**, by Ynnerman and colleagues. It essentially addresses wide-spread clinical practice adoption of 3D visualizations by trying to understand how multi-touch surfaces with patient-specific data have contributed to breaking this barrier. The main goal is to achieve intuitive interaction with patient-specific data captured by modern-day equipment, especially CT scan equipment, and graphics processing units (GPUs) with extensive capability, which has become widely available. In short, this chapter summarizes how patient-specific visualization on touch tables has the potential to find many uses in the medical workflow in a range of medical sub-disciplines.

The fourth chapter, Innovations in Microscopic Neurosurgery by Cherian et al., presents a new application called Hyperscope, a three-dimensional, ultra high definition camera mounted on a robotic arm compatible with Microsoft HoloLens. More concretely, it is "an endoscope-exoscope hybrid that can allow switching between the two using a foot control." The novelty stems from incorporating neuro-navigation and augmented reality to create a composite image of the inputs obtained. It can guide the surgeon through augmented reality-based neuronavigation performing point-matching between real and virtual anatomy.

Finally, the last chapter in this section approaches Cataracts, VR, and Digital Anatomy (Katharina Krösl). The author presents methods to simulate vision impairments in VR. Since at least 2.2 billion people were affected by vision impairments or blindness, this can be an impactful application area for VR. This work provides a solid foundation to determine the specific influence of vision impairments on perception and the effects of different lighting scenarios on perception by people with vision impairments.

1.4 Digital Anatomy as an Educational Tool

Teaching Anatomy requires students to imagine the complex spatial relations of the human body. Thus, they may benefit from experiencing immersion in the subject material. Also, integrating virtual and real information, e.g., muscles and bone overlaid on the user's body, is beneficial for imaging various anatomical structures. VR and AR systems for anatomy education compete with other media to support anatomy teachings, such as interactive 3D visualization and anatomy textbooks.

In this section of the book, we discuss the constraints of designing VR and AR systems to enable efficient knowledge transfer. **Patient-Specific Anatomy: the new area of anatomy based on 3D modeling**, by Soler et al., aims at overcoming the two main drawbacks of conventional visualizations: each voxel depicts density as grey levels, which are inadequate for human eye cones perception, and slicing the volumes makes any 3D mental representation of the real 3D anatomy of the patient highly complex. This chapter presents the significant advantages of using patient-specific 3D modeling and preoperative virtual planning compared to the usual anatomical definition using only image slices.

The second chapter in this section focuses on **Virtual and augmented reality for educational anatomy** (Preim et al.). Emphasizing the need for usability and simplicity in VR/AR systems aimed at educational anatomy, the authors present a series of constraints faced by modern approaches. Further, the authors advocate haptics as a novel sensory modality to support anatomy learning and add it to the visual sensations, among other factors.

Still inline with digital anatomy as an education tool, the third chapter, **The Road to Birth: Using Extended Reality to visualize pregnancy anatomy**, by Jones et al., makes a step forward. They focus on the design and use of digital technology for teaching pregnancy-related anatomy and physiology to undergraduate midwifery students. It has been deployed and tested amongst two international cohorts of undergraduate midwifery students, and the initial adoption has been quite successful, although more empirical evaluation is required.

The fourth chapter, titled **Towards Constructivist Approach Using Virtual Reality in Anatomy Education** (Seo et al.), presents Anatomy Builder VR and Muscle Action VR, examining how a VR system can support embodied learning in anatomy education. The authors investigated the relationship between direct manipulation in a VR environment and learning effectiveness.

Finally, the last chapter, **InNervate AR: Mobile Augmented Reality for Studying Motor Nerve Deficits in Anatomy Education** (Cook et al.), introduces educational AR for mobile systems. Authors expand the teaching of Anatomy beyond simple labeling of anatomical structures, layers, and simple identification interactions.

Future and Research Directions

In summary, Digital Anatomy and new imaging techniques are revolutionizing medicine. Different trends for future research directions emerge from the chapters in this new book. The advent of the "**personalized digital patient**" diagnosis and patient communication and **image-guided surgery** (IGS) are two new concepts that will also definitely revolutionize our surgical practices. Critical to advances in Image-Guided Surgery (IGS) is recording images from the different modalities and registering them with the patient. IGS's other key feature is monitoring instruments in real time during the operation and representing them as part of a practical operating volume model. Stereoscopic and virtual-reality techniques can usefully improve visualization. In the future, we argue that Digital Anatomy, coupled with progress in Artificial Intelligence and Augmented Reality, may enhance the performance of clinicians, surgeons and improve patient care in significant ways.

This book comes out during the most unfortunate and interesting times. Indeed, the Coronavirus (COVID-19) pandemic has not only been devastating for patients' health. However, it has also been extremely destructive in many fields, including medical education (Franchi 2020).

Nevertheless, many medical schools have used technical methods to mitigate medical education threats during this uncertain period. We argue that the digital anatomic atlases, coupled with advances in visualization tools, will eventually support real-time collaboration between local and remote surgeons while providing better visualizations of organs. Novel extended reality tools will populate the operating room of the future and assist surgeons in visualizing and communicating with the medical team before, during, and after critical procedures. New geometry reconstruction techniques from imaging data to dynamically match 3D anatomy will allow surgeons to visualize patient details better.

The convergence of AI, VR/AR, Computer Vision, Computer Graphics, and Robotic Surgery will open up new horizons for science, technology, and society. In this way, Digital Anatomy will impact surgical procedures, improve medical communication, and allow surgeons to make better decisions. It will reduce errors and increase efficiency by improving surgical team communication, allowing remote specialists to provide suggestions, offering expertise, and discussing alternatives in real time, obviating the need for air travel.

We hope this new tome will inspire readers and sow the ground for fertile research exploring more innovative approaches. The converging domains of Anatomy, Virtual and Augmented Reality, and Artificial Intelligence will become ever more unified. This book illustrates how Medical Researchers and Computer Scientists can collaborate and dialogue to gestate new intellectual offspring. Further, we anticipate that cross-pollination of research results among both fields helps advance the state-of-the-art with a positive and significant impact on human quality of life and well-being.

Acknowledgements Joaquim Jorge is thankful for the financial support given by Portuguese Foundation for Science and Technology (FCT). This work was also partially supported by national funds through FCT with reference UID/CEC/50021/2019 and IT-MEDEXPTDC/EEI-SII/6038/2014. The author would also like to thank Pedro Campos, Daniel Lopes and especially Jean-François Uhl and the UNESCO Chair of Digital Anatomy, for their contributions.

References

Ayache N (2015) Des images médicales au patient numérique. Leçonsinaugurales du Collège de France. Collège de France/Fayard, France March

Ayache N (2016) Towards a personalized computational patient. IMIA, Yearbook of Medical Informatics. https://hal.inria.fr/hal-01320985

Franchi T (2020) The impact of the Covid-19 pandemic on current anatomy education and future careers: a student's perspective. Anat Sci Educ 13:312–315. https://doi.org/10.1002/ase.1966

Park J, Chung M, Hwang S, Shin B-S (2006) Visible Korean human: its techniques and applications (Apr 2006). Clin Anat 19(3):216–224. https://doi.org/10.1002/ca.20275

Pennec X (2008) Statistical computing on manifolds: from Riemannian geometry to computational anatomy. Nielsen, Frank. Emerging Trends in Visual Computing, vol 5416. Springer, LNCS, pp 347–386. 978–3–642–00825–2.

Peters T (February 2000) Image-guided surgery: from X-rays to virtual reality. Comput Methods Biomech Biomed Eng 4(1):27–57. https://doi.org/10.1080/10255840008907997

UNESCO Chair in Digital Anatomy (2015). https://www.anatomieunesco.org

Vesalius A. De humani corporis fabrica libri septem, 1543

Waldby C (2003) The visible human project: informatic bodies and posthuman medicine. Routledge, p 4. ISBN978–0–203–36063–7

Zhang S-X, Heng P-A, Liu Z-J (2006) Chinese Visible Human Project. Clin Anat 19(3):204–215. https://doi.org/10.1002/ca.20273

Chapter 2
From Anatomical to Digital Dissection: A Historical Perspective Since Antiquity Towards the Twenty-First Century

Vincent Delmas, Jean-François Uhl, Pedro F. Campos, Daniel Simões Lopes, and Joaquim Jorge

Abstract As the oldest medical craft, anatomy remains the core and foundational field of medicine. That is why anatomy is in perpetual advancement, thanks to the technical progress in exploring the human body through computer science and biomedical research. Knowledge of the human body is the basis of medicine. Classical cadaver dissection, the standard discovery tool for centuries, is both unique and destructive of the surrounding tissues. For many years, anatomists have sought to preserve the shape of dissected organs for reference, teaching, and further inspection through different methods. Wax models make a copy of selected dissections. Vessel or duct injection with resin is another dissection-preserving technique. However, all these anatomical objects are unique in time and frozen in place. In contrast, modern Digital Anatomy aims to preserve structures from dissection in flexible ways. Then deliver the results quickly, flexibly, reproducibly, and interactively via advanced digital tools. Thus, computer-aided anatomical dissection addresses the limitations of classical dissection. Through it, experienced anatomists recognize the structures previously segmented with dedicated software to create accurate 3D models from macro or microscopic slices. Its interactivity, flexibility, and endless reusability make digital dissection a perfect tool for educational anatomy. This chapter explores the history of anatomical studies from their inception to the twenty-first century related to the remainder of the book.

V. Delmas · J.-F. Uhl
Université Paris Descartes, Paris, France

P. F. Campos (✉)
ITI/LARSyS and University of Madeira, Funchal, Portugal
e-mail: pedro.campos.pt@gmail.com

D. S. Lopes
INESC-ID Lisboa, Instituto Superior Técnico, Universidade de Lisboa, Lisbon, Portugal

Joaquim Jorge
INESC-ID/Universidade de Lisboa, Lisbon, Portugal

© The Author(s), under exclusive license to Springer Nature Switzerland AG 2021
J.-F. Uhl et al. (eds.), *Digital Anatomy*, Human–Computer Interaction Series,
https://doi.org/10.1007/978-3-030-61905-3_2

11

2.1   The Birth of Anatomy: The Knowledge of the Human
Body from Antiquity to the Middle Ages

Can the medical practice be performed without human body knowledge? Such a timeless question finds its answer in dissection. Scholars have performed dissections since thousands of years ago. Historically, anatomy was the first discipline directly hardwired to medicine, whatever the civilization: we can find schematic anatomical drawings in ancient Chinese (Unschuld1985, Tibetan (Walsh 1910), and Indian medical texts (Wujastyk 2009).

In Western civilization, Hippocrates is considered the father of occidental medicine. Based on observational medicine, Hippocrates created the medical science of diagnosis and prognosis, including the patient's questioning, exam, and symptoms analyses. At this time, anatomy was a relatively simple discipline. Based on the examination of viscera, it discarded the connection between physiology and disease. The disease was merely considered a natural disruption of one or more of the four "humors" influencing the body and emotions. As for treatment, all invasive acts required to deal with "humor" disruption were performed by specialized technicians (e.g., removal of bladder stones), not by medical physicians. Indeed, Hippocrates' oath specified that "*I will not use the knife, not even, verily, on sufferers from stone, but I will give place to such as are craftsmen therein*".

Under a purely homocentric perspective, the knowledge of man is a distinct characteristic of the Greek civilization. As demonstrated in his treatise about "History of animals", Aristotle searched in anatomy the characteristics of man, leading this philosopher to the following concise conclusion: "man is bimanual, erected, thinking and speaking animal". The Greek school developed further connections between medicine and anatomy after Alexander the Great. In Alexandria, during the second century BC under the patronage of King Ptolemaeus (Ist), a fantastic study of human anatomy began with contributions from Herophilus, Erasistrate or Diodore, to name a few of the great ancient intellectuals that contributed to Anatomy at this time. However, their explicit descriptions disappeared with thousands of ancient books when the Great Library of Alexandria was destroyed.

The importance of dissection as a procedure to deepen anatomical knowledge would surface in the second century AC with Galen of Pergamon (Fig. 2.1), doctor of the Roman emperor Marcus Aurelius (Singer 1956). Galen performed dissections on monkeys and pigs, described numerous anatomical structures in neuroanatomy and functional anatomy. In medical history, Galen is considered the heir of Hippocrates. He perpetuated the theory of "humors" but included drugs and their doses, hence the term "galenic" related to Galen himself, his drugs, or his methods.

The knowledge gathered by Galen made it through the Middle Ages, with the passing of Galen's anatomy to Persian and Arabian doctors such as *Albucasis* and *Avicenne*. This body of work was then amassed and transmitted later into Europe. However, during the Middle Ages, the daily need for anatomical knowledge was essentially for surgery practice. By then, surgeons required permission to perform human cadaver dissection before performing any surgery on living

Fig. 2.1 Bust of Galen

companions or compatriots. At the Faculty of Medicine in Montpellier, Guy de Chauliac (1298–1368), doctor of Pope Clement VI (Avignon), wrote in 1363 the "*InventariumsiveChirurgia magna*" (de Chauliac). In this *magnum opus* of surgery and medical practice, Guy de Chauliac described the anatomical knowledge and requirements for those needing to perform surgical procedures (Fig. 2.2).

2.2 The Anatomical Evolution: Scientific Dissections During Renaissance and Enlightenment

With the Renaissance came the printing press (Gutenberg, Mainz, 1452) that enabled the revolutionary diffusion of ancient books throughout the Western world and its colonies. Unsurprisingly, Galen's work on medicine and anatomy was among those re-prints. During the Renaissance, anatomical dissections became allowed, and even Leonardo Da Vinci (Fig. 2.3) was preparing a book (around 1496–1502) on personal

Fig. 2.2 *"Chirurgia magna"* manuscript, in 1363, by Guy de Chauliac. Anatomical dissections were presented for surgeons and medical practice. It was a very popular document that was also known as the *"Guidon"*, a familiar handbook for surgeons of this time and beyond, having been also printed in the Renaissance (BIUUniversité de Montpellier)

dissections that he conducted in Florence. Such a book would include original drawings and unusual representations of the human body. Unfortunately, this book was never published. Still, Da Vinci's anatomical drawings persisted over time (Vinci 1983) and are now in the Royal Collection at Windsor Castle (UK).

Besides Da Vinci, the Renaissance would see the rise of two famous names due to their contributions to the anatomical zeitgeist: Andreas Vesalius and AmbroiseParé. Andreas Vesalius (1514–1564) became a professor of anatomy at the University of Padua after studying in Louvain, Paris, and Montpelier, and was the official doctor of Emperor Charles V. Vesalius introduced a new mindset into the anatomy realm: he thought and practiced anatomy scientifically. In strict conflict with the masters of his time, who faithfully repeated Galen, Vesalius practiced dissection himself while comparing his findings with the descriptions of the ancient anatomists, namely those written by Galen. Remarkably, Vesalius publishes the first book on human anatomy, *De humani corporis fabrica*(Fig. 2.4a–c) a treatise in seven chapters (Vesalius 1543) describing the human body's major morphological systems. He was also a true pioneer as Vesalius' book became the first anatomical treatise featuring drawings—a genuine revolution in transmitting anatomical knowledge that was harshly criticized by his contemporary peers.

Fig. 2.3 Medallion of Leonardo Da Vinci performing a dissection by Terroir (1950), placed in front of the Faculté de Médecine, rue des Saints Pères at Paris. It evokes Leonardo Da Vinci as a precursor of anatomy. Da Vinci's drawings were based on dissections performed by himself. Leonardo was also the first to render anatomical slices in different 3D planes

AmbroiseParé (1510–1590) was the surgeon of France's kings. He is considered the father of modern surgery. Paré became notorious as he was considered the first to rely on anatomy as the foundation for surgical planning and practice. For example, Paré highlighted a specific ligation of arteries (Paré 1585), a feat that required detailed anatomical knowledge.

During the Renaissance, the notion emerged that surgery heavily relies on basic anatomical understanding. Amphitheaters of anatomy started to appear near the college of surgery (Delmas 1978). The same era saw anatomical theatres purposely built for dissections: Padua (1594), Leyden (1597), Bologna (1637), Strasbourg (1670), Academy of Surgery of Paris (1694) (Fig. 2.5a, b) or that of the Faculty of Medicine in Paris built by Winslow (1745).

As dissection became the basis of surgery learning, surgeons started to consider anatomical studies to directly prepare the surgical act. Indeed, brilliant masters in the Faculty of Medicine in Paris documented the link between anatomy and surgery. Namely, Jean Riolan, father and son (seventeenth century) (Fig. 2.6), and Jacques-Bénigne Winslow (eighteenth century) wrote scholarly treatises to teach anatomy.

Fig. 2.4 Engravings from the Vesalius book. **a** Vesalius dissecting a forearm. **b** Front page of the "*De homini corporis fabrica*"—note the context of a dissection in an amphitheater with many participants. **c** A more artistic than medical rendering of a contemplating skeleton, yet accurate drawn

Fig. 2.4 (continued)

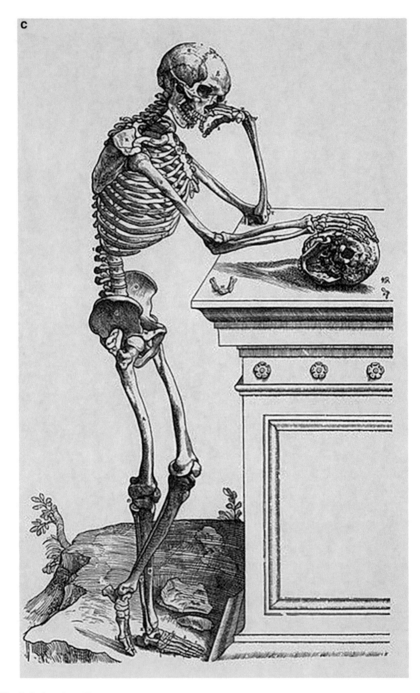

Fig. 2.4 (continued)

Fig. 2.5 a Anatomical dissection by Dionis in the amphitheater of surgery at Saint Côme in Paris (1707). **b** View of the Saint-Côme's theatre of anatomy in Paris held by the Academy of Surgery (1750)

It is important to remark that, in those days, anatomy and toxicology, just as chemistry at the time, were basic scientific disciplines with very few clinical applications. During the Renaissance, clinical medicine was mostly a medicine of "humors", strictly following the Hippocratic tradition. Curiously, at the same time, scientific anatomy progressed fantastically: William Harvey described blood circulation in 1628, and Nicolas Stenon described brain structure in 1665, to list a few. The seventeenth and eighteenth centuries are considered the *Golden Age* of anatomical dissection. So "golden" that celebrated paintings depicted dissections (the most famous is the anatomical lesson of Dr. Tulp painted by Rembrandt in 1632) and inspired dramaturgy (sometimes ironically in Molière's plays). Indeed, dissections were often considered performances in a purely artistic sense (Fig. 2.7).

Anatomy widened its field of interest to physiology through Albrecht von Haller's (1708–1777) pioneering works on taxonomy (Haller 1755; ElementaPhysiologiae Corporis Humani 1757–1766). He developed an anatomical thesaurus before naming anatomical structures became relevant to clinical practice.

The act of dissecting destroys tissues and the surrounding structures, thus producing unique but perishable, non-reproducible results that ultimately decay or decompose in a relatively short time. Therefore, Renaissance anatomists devised

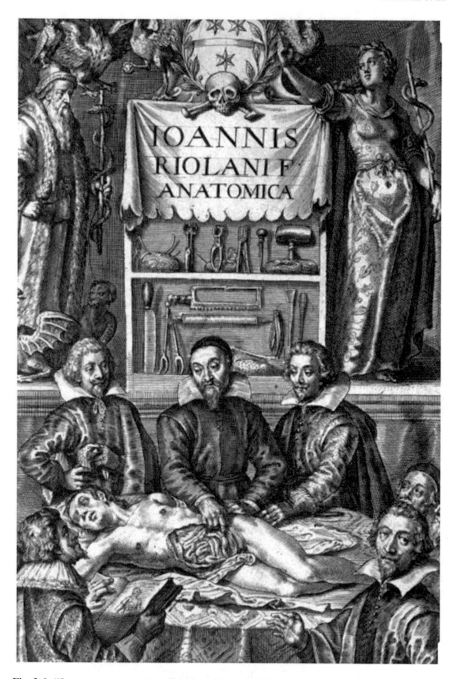

Fig. 2.6 *"Les oeuvres anatomiques"* de Jean Riolan (1628)

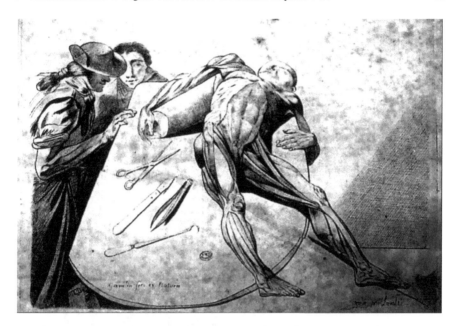

Fig. 2.7 Artistic anatomy in the eighteenth century: a surgeon and an artist look at a dissection. Print by Jacques Gamelin (1779) (BIUM Paris). In this century, anatomy is widespread and appreciated more in its artistic view of the man rather than in a more medical/clinical interest

techniques and media to preserve the morphological structures revealed during a dissection. Wax models became popular during the eighteenth century, as they formed real structures in this natural polymeric material. It was the first approach to permanent materialization in anatomy, the first 3D anatomical models. Reputed masterpieces, these wax molds reside today in museums and academic collections, but they are very fragile and often unique (Fig. 2.8a, b). Another preservation technique was cross-section or slices that preserve structures in situ. Due to technical limitations, the first slices were realized at the brain level and fixed by alcohol by Félix Vicqd'Azyr in 1786 (Fig. 2.9).

2.3   The Clinic Meets Anatomy: Expansion of Anatomy in the Eighteenth and Nineteenth Centuries

At the end of the eighteenth century, a new radical concept appeared within anatomy: diseases were a direct consequence of organ pathology and not caused by unsettled "humors". Giovanni Batista Morgagni (1682–1771) from the University of Padua was the first to demonstrate a link between clinical symptoms and autopsy findings. Known as the founder of pathology, Giovani Battista Morgagni wrote "*De sedibus et causismorborum per anatomenindagati*" (Giovani Battista Morgagni

Fig. 2.8 **a** Wax model of head of Gaetano GuilioZumbo (1696). **b** Wax model of head and neck.
(Muséed'anatomieDelmas-Orfila-Rouvière now in Montpellier museum)

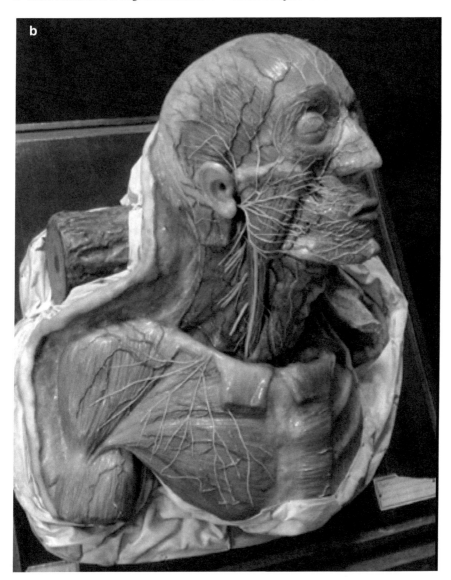

Fig. 2.8 (continued)

1761). However, pathology and clinical anatomy, as new anatomical paradigms, began as a discipline in Paris with Xavier Bichat (1771–1802) with the publication of the treatise "*Anatomiegénéraleappliquée à la physiologie et à la médecine*" (Bichat 1801).

Thanks to progress in other domains, the discipline evolved rapidly in the nineteenth century. The treatise on "*Anatomiepathologique du corps humain*" (Jean

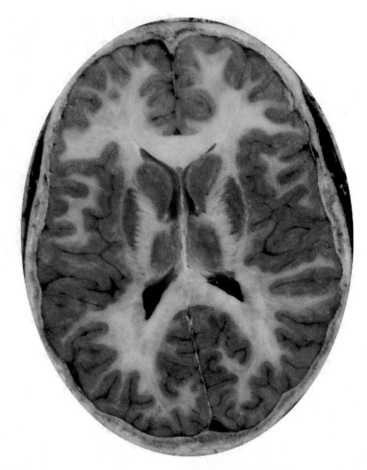

Fig. 2.9 Transversal slice of a human brain: grey and white matters are clearly visible; also identifiable are the cortex grey peripheral matter and the central grey nuclei (ex-Muséed'anatomieDelmas-Orfila-Rouvière)

CruveilhierAnatomiepathologique du corps humain 2020) by Jean Cruveilhier (1829–1842) was a macroscopic dissection of pathologic anatomy. Thanks to the technical progress in microscopy, Rudolf Virchow (1821–1902) created microscopic pathology, while Charles Robin (1874) developed histology. Nicolas Pirogoff of Saint Petersbourg introduced a new technique to study human anatomy using macroscopic slices in 1853. These parts preserved the organ's topography as each slice was made by freezing sectioned samples and immersing them in alcohol. Such macroscopic slices proved a powerful educational tool for learning anatomy. This dissection technique's most significant limitations were the slicing procedure's destructive nature and can be made possible only on dead bodies. However, cutting body parts in thin sections did bring about volumetric reconstruction based on slices as a novel concept.

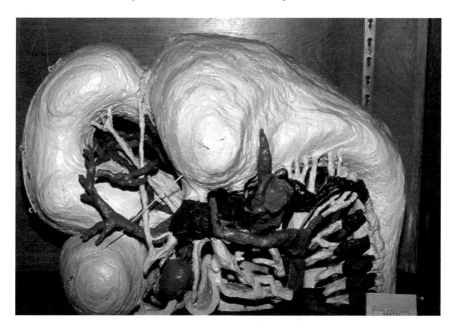

Fig. 2.10 Enlarged 3D reconstruction of the brain of an 9 mm embryo by Eyriès using the Born method in 1954 (collectiond'organogénèse, ex-MuséeDelmas-Orfila-Rouvière)

Just as wax models, such sets of slices orderly placed in 3D space consisted of another step towards the materialization of anatomy.

The late nineteenth century saw other anatomy materializations emerge. Born Gustav Jacob, a German embryologist (1851–1900), developed a method to reconstruct enlarged wax models of embryos (Fig. 2.10) or microscopic structures. Louis Auzoux (1825) introduced painted plaster models, leading to the impressive collection gathered by A. Nicolas, M. Augier, and Roux at the Faculty of Medicine of Paris. These display the human body's regions dissected in actual size and the whole body in a "clastic man" made of fully removable parts (Fig. 2.11).

It was also during the eighteenth and nineteenth centuries that a new medical paradigm arose: clinical anatomy. As anatomical knowledge became essential in diagnosis and treatment, the discipline became an integral part of medicine, with clinical applications and practical interest, contributing to surgery and the emergence of medical imaging. In particular, surgical anatomy knew a prodigious development thanks to anesthesia (Morton of Boston by ether 1846 and Simpson of Edinburgh by chloroform 1847), organ ablation, and reconstructive surgery. These three surgical procedures required in-depth anatomical knowledge for any surgeon to operate "quick and well". This era saw excellent anatomy textbooks such as the "*Traitéd'anatomie*" by Paul Poirieret Adrien Charpy (1892) in France (Poirier and Charpy 1911), "*Anatomie des Menschen*" by Von Bardeleben (1892) in Germany, or "*Anatomy*" by Mall and Gray (1900) (Gray's 1995) in the USA. Surgeons applied

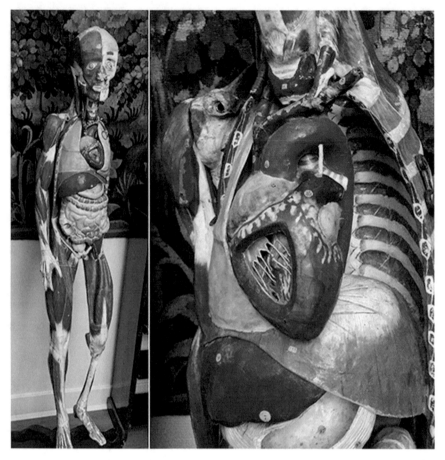

Fig. 2.11 "Clastic man" by Dr. Auzoux. Anatomical model painted in maché paper and composed of removable parts (which can be removed one by one, as in a real anatomical dissection). (private collection)

their anatomical knowledge for removing the sick organs electively without excessive bleeding while preserving the healthy ones around them.

2.4   New Perspectives: Anatomical Knowledge and Technical Evolution in the Twentieth Century

Several anatomical milestones and technical advancements marked the twentieth century. Vast new anatomical knowledge resulted in new anatomical disciplines while widening clinical and surgical anatomy to more complex, precise, and accurate domains. Specific surgical procedures became so specialized that needed textbooks dedicated to particular anatomical regions, organs, or tissues.

It was also the age that generated a new anatomical era for selective injections of vessels and ducts with resin, latex, or contrast agents in radiology. Such techniques are crucial for conservative surgery with partial ablation of an organ requiring precise knowledge of vascular territories that delimit segmental anatomy. Selective injection of vessels and ducts facilitated analysis of individual and unique structures of medical interest, namely, kidney segmentation (Max Brödel 1901; Marius Augier 1922; Augier 1923; André Dufour 1951, Graves 1954) (Fig. 2.12), liver segmentation (Couinaud 1957; Fig. 2.13) and lung segmentation (Gaston Cordier, Christian Cabrol; Fig. 2.14).

Novel surgical disciplines also arose with the advancement of anatomical knowledge. For instance, in the 1980s, microsurgery appeared as a new field, gathering surgical and anatomical research on vascular and nervous micropedicules. Microsurgery allowed micro transplants or reimplantation of small organs for preservative surgery. Fundamental to the discipline was infra optic anatomy settlement, as specific surgical procedures now required magnifying lens, microscopes, and microscopic anatomical knowledge with immunomarkers (Fig. 2.15). Hence, medicine now started to address both the visible anatomy and "invisible to the naked eye" anatomy.

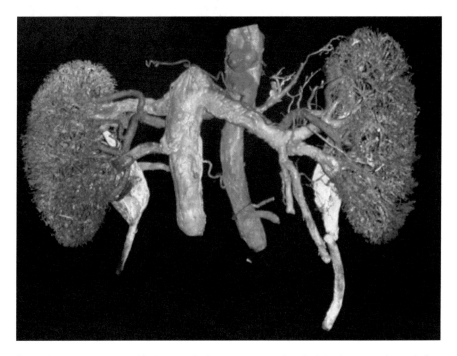

Fig. 2.12 Vascularization of kidneys and urinary tract seen after the injection-corrosion technique leading to vascular or urinary territories (ex-Muséed'anatomieDelmas-Orfila-Rouvière)

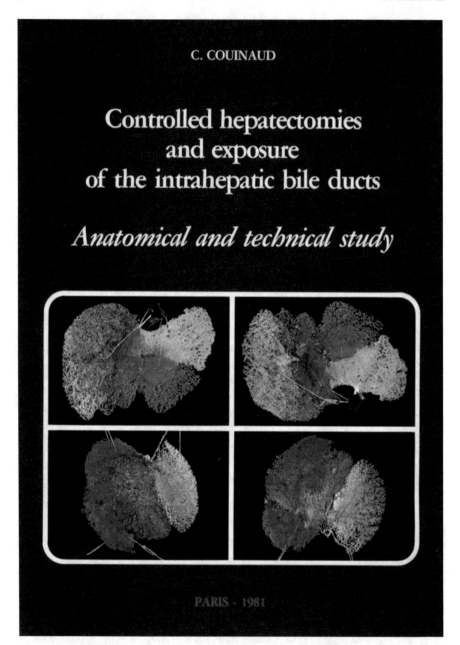

Fig. 2.13 Anatomical liver segmentation by Couinaud, first published in 1957 revisited in 1981, led to controlled hepatectomies (hepatic segmentectomies) and liver transplantation of part of the liver withdrawn from a living donor

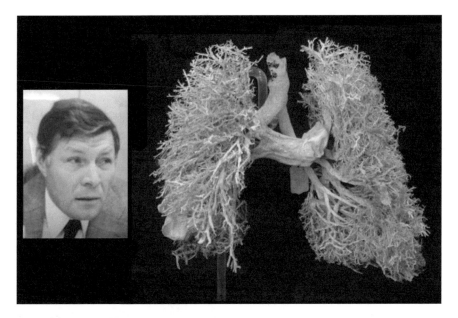

Fig. 2.14 Anatomical lung segmentation by C. Cabrol (injection-corrosion of the bronchial tree and vessels)

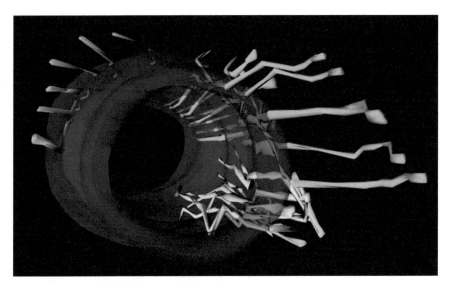

Fig. 2.15 3D reconstruction of the male urethra nerves recognized by immunohistochemistry (unmyelinated nervous fibers in yellow, myelinated nervous fibers in green). This image represents the superior view of a 3D reconstruction with bladder neck and proximal urethra, nerves reach the posterolateral grove along the urethra (Karam*EurUrol* 2005)

By the end of the nineteenth century, radiologic anatomy emerged, in simpler words, anatomy without skin incisions. In December 1895, Wilhelm Roentgen displayed living human bones for the first time without dissecting a single tissue (Fig. 2.16).

Combined with contrast agent injection, radiological anatomy could enable clinicians to visualize hollow organs and ducts. This originated novel approaches, including intravenous urography 1923, and the digestive system. Contrast agents could now reveal whole vessel systems under angiography, arteriography, phlebography, and lymphography. These in vivo methods can uncover the form of pathologic organs without opening the body. These emerging "virtual" anatomy fields required the precise knowledge of "standard" anatomy with its normal variations to be effective.

Two of the most revolutionary and popular anatomical technologies also appeared in the twentieth century: the X-ray-based computerized tomography (CT) invented by Hounsfield (1970) and the non-irradiating technique magnetic resonance imaging

Fig. 2.16 First radiography obtained by Roentgen in 1895 representing the hand of his wife, Anna Bertha. Higher density materials (bones and metallic ring) appear in darker tones

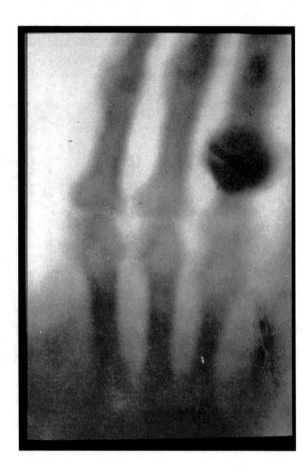

(MRI) later invented by Lauterbur-Mansfield (1975). Both technologies allow visualizing subject-specific anatomy through multiple slices or by reconstructing anatomical structures in 3D. Through these approaches, dissection becomes virtual anatomy. Interestingly, CT and MRI gave greater relevance to anatomical atlases that portrayed real anatomical slices published for pedagogical purposes. These include Nicolas IvnovichPirogoff's (Saint Petersburg) "*Anatomiatopographicasectionibus, per corpus humanumcongelatumtriplicedirectioneductis, illustrata*" (Nicolas Ivanovich-Pirogoff 2020) consisting of four volumes with 224 illustrations (1852–1859). Eugène Doyen's (Paris) "*Atlas d'anatomietopographique*" (1911) (Doyen and Bouchon 1911), Eycleshymer and Shoemaker's (New York) "*A cross-section anatomy*" (EycleshymerA and Shoemaker 1911), and Raymond Roy Camille's (Paris) "*Atlas de coupes anatomiques du tronc*" (Roy-Camille 1959) are still anatomical references to this day.

Another outstanding technique developed in the twentieth century was plastination: a process designed to preserve real tissues, organs, and entire systems for educational and instructional purposes. Plastinates are dry, odorless, durable, and incredibly valuable educational tools. They were invented by Gunther von Hagens (Heidelberg University, Germany) in 1977 (Gunther von Hagens 1977). Plastination consists of dehydrating the anatomical pieces, then injecting them with silicone to make them rot-proof (Fig. 2.17).

2.5 Modern Surgery: Twenty-First Century Anatomy in the Operating Theater

Functional anatomy leads to a more specific clinical exam; it is based on minimally-invasive surgery. Indeed, function sparing surgery leads to a refinement in anatomical knowledge. All vascular or nervous regions have a functional meaning; for example, tracing the erectile nerves' route is a fundamental step during radical prostatectomy to understand how to preserve them. The immunohistochemistry allows identification of the vegetative nerves (parasympathetic, sympathetic) and the somatic ones. Thus, it becomes possible to follow the nerves' route and location (vegetative, somitic) towards the urethral wall and its distribution (Fig. 2.16).

Minimally invasive surgery requires detailed and precise anatomical expertise. This is true for microscope surgery when reconstructing vessels and delicate structures with centimetric or millimetric precision. This is also true to make ablative surgery as non-invasive as possible. It is even truer for reconstructive surgery, especially in the study of pedicled cutaneous flaps, musculo-osseous or osseous grafts, and transplant surgery. These procedures need identification of the functional value and anatomical situation of patients' innards. To this end, both cœlioscopy and robotic-assisted laparoscopy have an enlarged vision of the specific surgical field. These

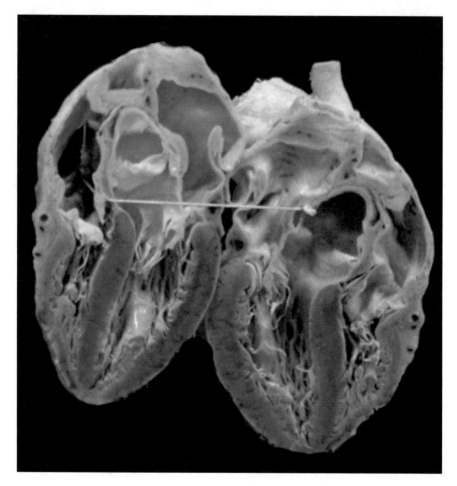

Fig. 2.17 Plastination of a human heart (ex-Muséed'anatomieDelmas-Orfila-Rouvière)

allow surgeons to visualize anatomical structures with greater precision. While visualizing minute anatomical details, these structures must have clearly assigned functional semantics, e.g., to accomplish nerve-sparing surgery with the cavernous nerves during radical prostatectomy.

2.6   Anatomy 2020: What is the Place of Anatomy Today?

To give it a name: anatomy provides a lexical meaning to each structure accompanied with an adequate semantic to the structure's definition. Peering under the skin has motivated a growing need for taxonomy. Why and how do we find such variable structures? What are people made of? After the answers of Greek anatomists and Galen's explanations, Vesalius provided a scientific approach to anatomy. Nowadays, the standard international anatomical nomenclature is in Latin, "*Parisiensa Nomina Anatomica*" (1955). It became the "*Terminologia Anatomica*" (Anatomica 1998), providing translations from Latin to other languages. Anatomy has evolved into a multifaceted tool for different purposes.

To guide surgery: all the progress in surgery geared to functional preservation leads to an increased and comprehensive anatomical background. Future surgery needs more **pre**-(using VR), **per**-(using Augmented Reality), and **post-operative** images (using Mixed Reality) to support a better surgical operation. Providing these images and the necessary support, image registration, and interactive techniques lies at the core of a novel emerging field, **image-guided surgery**.

To make clinicians aware of the real body: histology or *tissue anatomy*, or *general anatomy* as defined by Bichat (1801) in the eighteenth century, has made the transition from pure humoral therapy to an organ- and tissue-based medical practice. Anatomical pathology plays a central role in diagnosis and clinical practice since then. Understanding the specificities and variability of internal morphological structures is one of the significant challenges facing big data, visualization, and anatomical studies today.

To read inside the human body: Medical imaging allows in vivo anatomy. Classical cadaver dissection, while highlighting fundamental structures, destroys tissues and modifies their surroundings. Anatomists have made slices to preserve morphological arrangements, at least partially. Slice imaging, CT scan, and MRI allow viewing the anatomical structures again from different perspectives and serve as a fantastic interpretation tool. Thanks to progress in digital imaging techniques, CT scans in the axial or transverse planes, MRI in the three spatial planes enables a computerized reconstruction of structures seen in isolation as virtual dissections.

To communicate anatomical knowledge: there is a new approach to visualize and understand the human body developing today; it gives its entire meaning for educational anatomy. Fueled by progress in Computer Science, Imaging, Computer Graphics plus Extended Reality techniques, in synergy with the growing body of anatomical knowledge, excellent new tools are emerging to convey anatomical knowledge and further morphological research. Since the first images by Andreas Vesalius, physiology is rising to a new level, digital anatomy based on 3D reconstruction. This new representation of the human body is essential for medical doctors and, more generally, for biomedical scientists and health professionals.

2.7   Anatomy in the Age of Virtual, Mixed, Augmented Reality: Twenty-First Century Digital Dissections

Human anatomy is the basis of our knowledge in medicine and essential for diagnostic imaging and surgical treatments. However, learning human anatomy is a complex and fastidious process (Preim and Saalfeld 2018). Students must identify features of the human body and understand the relationships among these features. Ultimately, it depends on each student's ability to figure out the anatomical structures in 3D space and its implications for diagnostic or therapy. This is a tedious task when resorting to classical learning methods based on 2D images or illustrations. However, learning human anatomy cannot depend solely on imagery as students need to see, touch, and dissect real anatomy. That is why cadaver dissection remains the gold standard teaching process for anatomical education (Singer 1956). Classical dissection course settings place students within specialized facilities where anatomists produce unique materials for medical education, training, and research. The downside to classical dissection classes is that the results become irreversible once dissected since the surrounding structures get damaged after underlining the target structure. Another drawback of classical dissection classes is that there is a global shortage of cadavers in medical schools for training students and surgeons (Preim and Saalfeld 2018; Papa and Vaccarezza 2013).

Three-dimensional modeling of the human anatomy, also known as "digital anatomy", consists of building 3D representations from a series of related 2D medical images. Slices are stacked on top of each other and aligned neighboring slices. This way, anatomical structures can be reconstructed in three dimensions to portray realistic, subject-specific depictions of individual morphology or systems in relation to other organs with high accuracy. To not put such anatomical accuracy at waste, rendering techniques must generate precise results for 3D visualization. While these depictions are three-dimensional and subject-specific, they represent anatomy and are not subject to real dissections' limitations.

Modeling the three-dimensional, geometric nature of human anatomy requires virtual dissection tools, as those presented in this book. Such digital anatomy tools are adequate for modern medical education (Preim and Saalfeld 2018; Zorzal et al. 2019), but are they also useful for diagnostic (Papa and Vaccarezza 2013; Sousa et al. 2017), surgical training (Lopes et al. 2018), and surgical planning (Mendes et al. 2020) (Fig. 2.18). The size of organs, the exact shape a pathology may take, or the relative position between anatomical landmarks and vital structures is unique to each individual. That is why digital anatomy tools match specialists' requirements in radiology, surgery, and neurology as these professionals request specific geometric knowledge of a patient under diagnostic or treatment. Nevertheless, image-based anatomical accuracy, rendering precision, and interactive tools are the combination that advances the understanding of human anatomy (Wen et al. 2013).

To alleviate classical dissection education's significant problems, anatomists and students need to continue dissecting but digitally. In the digital realm, subject-specific anatomy can be represented in two major formats, either in two-dimensional (2D) medical images or three-dimensional (3D) anatomical models. There are many digital tools available to perform digital dissections. However, all yield the same result: a 3D reconstruction of anatomical information within a stack of 2D medical images, allowing users to create digital representations of subject-specific anatomical structures. In order to conduct a digital dissection, experts and students have to perform three fundamental tasks (Papa and Vaccarezza 2013):

**Mixed Reality
Dissection Table**

**Tablet and Stylus for Image
Segmentation**

(a)

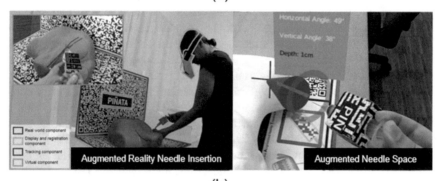

Augmented Reality Needle Insertion

Augmented Needle Space

(b)

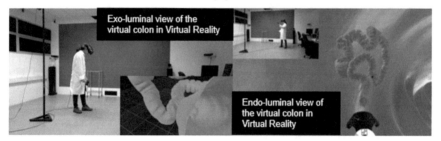

**Exo-luminal view of the
virtual colon in Virtual Reality**

**Endo-luminal view of
the virtual colon in
Virtual Reality**

(c)

Fig. 2.18 Modern examples of **a** Mixed, **b** Augmented and **c** Virtual Reality applications in Digital Anatomy for medical education (Zorzal et al. 2019), surgical training (Mendes et al. 2020) and diagnostics (Lopes et al. 2018), respectively. ((**a**) Adapted from (Zorzal et al. 2019) with Elsevier permission; (**b**) Adapted from (Mendes et al. 2020) with Elsevier permission; (**c**) Adapted from (Lopes et al. 2018) with author's permission)

1. Browse through large sequences of images (e.g., MRI, CT, cryosections) using slice-by-slice navigation tools to reach and identify relevant anatomical details.
2. Manually segment the geometric *loci* of the anatomical structures of interest to reveal relationships between neighboring organs.

3. Iteratively explore the 3D reconstructed model by panning, zooming, and rotating the 3D model and/or a virtual camera.

These tasks are naturally laborious, demand a high cognitive load, and are even error-prone but are mandatory to guarantee high-quality digital dissections. Such digital tools are a complementary medium to live dissection, not their replacement. Thus, digital dissection complements classical dissection and not the other way around.

By default, digital reconstruction tools are designed for slice-by-slice navigation and force users to trace contours around anatomical structures throughout many slices. Once a set of segmented curves is assembled, it is possible to reconstruct an isolated 3D digital organ. However, 3D reconstruction can supervene not only by tracing organs from a sequence of images but also by generating a 2D projection of 3D voxel data through volume rendering techniques. These digital representations materialize in the form of 3D meshes—a set of vertices, edges, and faces that form a geometrical arrangement of tiny triangles. Or 2D images generated by projecting 3D voxel data into a 2D screen (e.g., Maximum Intensity Projection, Direct Volume Rendering, or more recently, Cinematic Rendering (Lopes and Jorge 2019; Paladini et al. 2015; Engel 2016)). Meshes can even be used to feed a 3D printer and fabricate real-sized subject-specific anatomy, leading to digitally 3D reconstructed objects' materialization. Such 3D meshes, volume rendering images, and 3D printed models can then be used as source material for education and training purposes.

In summary, 3D reconstruction from medical images is the ultimate goal of virtual dissection. When performing a 3D reconstruction, users are dissecting pixels, voxels, vertices, edges, and triangles. Yet, these digital entities encode anatomical data and anatomical information that will be perceived and processed by the user into anatomical knowledge. Each digital dissection results in accurate, subject or patient-specific models that can be used to analyze specific structures, their functionality, and relationships with neighboring structures. Given its digital nature, all the reconstruction-related tasks are performed interactively through a user interface that provides feedback (mostly visual) in real time. Another plus of the digital format is that anatomical data can be shared with many people while promoting collaborative anatomical dissection, thus, enriching teaching and learning experiences.

2.8 Future Directions in Digital Anatomy: Twenty-First Century as a Stage of Innovative Virtual Dissections

Since the Visible Human Project (Ackerman 1999), many digital dissection tools have been developed. However, during the time-lapse that stretches from the late 1980s until the early 2000s, the Windows, Icons, Menus, and Pointer (WIMP) paradigm prevailed and became standard (Olsen et al. 2009). Such an interaction paradigm lacks direct spatial input, affords limited exploration control, prescribes timely slice-by-slice segmentation, forces users to build a 3D mental image from a set of 2D cryosections, and promotes single-user interaction. For instance, the Korean Visible Human took 8 years to segment using WIMP interfaces and mouse input. Nevertheless, virtual dissection presents a challenging conundrum. There is a need to speed up segmentation without discarding manual intervention. Anatomists want to retain the control to produce accurate and informed contours manually, overcoming the limitations of fully automatic algorithms.

The commonly used digital dissection tools still present limited deployment, besides underdeveloped usage of mature technologies, as the vast majority of tools are stuck with WIMP interfaces. Therefore, virtual dissection of the **twenty-first** century needs more natural and familiar interfaces, along with tools that are more immersive, spatial, tangible, and encourages closely coupled collaborations and group discussion. Such interfaces and tools are more than welcome. They should become a standard.

In recent years, a real technological revolution surfaced due to the growing power of computers, graphic processors, and machine learning: the Renaissance of VR, Augmented Reality (AR), and Mixed Reality (MR). Suddenly, a myriad of interactive technologies that are tangible, spatial, stereoscopic, immersive, and mobile appeared as off-the-shelf consumer products with relatively affordable price tags. This represents a golden opportunity for Digital Anatomy as a whole. By updating the discipline of Digital Anatomy with more interesting interaction paradigms, namely, interactive surfaces, tangible user interfaces, VR, AR, and MR, it is possible to design new and more exciting tools for 3D reconstruction anatomical structures. Compared to conventional WIMP interfaces, VR, AR, and MR promote greater freedom of movement, superior camera control (e.g., head, hands, arms, feet, full-body). These allow more flexible control of users' perspectives towards the anatomical content or allow more natural ways to manipulate data in 3D space. Immersion also promotes a greater visual bandwidth and improves the 3D perception that combined can improve the effectiveness when studying medical data (Laha et al. 2013) and scientific data sets (Kuhlen and Hentschel 2014).

Such an update to redesign conventional interfaces and interaction techniques is welcome. Such a redesign can ease virtual dissection processes, e.g., manual tracing, semi-automatic segmentation, or thresholding opacity maps for volume rendering. These semi-automatic techniques promote expeditious exploration of rich and complex volumetric medical data sets and facilitate 3D model navigation. Moreover, how users explore, teach and learn human anatomy can leap forward thanks to stereoscopic (semi-)immersive 3D visualization, body tracking devices, and full-body movement, all features that characterize so well VR, AR, and MR.

Several studies have demonstrated that more immersive, spatial, and interactive tools can boost digital dissection tasks and attenuate dissection workload. It is important to remark that current digital dissection tools promote single-user slice navigation and mouse-based input to create content through manual segmentation, which is often performed using a single flat display and mouse-based systems, forcing multiple scrolling and pinpointing mouse clicks. On the contrary, interactive surfaces, tangible user interfaces, Virtual Reality (VR), Augmented Reality (AR), and Mixed Reality (MR) are more prone to promote the design of collaborative interfaces for digital dissection. Also, several studies have shown that VR, AR, and MR applied to teach anatomy are more effective than conventional techniques (Couinaud 1957; EycleshymerA and Shoemaker 1911; Papa and Vaccarezza 2013).

Acknowledgements All authors are thankful to the Unesco Chair of Digital Anatomy (Paris Descartes University) participating in this book with the common aim to promote Digital anatomy worldwide.

References

Ackerman MJ (1999) The visible human project: a resource for education. Acad Med J Assoc Am Med Coll 74(6):667–670. https://doi.org/10.1097/00001888-199906000-00012

Anatomica T (1998) International anatomical terminology. Federative Committee on anatomical Terminology, Stuttgart, Thieme

Augier M (1923) Anatomie du rein. In: Poirier & Charpy, Traitéd'anatomiehumaine. T VI, Paris Masson

Bichat X (1801) Anatomiegénéraleappliquée à la physiologie et à la médecine. Brosson et Gabon, Paris

Couinaud C (1957) Le foie. Étudesanatomiquesetchirurgicales, Paris, Masson

Da Vinci L (1983) Leonardo on the human body. Dover pub, New York

de Chauliac G. In: McVaugh MR (ed) Inventarium sive Chirurgia magna, series: studies in ancient medicine, vol 14/1

Delmas A (1978) Histoire de l'anatomie. In: Histoire de la médecine, de la pharmacie, de l'artdentaire et de l'artvétérinaire, Soc Fr d'étudesprofessionnellesmédicales et scientifiques, Paris, Albin Michel-Laffont-Tchou, T. 3, pp 71–129

Doyen E, Bouchon JP (1911) Atlas d'anatomietopographique, 12 fascicules en 3 vol. Maloine, Paris, Paris

Engel K (2016) Real-Time Monte-Carlo path tracing of medical volume data, GPU technology conference, 4–7 Apr 2016. San Jose Convention Center, CA, USA

EycleshymerA C, Shoemaker DM (1911) A cross-section anatomy. D. Appleton & Co, New York

Gray's A (1995) 38th edn. New York, Churchill Livingstone

Jean CruveilhierAnatomiepathologique du corps humain (2020) Descriptions avec figures lithographiéesetcolorisées des diversesaltérationsmorbidesdont le corps humain est susceptible. Paris, Chez J B Bailliére, 1829–1835. https://anatomia.library.utoronto.ca/islandora/object/ana tomia:RBAI072. Accessed 28 Sep 2020

Kuhlen TW, Hentschel B (2014) Quo vadis CAVE: does immersive visualization still matter? IEEE Comput Graph Appl. 34(5):14–21. https://doi.org/10.1109/MCG.2014.97

Laha B, Bowman DA, Schiffbauer JD (2013) Validation of the MR simulation approach for evaluating the effects of immersion on visual analysis of volume data. IEEE Trans Vis Comput Gr 19(4):529–538

Lopes DS, Jorge JA (2019) Extending medical interfaces towards virtual reality and augmented reality. Ann Med 51(sup1):29. https://doi.org/10.1080/07853890.2018.1560068

Lopes DS, Medeiros D, Paulo SF, Borges PB, Nunes V, Mascarenhas V, Veiga M, Jorge JA (2018) Interaction techniques for immersive CT colonography: a professional assessment. In: Frangi AF, Schnabel JA, Davatzikos C, Alberola-Lopez C, Fichtinger G (eds) Medical image computing and computer assisted intervention—MICCAI 2018. Lecture notes in computer science, vol 11071, pp 629–637. Springer, Cham. https://doi.org/10.1007/978-3-030-00934-2_70

Mendes HCM, Costa CIAB, da Silva NA, Leite FP, Esteves A, Lopes DS (2020) PIÑATA: pinpoint insertion of intravenous needles via augmented reality training assistance. Comput Med Imaging Graph 82:101731. https://doi.org/10.1016/j.compmedimag.2020.101731

Morgagni GB (1761) de Sedibus Et Causis Morborum Per AnatomenIndagatis: Dissectiones Et Animadversiones, Nunc Primum Editas, ComplectunturPropemodumInnumeras, Medicis, Chirurgis, AnatomicisProfuturas, Ex. TypographiaRemondiniana, Venice

Nicolas Ivanovich Pirogoff (2020) Anatomia topographica sectionibus, per corpus humanum congelatum triplice directione ductis, illustrata (1852–1859). National Library of Medicine at https://resource.nlm.nih.gov/61120970RX6. Accessed 28 Sep 2020

Olsen L, Samavati FF, Sousa MC, Jorge JA (2009) Sketch-based modeling: a survey. Comput Gr 33(1):85–103. https://doi.org/10.1016/j.cag.2008.09.013

Paladini G, Petkov K, Paulus J, Engel K (2015) Optimization techniques for cloud based interactive volumetric monte carlo path tracing. Industrial Talk, EG/VGTCEuroVis

Papa V, Vaccarezza M (2013) Teaching anatomy in the XXI century: new aspects and pitfalls. Sci World J, (Article ID 310348), 5. https://doi.org/10.1155/2013/310348

Paré A (1585) De l'anatomie, livres 3–6. In: Oeuvres, Paris, Buon

Poirier P, Charpy A (1911) Traitéd'anatomiehumaine, 5 tomes. Masson, Paris

Preim B, Saalfeld P (2018) A survey of virtual human anatomy education systems. Comput & Graph 71:132–153

Roy-Camille R (1959) Coupes horizontales du tronc. Atlas anatomique etradiologique à l'usage des chirurgiens et des radiologistes, Paris, Masson

Shaw V, Diogo R, Winder IC (2020) Hiding in Plain Sight-ancient Chinese anatomy. Anat Rec 1–14. https://doi.org/10.1002/ar.24503

Singer C (1956) Galen On anatomical procedures. London, Oxford University Press for the Wellcome Historical Medical Museum, de Anatomicisadministrationibus

Sousa M, Mendes D, Paulo S, Matela N, Jorge J, Lopes DS (2017) VRRRRoom: virtual reality for radiologists in the reading room. In: Proceedings of the 35th annual ACM conference on human factors in computing systems (CHI 2017). ACM Press, New York. https://doi.org/10.1145/3025453.3025566

Unschuld P (1985) Medicine in China, a history of ideas. University of California Press, Berkeley, CA

Vesalius A (1543) De humani corporis fabricalibriseptem, Basel, Ex officina JoannisOporini

Von Bardeleben K, des Menschen A (1913) Leipzig

von Hagens G (1977) US Patent 4,205,059 animal and vegetal tissues permanently preserved by synthetic resin impregnation, filed Nov 1977, issued May 1988

von Haller A (1755) Onomatologia medica completa. Gaum, Ulm

von Haller A (1757–1766) ElementaPhysiologiae Corporis Humani, vol 8

Walsh E (1910) The Tibetan anatomical system. J R Asiat Soc Great Br Ireland 1215–1245. https://www.jstor.org/stable/25189785. Accessed 28 Sept 2020

Wen R, Nguyen BP, Chng C-B, Chui C-K (2013) In situ spatial AR surgical planning using projector-Kinect system. In Proceedings of the fourth symposium on information and communication technology (SoICT '13). Association for Computing Machinery, New York, NY, USA, pp 164–171. https://doi.org/10.1145/2542050.2542060

Wujastyk D (2009) A body of knowledge: The WellcomeAyurvedic anatomical man and his Sanskrit context, Asian medicine. Brill, Leiden, The Netherlands. https://doi.org/10.1163/157342109X423793

Zorzal ER, Sousa M, Mendes D, dos Anjos RF, Medeiros D, Paulo SF, Rodrigues P, Mendes JJ, Delmas V, Uhl J-F, Mogorrón J, Jorge JA, Lopes DS (2019) Anatomy studio: a tool for virtual dissection through augmented 3D reconstruction. Comput & Graph 85:74–84. https://doi.org/10.1016/j.cag.2019.09.006

Chapter 3
A Tool for Collaborative Anatomical Dissection

**Ezequiel Roberto Zorzal, Maurício Sousa, Daniel Mendes,
Soraia Figueiredo Paulo, Pedro Rodrigues, Joaquim Jorge,
and Daniel Simões Lopes**

Abstract 3D reconstruction from anatomical slices permits anatomists to create three-dimensional depictions of real structures by tracing organs from sequences of cryosections. A wide variety of tools for 3D reconstruction from anatomical slices are becoming available for use in training and study. In this chapter, we present Anatomy Studio, a collaborative Mixed Reality tool for virtual dissection that combines tablets with styli and see-through head-mounted displays to assist anatomists by easing manual tracing and exploring cryosection images. By using mid-air interactions and interactive surfaces, anatomists can easily access any cryosection and edit contours, while following other user's contributions. A user study including experienced anatomists and medical professionals, conducted in real working sessions, demonstrates that Anatomy Studio is appropriate and useful for 3D reconstruction. Results indicate that Anatomy Studio encourages closely coupled collaborations and group discussion, to achieve deeper insights.

E. Roberto Zorzal (✉)
ICT/UNIFESP, Institute of Science and Technology, Federal University of São Paulo, São Paulo, Brazil
e-mail: ezorzal@unifesp.br

M. Sousa · D. Mendes · S. Figueiredo Paulo · J. Jorge · D. S. Lopes
INESC-ID Lisboa, Instituto Superior Técnico, Universidade de Lisboa, Lisbon, Portugal
e-mail: antonio.sousa@tecnico.ulisboa.pt

D. Mendes
e-mail: danielmendes@tecnico.ulisboa.pt

S. Figueiredo Paulo
e-mail: soraiafpaulo@inesc-id.pt

J. Jorge
e-mail: jorgej@tecnico.ulisboa.pt

D. S. Lopes
e-mail: daniel.lopes@inesc-id.pt

P. Rodrigues
Clinical Research Unit (CRU), CiiEM, IUEM, Almada, Portugal
e-mail: prodrigues@egasmoniz.edu.pt

© The Author(s), under exclusive license to Springer Nature Switzerland AG 2021
J.-F. Uhl et al. (eds.), *Digital Anatomy*, Human–Computer Interaction Series,
https://doi.org/10.1007/978-3-030-61905-3_3

3.1 Introduction

The traditional methods for anatomy education involve lectures, text books, atlases, and cadaveric dissections Preim and Saalfeld (2018). Cadaveric dissection plays an essential role for the training of manual dexterity and communication skills (Brenton et al. 2007; Preim and Saalfeld 2018). Also, according to Shaikh et al. (2015), the practice of cadaveric dissection helps students to grasp the three-dimensional anatomy and concept of innumerable variations.

Cadaveric dissection is considered a tool for studying the structural details of the body and the source of teaching material for anatomical education. However, the cadaveric dissection for teaching and training purposes is surrounded by ethical uncertainties (McLachlan et al. 2004; Shaikh et al. 2015). Also, once dissected, the results become irreversible since the surrounding structures are damaged for underlining the target structure. Furthermore, there is a global shortage of cadavers in medical schools for training students and surgeons. According to Shaikh et al. (2015), because of problems related to the use of cadavers, many curricula in anatomy have introduced a shift toward greater use of alternative modalities of teaching involving cadaveric plastination, non-cadaveric models, and computer-based imaging Kerby et al. (2010). To alleviate these problems, innovative technologies, such as 3D printing, Virtual Reality (VR) and Mixed Reality (MR), are becoming available for use. According to Burdea et al. (1996), VR is a high-end user-computer interface that involves real-time simulation and interactions through multiple sensorial channels. Rather than compositing virtual objects and a real scene, VR technology creates a virtual environment presented to our senses in such a way that we experience it as if we were really there. On the other hand, MR refers to the incorporation of virtual objects into a real three-dimensional scene, or alternatively the inclusion of real-world real objects into a virtual environment. VR and MR have been proposed as a technological advance that holds the power to facilitate learning Pan et al. (2006). Also, anatomists and students rely on a wide variety of tools for 3D Reconstruction from Anatomical Slices (3DRAS) from these technologies. These tools suit several purposes: promote novel educational methods (Papa and Vaccarezza 2013; Chung et al. 2016; Zilverschoon et al. 2017), allow statistical analysis of anatomical variability Shepherd et al. (2012), and support clinical practice to optimize decisions Malmberg et al. (2017). It should be noted that 3DRAS tools are a complementary medium to live dissection, not their replacement (Ackerman 1999; Park et al. 2005; Pflesser et al. 2001; Uhl et al. 2006).

3DRAS make possible the virtual dissection resulting in accurate and interactive 3D anatomical models. Due to its digital nature, 3DRAS promote new ways to share anatomical knowledge and, more importantly, produces accurate subject-specific models that can be used to analyze a specific structure, its functionality, and relationships with neighboring structures Uhl et al. (2006).

By default, 3DRAS tools are designed for laborious manual segmentation forcing an expert to trace contours around anatomical structures throughout many sections. Once a set of segmented curves is assembled, it is then possible to reconstruct a

3D organ. Again, we remark that current 3DRAS tools promote single-user slice navigation and manual segmentation. These tasks are often performed using single flat display and mouse-based systems, forcing multiple scrolling and pinpointing mouse clicks. Such limited deployment is the foundation for the work presented in this chapter.

Clearly, this specific application domain presents a situation of limited deployment and underdeveloped usage of mature technologies, namely interactive surfaces and MR that bring high potential benefits. Therefore, we hypothesize that group interaction conveyed through spatial input and interactive surfaces can boost 3DRAS related tasks and attenuate dissection workload. According to Xiang Cao et al. (2008), interactive surfaces allow users to manipulate information by directly touching them, thus enabling natural interaction styles and applications. The ability to interact directly with virtual objects presented on an interactive surface suggests models of interaction based on how we interact with objects in the real world.

In this chapter, we present Anatomy Studio Zorzal et al. (2019), a collaborative MR dissection table approach where one or more anatomists can explore a whole anatomical data set and carry out manual 3D reconstructions. As stated in Mahlasela et al. (2016) and Zamanzadeh et al. (2014), collaboration is essential of work relationships in any profession, as it is through this continuous process that a common vision, common goals, and realities are developed and maintained. Collaboration in the workplace has become a popular research topic since it allows users to get

Fig. 3.1 Overview of Anatomy Studio, a collaborative MR dissection table approach where one or more anatomists can explore anatomical data sets and carry out manual 3D reconstructions using tablets and styli

involved in group activities that not only increase learning, but also produce other benefits, such as the development of relationships and social skills Garcia-Sanjuan et al. (2018).

Anatomy Studio mirrors a drafting table, where users are seated and equipped with head-mounted see-through displays, tablets, and styli. Our approach adopts a familiar drawing board metaphor since tablets are used as sketch-based interfaces to trace anatomical structures, while simple hand gestures are employed for 3D navigation on top of a table, as shown in Fig. 3.1. By using hand gestures combined with mobile touchscreens, the anatomists can easily access any cryosection or 2D contour and follow each user's contribution toward the overall 3D reconstructed model.

3.2 Related Work

Since Höhne and Hanson (1992) presented a pioneering work on computer-assisted anatomy education and the advent of the Visible Human Project Ackerman (1999), interactive solutions have been proposed for virtual dissection, yet still the Windows, Icons, Menus and Pointer (WIMP) paradigm prevails ecumenical for image segmentation within the 3DRAS community (Wu et al. 2013; Fang et al. 2017; Asensio Romero et al. 2018; Chung et al. 2018). More effective approaches are sorely needed as conventional WIMP interfaces are known to hamper 3D reconstruction tasks because they rely on mouse-based input and 2D displays (Olsen et al. 2009; Meyer-Spradow et al. 2009). Besides lacking direct spatial input and affording limited navigation control, WIMP approaches for 3DRAS also promote single-user interaction, even though several studies refer to the importance of collaborative drawing (Lyon et al. 2013; Alsaid 2016) such has not been performed for a strictly 3D reconstruction purpose.

Another serious limitation of WIMP is that they prescribe timely slice-by-slice segmentation. For instance, the Korean Visible Human took 8 years to segment using mouse input (Park et al. 2005; Chung et al. 2015). Clearly, there is a need to speedup the segmentation process without discarding manual operability, as anatomists feel more in control to produce meticulous and informed contours manually (Igarashi et al. 2016; Sanandaji et al. 2016). Another restriction consists of the limited 3D perception offered by WIMP interfaces, as this induces a greater cognitive load by forcing anatomists to build a 3D mental image from a set of 2D cryosections.

Other interaction paradigms have been proposed for 3DRAS, namely, Augmented Reality (AR) and Virtual Reality (VR) have been explored for medical visualization, since immersion can improve the effectiveness when studying medical data Laha et al. (2013). For instance, Ni et al. (2011) developed AnatOnMe, a prototype AR projection-based handheld system for enhancing information exchange in the current practice of physical therapy. AnatOnMe combines projection, photo, and video capture along with a pointing device for input, while projection can be done directly on the patient's body. Another related study proposed the introduction of AR above the Tabletop for the analysis of multidimensional data sets, as their approach facilitated collaboration, immersion with the data, and promoted fluid analyses of the data Butscher et al. (2018). Furthermore, a collaborative learning intervention using AR has been proposed for learning clinical anatomy system Barmaki et al. (2019). The system uses the AR magic mirror paradigm to superimpose anatomical visualizations over the user's body in a large display, creating the impression that she sees the relevant anatomic illustrations inside her own body.

Another advantage of AR and VR paradigms is that they promote expeditious navigation of volumetric data along complex medical data sets. To this regard, Hinckley et al. (1994) adopted two-handed interactions on a tangible object to navigate multiple cutting planes on a volumetric medical data set. Coffey et al. (2012) proposed a VR approach for volumetric medical data sets navigation using an interactive multitouch table and a large-scale stereoscopic display. Sousa et al. (2017) introduced a VR visualization tool for diagnostic radiology. The authors employed a touch-sensitive surface to allow radiologists to navigate through volumetric data sets. Lopes et al. (2018) explored the potential of immersion and freedom of movement afforded by VR to perform CT Colonograpy reading, allowing users to freely walk within a work space to analyze 3D colon data.

Furthermore, the combination of immersive technologies and sketch-based interfaces have been proposed for 3DRAS education and training, but not for accurate 3D reconstruction (Lundström et al. 2011; Teistler et al. 2014; Lu et al. 2017; Saalfeld et al. 2016). Immersive solutions usually place anatomical representations within a 3D virtual space Lu et al. (2017), similarly to plaster models used in the anatomical theater, or consider virtual representations of the dissection table (Lundström et al. 2011; Teistler et al. 2014) but often require dedicated and expensive hardware. Only recently have 3D or VR approaches been considered to assist the medical segmentation process (Heckel et al. 2011; Johnson et al. 2016; Jackson and Keefe 2016) but the resulting models continue to be rough representations of subject-specific anatomy. In turn, sketch-based interfaces have been reported to complement or even finish off automatic segmentation issues that rise during anatomical modeling (Olabarriaga and Smeulders 2001; Malmberg et al. 2017). Although tracing can be guided by simple edge-seeking algorithms or adjustable intensity thresholds, these often fail to produce sufficiently accurate results (Shepherd et al. 2012; van Heeswijk et al. 2016).

Given the size and complexity of the data set, coordinating 3D reconstruction with navigation can be difficult as such tasks demand users to maintain 3D context, by choosing different points of view toward the 3D content, while focusing on a subset of data materialized on a 2D medium. To assist the visualization task, head-tracked stereoscopic displays have proven to be useful due to the increased spatial understanding (Coffey et al. 2012; Hwaryoung Seo et al. Hwaryoung Seo et al.; Lopes et al. 2018). However, prior work has been primarily conducted within navigation scenarios and not for 3D reconstruction from medical images; thus, it is not clear if there are benefits of complementing 3D displays with 2D displays Tory et al. (2006).

Despite the many advancements in medical image segmentation, most semi- and automatic algorithms fail to deliver infallible contour tracing. That is why clinical practice in medical departments is still manual slice-by-slice segmentation, as users feel more in control and produce a more informed, meticulous 3D reconstruction (Igarashi et al. 2016; Sanandaji et al. 2016). Note that, segmentation of cryosections is a labeling problem in which a unique label that represents a tissue or organ is assigned to each pixel in an input image.

Tailored solutions for 3D reconstruction that rely on easily accessible, interactive, and ubiquitous hardware, besides guaranteeing qualified peer-reviewing, are welcomed by the Anatomy community. While using HMDs or tablets to interact with 2D and 3D data is not new, combining them for 3DRAS has not been studied. Much research focuses on VR-based navigation for surgical planning and radiodiagnosis. However, our approach addresses 3D reconstruction. Moreover, we specifically worked with anatomists and our interaction was purposely designed to combine a 2D sketch-based interface for expedite segmentation with spatial gestures for augmented visualization.

3.3   Anatomy Studio

Our approach, Anatomy Studio, combines sketching on a tablet with MR based visualization to perform 3D reconstruction of anatomic structures through contour drawing on 2D images of real cross-sections (i.e., cryosections). While the tablet's interactive surface offers a natural sketching experience, the 3D visualization provides an improved perception of the resulting reconstructed content over traditional desktop approaches. It is also possible to interact with Anatomy Studio using mid-air gestures in the MR visualization to browse throughout the slices. The combination of mid-air input with interactive surfaces allows us to exploit the advantages of each interaction paradigm, as most likely their synergistic combination should overcome the limitations of either modality in isolation, a result well-known from multi-modal interface research. Additionally, Anatomy Studio enables two or more experts to collaborate, showing in real time the modifications made to the contours by each other, and easing communication.

The main metaphor used in Anatomy Studio is the dissection table. Using MR, collaborators can visualize 3D reconstructed structures in real size above the table, as depicted in Fig. 3.1. The content becomes visible to all people around the virtual dissection table who are wearing MR glasses. Also, users can select slices from the common MR visualization to be displayed on their tablet device in order to perform tracing tasks.

In order to support tablet and MR glasses for each user and the collaboration between all participants, Anatomy Studio uses the distributed architecture illustrated in Fig. 3.2. Anatomy Studio was developed using Unity 3D (version 2018.3.8f1), C# programming language for scripting and Meta SDK 2.8. Two applications were developed to run on both device types: Windows-based ASUS T100HA tablets and Meta 2 headsets. The whole data set, comprised 12.2 gigabytes in high-resolution images, as well existing contours already traced, are stored in a Web Server, accessible

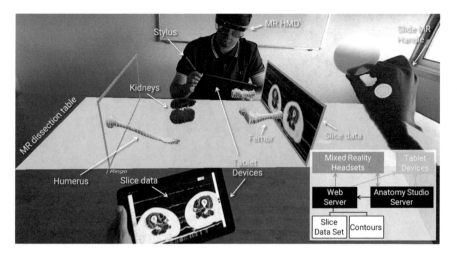

Fig. 3.2 Anatomy Studio's distributed architecture

by all devices in the session. However, to show immediate previews during slice navigation, each device displays thumbnails as slice previews, which consist in low-resolution images. All together, these thumbnails require only 36 megabytes.

Located on the same machine as the Web Server, is the Anatomy Studio server to which all devices connect. While only this server can make changes to the files in the Web Server, such as storing contours, all clients can read from it. The clients, both MR glasses and tablet devices, have an associated user ID so that they can be properly paired between each other. Every time a user changes his active slice or modifies a contour, the client device immediately notifies the server and all other clients through UDP messages.

Existing digitizations of sectioned bodies consist of thousands of slices, each of which with a thickness that can be less than 1mm. As such, Anatomy Studio offers two possible ways to browse the collection of slices: one fast and coarse, useful for going swiftly to a region of the body, and another that allows specific slice selection.

To perform a quick selection of a slice in a region of the body, Anatomy Studio resorts to mid-air gestures. Attached to the frame representing the current slice in the MR visualization, there is a sphere-shaped handle, as depicted in Fig. 3.1, which can be grabbed and dragged to access the desired slice. This allows to switch the current slice for a distant one efficiently. Slices selected by other collaborators are also represented by a similar frame, without the handle, with the corresponding name displayed next to it. To ease collaboration, when dragging the handle and approaching a collaborator's slice, it snaps to the same slice.

The very small thickness of each slice (≤ 1 mm) together with inherent precision challenges of mid-air object manipulation Mendes et al. (2016) makes it difficult to place the MR handle in a specific position to exactly select a desired slice. Thus, Anatomy Studio also provides a scrollable list of slices in the tablet device (Fig. 3.3)

Fig. 3.3 Tracing the contour of a kidney with the stylus on the tablet. On the left pane there is a scrollable list of slices, and the right pane shows the available structures

that only shows a very small subset of 20 slices around the currently selected one. This list is constantly synced with the MR handle and, after defining a region, users are able to unequivocally select a specific slice. Of course, due to the high number of slices, this scroll alone was not feasible to browse the whole data set, and needs to be used in conjunction with our Fast Region Navigation approach. In addition, slices' numbers are accompanied with the name of the collaborators that have them currently selected, which makes them reachable by a single tap. In Anatomy Studio only coarse slice selection is done in mid-air, as more precise slice selection is performed through the tablet device.

To provide a natural experience that fashions sketching on paper with a pen, Anatomy Studio offers anatomists a tablet device and a stylus.

To ease the tracing process, the image can be zoomed in and out, to provide both overall and detailed views, as well as translated and rotated, using the now commonplace Two-Point Rotation and Translation with scale approach Hancock et al. (2006). After each stroke is performed, either to create or erase contours, Anatomy Studio promptly propagates the changes to the MR visualization making them available to all collaborators. It also re-computes the structure's corresponding 3D structure according to the new information, offering a real-time 3D visualization of the structure being reconstructed.

We implemented a custom 3D reconstruction algorithm that uses the strokes created by the users to recreate an estimated three-dimensional mesh of a closed 3D model. Each time a user changes the drawing made on a certain slice, a localized reconstruction process is initiated that comprises 3 steps: (1) Contouring can be performed by inputting smaller strokes; (2) The algorithm then iterates through the line to find the extreme points, which will help iterate through the line during reconstruction; and (3) A mesh is finally created by connecting two closed lines from neighboring slices.

Therefore, each individual triangle is created so the normal vectors are coherently oriented to the outside of the final 3D model. By applying this simple process to each pair of neighboring lines, we can create a complete closed 3D model in real time, so alterations can be immediately reflected on the 3D augmented space.

3.4   Evaluation

Our main goal was to assess whether collaborative tools such as Anatomy Studio can provide viable alternatives to current methods, and whether these would be well received by the medical community, focusing on qualitative valuations rather than basic performance metrics.

To assess whether Anatomy Studio can be used as a mean to enable collaboration and aid in the process of anatomical 3D reconstruction, we conducted a user study with experienced anatomists and medical professionals. To this end, we resorted to a data set that consists of serial cryosection images of the whole female body from the Visible Korean Project Seok Park et al. (2015).

For testing our prototype we used two Meta2 optical see-through head-mounted displays to view the augmented content above the table. We used this device mainly because of its augmented 90 degree field of view, which facilitates the visualization and interaction with the augmented body being reconstructed. We used the Meta2 headsets to perform the interaction in the environment, as they possess an embedded depth camera similar to the Microsoft Kinect or the Leap Motion that, besides tracking the headset position and orientation, also track users' hands and fingers, detecting their position, orientation, and pose. Each of the MR glasses was linked to a PC with dedicated graphics card. We also used one Windows-based ASUS T100HA tablet with a 10 inch touch screen and an Adonit Jot Pro stylus for each participant. An additional Microsoft Kinect DK2 was used recording video and audio of the test session for further evaluation.

Participants were grouped in pairs, seating at a table, facing each other as shown in Fig. 3.4. Each was equipped with an optical see-through head-mounted display, a tablet, and a stylus. Firstly, researchers outlined the goals of the session and provided an introduction to the prototype. Prior to start, participants were asked to fill a demographic questionnaire, regarding their profile information and previous experience with the tested technologies (MR glasses, virtual dissection applications, and multi-touch devices), as well as an informed consent. A calibration process was performed to enable each headset to locate the virtual objects in real space.

Then, both participants were instructed to perform a training task, individually, where they were free to interrupt and ask questions whenever they deemed necessary. This was followed by the test task, in which participants where asked to collaborate to achieve the final result. Both tasks were based on reconstructing different anatomical

Fig. 3.4 A pair of participants during a user evaluation session

structures using sketches. To prevent excessively long sessions, both the solo training task and the collaborative test task were limited to 15 min. Participants were then asked to fulfill a questionnaire about their user experience. Finally, we conducted a semi-structured interview in order to gather participants opinions, suggestions and to clarify the answers obtained from the questionnaires.

We conducted usability testing and evaluated our prototype with ten participants (one female), eight of which were medical professionals and two were medical students, recruited during an international congress on Digital Anatomy using a convenience sampling strategy. Participants' ages varied between 23 and 69 years old ($\overline{x} = 43.6$, $s = 19.5$). Having this particular sample size also ensured that we met recommended minimum criteria for usability evaluation of the intervention. According to Faulkner (2003), in a group of ten people, 82–94,6% of usability problems will be found. Participants who have evaluated our prototype are domain experts, have worked for a long time and have many years of experience. Because of this expertise, the expert is a trusted source of valuable information about the topic and the domain (Costabile t al. 2003; Caine 2016; Sousa et al. 2017; Akram Hassan et al. 2019).

Among the professionals, four were radiologists (with an average of five years of experience), one neurologists, one surgeon, one dental surgeon, and one internist with 27 years of experience.

The majority (80%) were familiarized with touch screen devices, but 70% reported having no prior experience with optical see-through MR technology. Five participants stated to perform virtual dissections, four of them on a daily basis.

3.5   Results and Discussion

We reviewed the usability testing videos and observed that users behaved in three ways when they were focusing on the MR environment. Figure 3.5 shows the total time used in each session and the general percentage of interaction modes identified. We assign a tablet time for tablet interaction mode when a user focuses on device usage. MR interaction mode when a user focuses or explores in the MR environment. As well, when a user interacts with the MR environment using his hands. Finally, a collaboration time for collaboration when a user interacts with other participants through conversation.

Also, we assessed user preferences and experience through a questionnaire with a list of statements for participants to score on a 6-point Likert Scale (6 indicates full agreement). Our evaluation with medical experts suggests that MR combined with tablets can be a viable approach to overcome existing 3DRAS issues. This chapter presents the summarized evaluation results, more details are available in Zorzal et al. (2019).

Regarding the overall prototype, the participants found it easy to use and, in particular, considered the combination of MR and tablet sliders to function well together. They also considered that the tablet's dimensions were appropriate for the tasks performed, and that contouring using a stylus was an expedite operation. Participants that perform virtual dissections professionally found it easier to segment slices using Anatomy Studio when compared to the mouse-based interface they are acquainted to. All participants remarked that Anatomy Studio is a viable alternative to conventional virtual dissection systems. Using AR, we are able to show that a virtual surface on top of each body's reconstructed structures are rendered volumetrically in full size, as depicted in Fig. 3.6, visible for all collaborators around it, provided that they are properly equipped with AR glasses. Also, users can choose slices in the AR visualization, in order for them to be shown on the tablet device and to be sketched

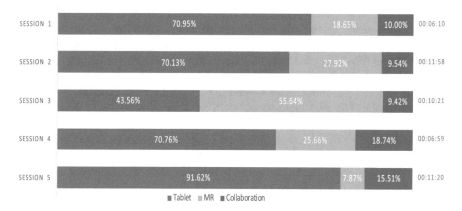

Fig. 3.5 Total time used in each session and the general percentage of interaction modes

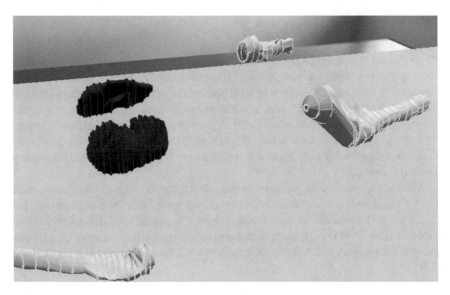

Fig. 3.6 Detail of the AR volumetric rendering above the table, showing lines and the corresponding reconstructed volumes of two kidneys (in burgundy) and three bones (in beige)

upon. They also noted that the visual representations of the 3D model and the slices above the virtual table are appropriate for anatomical study. The participants agreed that the 3D model overview allowed them to rapidly identify and reach anatomical locations. Furthermore, the augmented 3D space created a shared understanding of the dissection tasks and promoted closely coupled collaboration and face-to-face interactions.

We also gathered observational notes taken during evaluation sessions and transcripts of recorded semi-structured interviews, in order to obtain participants' opinions, suggestions, and to clarify the answers from the questionnaires. Participants stated that Anatomy Studio is adequate to "distinguish the several structures" and "understand the spatial relation between [them]". Therefore, "[with tools like Anatomy Studio] we do not need a corpse to learn anatomy". Notwithstanding, "virtual is different from cadaveric material, because we do not have the feeling of cutting tissue". Lastly, the collaborative capabilities of Anatomy Studio were praised, since "working in groups is more effective because, as medics, the experience counts a lot to do a better job, and there should be a mixture of experiences during these sections".

Overall, participants work daily alone and rarely collaborate. Participants said that collaboration offered an equal opportunity to share ideas. Assisted in understanding and respecting diversity better, make team-focused decisions leading the team to a swift achievement of a common goal. The most observed benefit of collaboration was of the less time spent to complete a task.

Also, the participants mentioned some challenges. Two participants said that the stylus contour was very thick and made it difficult for the task. Another mentioned that they had to adapt to the orientation of the drawing presented on the tablet, because the orientation in the computed tomography image is so that the anterior is on top, posterior is bottom, left of the patient is on the right side of the image and the right is on the left side of the image. One participant reported that initially, Anatomy Studio seemed complex because it has many gadgets. Another suggestion mentioned by two participants is the need for prior training to get accustomed to the environment of MR. Another participant mentioned with although the virtual does provide a good interaction, the experience is not identical to that of the real body. In a real body we can feel the difference through touch and cutting the tissues.

The advantage of using technological tools for teaching anatomy is that, in addition to the static figure, one can also understand and demonstrate the dynamics of movement. However, there are challenges to be explored. These challenges limit the actual use of these applications in the routine of health professionals and the transfer of this technology to the productive sector; on the other hand, these challenges create opportunities for research and development.

A significant challenge in the area is to make applications that offer realistic simulations of anatomical features. It is interesting to develop techniques that improve user perception, tactile sensitivity and spatial correlation between physical and virtual objects. Furthermore, Periya and Moro (2019) expressive finger-gestures may assist in identifying comparisons between scans, or unique anatomical variations and features when compared to using a mouse-and-keyboard approach. Also, introducing new teaching approaches in traditional culture is a current challenge for the applications that work in the area of health education.

3.6 Lessons Learned

The lessons learned serve as a valuable tool for use by other researchers and developers who are assigned related projects. These lessons may be used as part of new project planning in order to present what main guidelines for the development of tool for collaborative anatomy. The following lists the lessons learned for the Anatomy Studio.

- **Combined approaches**: Mobile devices such as tablets bring the potential of MR into every learning and collaborative environment Birt et al. (2018). The self-directed approach allowed by MR can enhance experiential learning, engagement, while tackling challenging content in both medical practice and health sciences. In addition, previous research Iannessi et al. (2018) reported that MR allows for better visualization of 3D volumes regarding the perception of depth, distances, and relations between different structures. Accordingly, we chose to follow these approaches, because when comparing MR through an HMD with a virtual window through a tablet, the first is more practical and natural, provides stereoscopic visualization, and can be easily combined with a tablet for 2D tasks, where these devices excel.

- **Transmission and display improvements**: Although medical applications require high-resolution images with realistic features, good practice in collaborative applications is to display instant thumbnails during slice navigation. Also, to avoid network congestion and even decrease throughput, shared virtual environments should change only when there is a change considered significant. We consider the UDP protocol a suitable choice for real-time data streams since no connection is created and only direct data is sent.
- **Slice browsing**: Due to the thickness of the slices and the precision challenges inherent in handling objects in the air, we consider two possible ways to browse the collection of slices: fast region navigation, useful forgoing swiftly to a region of the body, and another precise slice selection that allows specific slice selection.
- **Real-time 3D visualization**: After each stroke is performed, either to create or erase contours, Anatomy Studio promptly propagates the changes to the MR visualization making them available to all collaborators. It also re-computes the structure's corresponding 3D structure according to the new information, offering a real-time 3D visualization of the structure being reconstructed.
- **Interaction and collaboration**: Users behaved in three ways when they were focusing on the MR environment: (i) MR preview when the user raised his head and looked at the environment, (ii) MR exploration when the user analyzed the environment moving the head or body to different directions and kept a fixed eye on the environment of MR content, and (iii) MR interaction when the user interacted with the environment using his hands. Also, participants did use collaborative conversation to complete the task. This ability is an outcome-driven conversation aimed at building on each other's ideas and a solution to a shared problem.

3.7 Conclusions

In this chapter, we presented a collaborative MR dissection table where one or more anatomists can explore large data sets and perform expedite manual segmentation.

Collaborative MR systems are either visualization-based or systems in which users can create and modify a 3D model collaboratively in a 3D space. Our results show that collaborative virtual dissection is feasible supporting two tablets, and has the potential to scale to more simultaneous collaborators, whereby users that can choose the slice to trace on simultaneously, thus contributing to mitigating the reconstruction workload. Moreover, our approach provides for a portable and cost-effective 3DRAS tool to build anatomically accurate 3D reconstructions even for institutions that do not have the possibility to perform actual dissections on real cadavers. Our results illustrate the perceived potential of the approach, and its potential to motivate novel developments. Furthermore, all test sessions involved real drawing tasks, in a realistic setting, where participants were asked to build a 3D reconstruction of an anatomical structure as best as they (anatomists) could.

While the work presented in this chapter represents a first step toward MR for virtual dissection, as future work, we intend to conduct a comprehensive user evaluation with non-experienced students, to compare the learning curve and the ease of use of an iterated version of Anatomy Studio against the most common approaches to 3DRAS.

According to de Souza Cardoso et al. (2020), a crucial aspect to consider is that the selected visualization device should be ergonomic and should not limit or increase the necessary movement required to execute the main activity. Although fatigue, stress tests and cognitive load are important variables to understand the limitations of the proposed system, they were not considered in this chapter, as the focus of our work was to explore the potential of Anatomy Studio as an MR system to perform virtual dissection through sketches by enabling collaboration between multiple users. We intend to study such variables in the near future. While the work presented is exploratory, we see it as the precursor to a new generation of collaborative tools for anatomical applications.

Acknowledgements The first author would like to thank the São Paulo Research Foundation (FAPESP) for support of this research: grant#2018/20358-0, FAPESP. This work was supported by national funds through FCT, Fundação para a Ciência e a Tecnologia, under project UIDB/50021/2020. The authors would like to thank the Champalimaud Foundation for its collaboration and support in the development of the study.

References

Ackerman MJ (1999) The visible human project: a resource for education. Acad Medic: J Assoc Amer Med Colleges 74(6):667–670

Alsaid B (2016) Slide shows vs graphic tablet live drawing for anatomy teaching. Morphologie 100(331):210–215

Asensio Romero L, Asensio Gómez M, Prats-Galino A, Juanes Méndez JA (2018) 3d models of female pelvis structures reconstructed and represented in combination with anatomical and radiological sections. J Med Syst 42(3):37

Barmaki R, Kevin Yu, Pearlman R, Shingles R, Bork F, Osgood GM, Navab N (2019) Enhancement of anatomical education using augmented reality: an empirical study of body painting. Anatomical Sci Educ 12(6):599–609

Birt J, Stromberga Z, Cowling M, Moro C (2018) Mobile mixed reality for experiential learning and simulation in medical and health sciences education. Information 9(2):1–14

Brenton H, Hernandez J, Bello F, Strutton P, Purkayastha S, Firth T, Darzi A (2007). Using multimedia and web3d to enhance anatomy teaching. Comput Educ 49(1):32–53. Web3D Technologies in Learning, Education and Training

Burdea G, Richard P, Coiffet P (1996) Multimodal virtual reality: input-output devices, system integration, and human factors. Int J Human-Comput Inter 8(1):5–24

Butscher S, Hubenschmid S, Müller J, Fuchs J, Reiterer H (2018) Clusters, trends, and outliers: how immersive technologies can facilitate the collaborative analysis of multidimensional data. In: Proceedings of the 2018 CHI conference on human factors in computing systems, p 90. ACM

Caine K (2016) Local standards for sample size at chi. In: Proceedings of the 2016 CHI conference on human factors in computing systems, CHI '16, pp 981–992, New York, NY, USA. ACM

Cao X, Wilson AD, Balakrishnan R, Hinckley K, Hudson SE (2008) Shapetouch: Leveraging contact shape on interactive surfaces. In: 2008 3rd IEEE international workshop on horizontal interactive human computer systems, pp 129–136

Chung BS, Chung MS, Park HS, Shin B-S, Kwon K (2016) Colonoscopy tutorial software made with a cadaver's sectioned images. Ann Anatomy - Anatomischer Anzeiger, 19–23

Chung BS, Chung MS, Shin B-S, Kwon K (2018) Three software tools for viewing sectional planes, volume models, and surface models of a cadaver hand. J Korean Med Sci 33(8):e64

Chung BS, Shin DS, Brown P, Choi J, Chung MS (2015) Virtual dissection table including the visible korean images, complemented by free software of the same data. Int J Morphol 33(2):440–445

Coffey D, Malbraaten N, Le TB, Borazjani I, Sotiropoulos F, Erdman AG, Keefe DF (2012) Interactive slice wim: Navigating and interrogating volume data sets using a multisurface, multitouch vr interface. IEEE Trans Visualizat Comput Graphics 18(10):1614–1626

de Souza Cardoso LF, Mariano FCQ, Zorzal ER (2020) A survey of industrial augmented reality. Computers Industr Eng 139:106159

Fang B, Yi W, Chu C, Li Y, Luo N, Liu K, Tan L, Zhang S (2017) Creation of a virtual anatomy system based on chinese visible human data sets. Surgical Radiol Anatomy 39(4):441–449

Faulkner L (2003) Beyond the five-user assumption: benefits of increased sample sizes in usability testing. Behav Res Methods Instrum Compute 35(3):379–383

Garcia-Sanjuan F, Jurdi S, Jaen J, Nacher V (2018) Evaluating a tactile and a tangible multi-tablet gamified quiz system for collaborative learning in primary education. Comput Educ 123:65–84

Hancock MS, Carpendale MST, Vernier F, Wigdor D, Shen C (2006) Rotation and translation mechanisms for tabletop interaction. In: Tabletop, pp 79–88

Hassan KA, Liu Y, Besanãon L, Johansson J, Rännberg N (2019) A study on visual representations for active plant wall data analysis. Data 4(2)

Heckel F, Konrad O, Hahn HK, Peitgen H-O (2011) Interactive 3d medical image segmentation with energy-minimizing implicit functions. Comput Graph 35(2):275–287

Hinckley K, Pausch R, Goble JC, Kassell NF (1994) Passive real-world interface props for neurosurgical visualization. In: Proceedings of the SIGCHI conference on Human factors in computing systems, pp 452–458. ACM

Höhne KH, Hanson WA (1992) Interactive 3d segmentation of mri and ct volumes using morphological operations. J Comput Assis Tomogr 16(2):285–294

Iannessi A, Marcy P-Y, Clatz O, Bertrand A-S, Sugimoto M (2018) A review of existing and potential computer user interfaces for modern radiology. Insights into Imaging 9(4):599–609

Igarashi T, Shono N, Kin T, Saito T (2016) Interactive volume segmentation with threshold field painting. In: Proceedings of the 29th Annual symposium on user interface software and technology, UIST '16, pp 403–413, New York, NY, USA. ACM

Jackson B, Keefe DF (2016) Lift-off: using reference imagery and freehand sketching to create 3d models in vr. IEEE Trans Visualizat Comput Graph 22(4):1442–1451

Johnson S, Jackson B, Tourek B, Molina M, Erdman AG, Keefe DF (2016) Immersive analytics for medicine: Hybrid 2d/3d sketch-based interfaces for annotating medical data and designing medical devices. In: Proceedings of the 2016 ACM companion on interactive surfaces and spaces, ISS Companion '16, pp 107–113, New York, NY, USA. ACM

Kerby J, Shukur ZN, Shalhoub J (2010) The relationships between learning outcomes and methods of teaching anatomy as perceived by medical students. Clinical Anatomy 24(4):489–497

Laha B, Bowman DA, Schiffbauer JD (2013) Validation of the mr simulation approach for evaluating the effects of immersion on visual analysis of volume data. IEEE Trans Visualiz Comput Graph 19(4):529–538

Lopes DS, Medeiros S, Paulo SF, Borges PB, Nunes V, Mascarenhas V, Veiga M, Jorge JA (2018). Interaction techniques for immersive ct colonography: a professional assessment. In: Frangi AF, Schnabel JA, Davatzikos C, Alberola-López C, Fichtinger G (eds) Medical image computing and computer assisted intervention—MICCAI 2018, pp 629–637, Cham. Springer International Publishing

Lundström C, Rydell T, Forsell C, Persson A, Ynnerman A (2011) Multi-touch table system for medical visualization: application to orthopedic surgery planning. IEEE Trans Visualiz Comput Graph 17(12):1775–1784

Lu W, Pillai S, Rajendran K, Kitamura Y, Yen C-C, Yi-Luen Do E (2017) Virtual interactive human anatomy: Dissecting the domain, navigating the politics, creating the impossible. In: Proceedings of the 2017 CHI conference extended abstracts on human factors in computing systems, CHI EA '17, pp 429–432, New York, NY, USA. ACM

Lyon P, Letschka P, Ainsworth T, Haq I (2013) An exploratory study of the potential learning benefits for medical students in collaborative drawing: creativity, reflection and 'critical looking'. BMC Med Educ 13(1):86

Malmberg F, Nordenskjöld R, Strand R, Kullberg J (2017) Smartpaint: a tool for interactive segmentation of medical volume images. Comput Methods Biomech Biomed Eng Imaging Visualiz 5(1):36–44

McLachlan JC, Bligh J, Bradley P, Searle J (2004) Teaching anatomy without cadavers. Med Educ 38(4):418–424

Mendes D, Relvas F, Ferreira A, Jorge J (2016) The benefits of dof separation in mid-air 3d object manipulation. In: Proceedings of the 22nd ACM conference on virtual reality software and technology, pp 261–268. ACM

Meyer-Spradow J, Ropinski T, Mensmann J, Hinrichs K (2009) Voreen: a rapid-prototyping environment for ray-casting-based volume visualizations. IEEE Comput Graph Appl 29(6):6–13

M-Francesca Costabile D, Letondal FC, Mussio P, Piccinno A (2003) Users domain-expert, their needs of software development. In: HCI, (2003) End user development session, proceedings of the HCI 2003 end user development session. Crète, Greece

Ni T, Karlson AK, Wigdor D (2011) Anatonme: facilitating doctor-patient communication using a projection-based handheld device. In: Proceedings of the SIGCHI conference on human factors in computing systems, pp 3333–3342. ACM

Olabarriaga SD, Smeulders AWM (2001) Interaction in the segmentation of medical images: a survey. Medical Image Anal 5(2):127–142

Olsen L, Samavati FF, Sousa MC, Jorge JA (2009) Sketch-based modeling: a survey. Comput Graphi 33(1):85–103

Pan Z, Cheok AD, Yang H, Zhu J, Shi J (2006) Virtual reality and mixed reality for virtual learning environments. Comput Graph 30(1):20–28

Papa V, Vaccarezza M (2013) Teaching anatomy in the xxi century: New aspects and pitfalls. Scient World J (Article ID 310348):5

Park HS, Choi DH, Park JS (2015) Improved sectioned images and surface models of the whole female body. Int J Morphol 33:1323–1332

Park JS, Chung MS, Hwang SB, Lee YS, Har D-H, Park HS (2005) Visible korean human: Improved serially sectioned images of the entire body. IEEE Trans Med Imaging 24(3):352–360

Periya SN, Moro C (2019) Applied learning of anatomy and physiology: virtual dissectiontables within medical and health sciences education. Bangkok Medical J 15(1):121–127

Pflesser B, Petersik A, Pommert A, Riemer M, Schubert R, Tiede U, Hohne KH, Schumacher U, Richter E (2001) Exploring the visible human's inner organs with the voxel-man 3d navigator. Stud, Health Technol Inform pp 379–385

Preim B, Saalfeld P (2018) A survey of virtual human anatomy education systems. Comput Graph 71:132–153

Rakhudu Mahlasela A, Mashudu D-M, Ushanatefe U (2016) Concept analysis of collaboration in implementing problem-based learning in nursing education. Curationis 39(1)

Saalfeld P, Stojnic A, Preim B, Oeltze-Jafra S (2016) Semi-immersive 3d sketching of vascular structures for medical education. In: VCBM, pp 123–132

Sanandaji A, Grimm C, West R, Parola M (2016) Where do experts look while doing 3d image segmentation. In: Proceedings of the ninth biennial acm symposium on eye tracking research & applications, ETRA '16, pp 171–174, New York, NY, USA. ACM

Seo JH, Smith BM, Cook ME, Malone ER, Pine M, Leal S, Bai Z, Suh J (2017) Anatomy builder vr: Embodied vr anatomy learning program to promote constructionist learning. In: Proceedings of the 2017 CHI conference extended abstracts on human factors in computing systems, CHI EA '17, pp 2070–2075, New York, NY, USA. ACM

Shaikh ST (2015) Cadaver dissection in anatomy: tethical aspect. Anatomy Physiol s5

Shepherd T, Prince SJD, Alexander DC (2012) Interactive lesion segmentation with shape priors from offline and online learning. IEEE Trans Med Imaging 31(9):1698–1712

Sousa M, Mendes D, Paulo S, Matela N, Jorge J, Lopes DS (2017) Vrrrroom: Virtual reality for radiologists in the reading room. In: Proceedings of the 2017 CHI conference on human factors in computing systems, pp 4057–4062. ACM

Teistler M, Brunberg JA, Bott OJ, Breiman RS, Ebert LC, Ross SG, Dresing K (2014) Understanding spatial relationships in us: a computer-based training tool that utilizes inexpensive off-the-shelf game controllers. RadioGraphics 34(5):1334–1343 PMID: 25110963

Tory M, Kirkpatrick AE, Atkins MS, Moller T (2006) Visualization task performance with 2d, 3d, and combination displays. IEEE Trans Visualiz Comput Graph 12(1):2–13

Uhl JF, Plaisant O, Ami O, Delmas V (2006) La modelisation tridimensionnelle en morphologie: Méthodes, intérêt et resultants. Morphologie 90(288):5–20

van Heeswijk MM, Lambregts DMJ, van Griethuysen JJM, Oei S, Rao S-X, de Graaff CAM, Vliegen RFA, Beets GL, Papanikolaou N, Beets-Tan RGH (2016) Automated and semiautomated segmentation of rectal tumor volumes on diffusion-weighted mri: Can it replace manual volumetry? Int J Radiat Oncol Biol Phys 94(4):824–831

Wu Y, Luo N, Tan L, Fang B, Li Y, Xie B, Liu K, Chu C, Li M, Zhang S (2013) Three-dimensional reconstruction of thoracic structures: based on chinese visible human. Comput Math Methods Medic 2013(Article ID 795650):7

Zamanzadeh V, Irajpour A, Valizadeh L, Shohani M (2014) The meaning of collaboration, from the perspective of iranian nurses: a qualitative study. Scient World J 1–9:2014

Zilverschoon M, Vincken KL, Bleys RLAW (2017) Creating highly detailed anatomy models for educational purposes. The virtual dissecting room. J Biomed Inform 65:58–75

Zorzal ER, Sousa M, Mendes D, dos Anjos RK, Medeiros D, Paulo SF, Rodrigues P, Mendes J, Delmas V, Uhl J-F, Mogorrón J, Jorge JA, Lopes DS (2019) Anatomy studio: a tool for virtual dissection through augmented 3d reconstruction. Comput Graph

Chapter 4
3D Modeling from Anatomical and Histological Slices: Methodology and Results of Computer-Assisted Anatomical Dissection

Jean-François Uhl and Maxime Chahim

Abstract Computer-Assisted Anatomical Dissection (CAAD) is a new method of 3D reconstruction of anatomical structures from histological or anatomical slices. It uses staining and immunomarking of the tissues for a more precise identification, in particular for the nerves and the vessels, leading to an easy morphological segmentation. Starting from a digitalized series of transverse histological sections, we perform a staining by immune-markers (protein S100, VAChT and D2-40), then alignment of the slices and finally a manual segmentation of the main anatomical structures by using the Winsurf® software version 3.5. This chapter shows the results of CAAD in embryology and for the pelvic nerves in adults. Its main interest is in the field of pelvic surgery for cancer, to improve the knowledge of the pelvic nervous anatomy and preserve the inferior hypogastric plexus. It is also an original method to provide 3D reconstruction of the human embryo, and so bring us an improved understanding of embryogenesis.

J.-F. Uhl (✉) · M. Chahim
Department of Anatomy, Paris University, Paris, France
e-mail: jeanfrancois.uhl@gmail.com

J.-F. Uhl
Scientific Director - Unesco Chair of Digital Anatomy - Paris University, Paris, France

M. Chahim
Service de médecinevasculaire HEGP, Paris 75015, France

© The Author(s), under exclusive license to Springer Nature Switzerland AG 2021   59
J.-F. Uhl et al. (eds.), *Digital Anatomy*, Human–Computer Interaction Series,
https://doi.org/10.1007/978-3-030-61905-3_4

4.1 Introduction

The principle of 3D reconstruction from a series of transverse slices has first been described by Born in 1883 (Born 1883; Pillet et al. 1995). Born, Gustav Jacob, is a German embryologist (1851–1900). The principle of the Born method is the making of three-dimensional models of structures from serial sections; it depends on the building up of a series of wax plates, cut out to scaled enlargements of the individual sections involved in the region to be reconstructed.

Figure 4.1 shows an example of 3D reconstruction of the inner ear and Fig. 4.2 an embryo's brain done by Eyries in 1954.

Modern 3D anatomical modeling uses the same principle but enhanced by the power of digital techniques and computerized reconstruction thanks to the CAD software.

In fact, clinical research techniques in morphology have the following limitations: the small size of the specimen (in embryology) and to the lack of determination of the nature of the nerve fibers (cholinergic or adrenergic).

The association of 3D vectorial modeling reconstruction and immunostaining of the nerve solves these limitations. The name of this original method is "computer assisted anatomical dissection" (CAAD). First published by Yucel in 2004 (Yucel and Baskin 2004), it has then been developed by our research unit (URDIAEA4465) in collaboration with the experimental laboratory of surgery, Paris XI University (Pr G. Benoît).

The aim of this chapter is to describe the methodology of the CAAD technique, its main results and its interest in the field of pelvic surgery, avoiding the main complications by the preservation of the nerves. This technique takes advantage

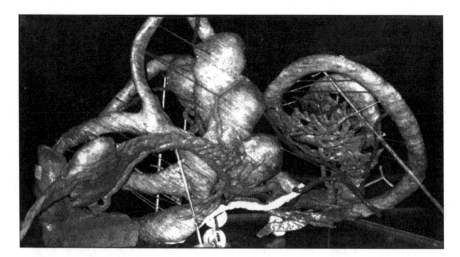

Fig. 4.1 3D Reconstruction by Born method of the inner ear of a 135 mm human embryo by Eyriès (X 200) from former museum Delmas-Orphila-Rouvière (Paris University)

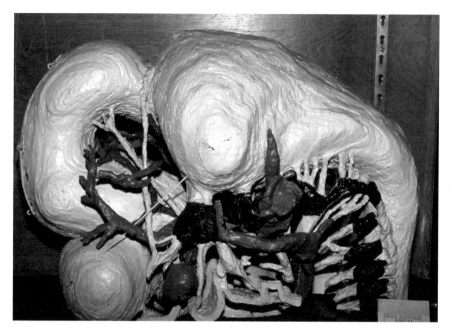

Fig. 4.2 3D Reconstruction by Born method of the brain of a 9 mm human embryo by Eyriès, 1954, n° 721 (X100) from former museum Delmas-Orphila-Rouvière (Paris University)

of advanced human–computer interface for building the 3D models as well as for displaying the complex segmented anatomy.

4.2   Materials and Methods

The four main steps of this technique are: performing thin slices, staining by immune-markers, alignment and segmentation. A series of transverse sections of anatomical structures of fetus or adult are at the basis of this method. The ideal situation is to make sections perpendicularly to the main axis of the reconstructed structures. The thickness of the sections is 5–10 μ after inclusion into a paraffin block of the histological sample, previously fixed with formaldehyde.

Then, the histological sections are stained by immunomarkers (Table 4.1).

This immunohistochemical staining technique is specific of the different tissues and makes possible an immediate identification of the different types of nerves and vessels, leading to an easy morphological segmentation. The *main markers* used are the following:

1. *Neuronal immuno-labelling* (Rabbit anti-cow protein S100): is a general immunomarker which is considered as a marker for Schwann cells (Stefansson et al. 1982). The antibody used (Code No. Z0311) is diluted to 1/400. An example

Table 4.1 Main markers used for the CAAD technique

Marker	Type	Ref	Tissue
HES	Classical staining	Hematin- Eosin- Safran	Bone muscles
Masson trichrome	Classical staining		Connective
S100	Immunomarker	Rabbit anti-cow protein	Neuronal
TH	Immunomarker	Rabbit anti-rat tyrosine hydroxylase	Adrenergic fiber
VAChT	Immunomarker	Rabbit anti-rat vesicular acetylcholine transporter	Cholinergic fibers
D2-40	Immunomarker		Endothelial cells vessels
Prox-1 &VEGFR3	Immunomarker	*Prox-1 and VEGFR3 Antibodies*	Lymphatics

of the obtained result is shown on a lower limb of a human embryo of 13 weeks (Fig. 4.3).

2. *Adrenergic fiber immunolabeling* (Rabbit anti-rat tyrosine hydroxylase (TH)): is employed as a marker for the adrenergic activity and has been considered as an adrenergic fiber marker (Lewis et al. 1987). The antibody used (Code No. ab112) is diluted to 1/750

3. *Cholinergic fiber immunolabeling* (Rabbit anti-rat vesicular acetylcholine transporter (VAChT)): was considered as specific marker for the cholinergic neurons and fibers (Usdin et al. 1995). The antibody used (Code No. V 5387) is diluted to 1/2000

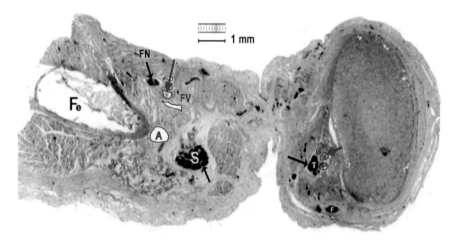

Fig. 4.3 Immunomarking of the nerves using protein S100 (slice # 535 from 15-week-old fetal right limb). The nerves are shown by black arrows. S = sciatic nerve, T = tibial nerve, F = fibular nerve. FN = femoral nerve (black arrow). Veins are outlined in blue. P = popliteal vein, FV = femoral vein plexus, A = axial vein. Arteries are shown by red arrows (outlined in red) Fe = femur bone

4. *Vascular immunolabeling* (*D240* specific of the vessels) D2-40 is a marker of endothelial cells (Ren et al. 2011), created by the production of antibodies in cloned mice (Code No. M3619, DAKO, Denmark after dilution 1/10 at pH6).
5. *Prox-1 and VEGFR3 Antibodies* are specific of the lymphatics (Eumenia Costa da Cunha Castro and CsabaGalambos 2011).
6. In addition, *classical staining* is also used for the surrounding tissues:
   Hematin-Eosin-Safran (HES): considered as reference section.
   Masson trichrome: is suited for connective tissue (Schaefer 1957).

These staining techniques are alternatively used on the series of thin slices (5 μ) for example we can divide the sections in blocks of 10. In each block of 10 slices, the section number one will be stained by Masson trichrome, the number 2 by HES, number 3 by ProtS100, number 4 by TH, number 5 by VaChT the number 6 by D240 and the number 7 by Prox-1 and VEGFR3 Antibodies. The three last slices are left blank for further studies.

In fact, the 3D reconstruction is possible because the structures are almost at the same level with a negligible interval of 5 μ between the slices. The main trick of this technique is to finally gather the different elements, segmented separately with the different markers, to obtain the full 3D model.

Then the important step is to align the stained sections. This is mandatory to allow an accurate 3D reconstruction of the anatomical elements.

This could be done manually but this is time consuming and tedious. The technique uses the transparent layers of Photoshop® software and the magnetic lasso tool to easily draw boundaries (Park et al. 2007, 2005). We could also use fiducial markers (like small brass wires) to be included into the paraffin block to help and automatic or semi-automatic alignment of the slices. But indeed the FIJI software is the best solution to automatize the alignment of a big number of slices. It is an open source Java image processing program inspired by NIH Image derivate from imageJ software (Schneider et al. 2012). It uses a special plugin for realignment: RVSS (Register Virtual Stack Slices).

The next step is to perform the manual segmentation of each anatomical stained structure. For this purpose, we used *Winsurf® software v 3.5* (Bardol et al. 2018; Lozanoff et al. 1988). This is done manually by outlining each anatomical structure slice by slice. The segmentation tools used in Winsurf software are shown in Fig. 4.4a. When finished, the software provides objects as a 3D mesh in a separate window (Fig. 4.4b)

Finally, we just have to gather the 3D objects of the different anatomical elements by importing them into the CAD software to obtain the whole reconstruction of the 3D anatomy.

The right window shows the 3D surface rendered reconstruction of the leg muscles.

– Winsurf software has no option for exportation of the whole 3D meshes, thus the exportation of the 3D mesh created by Winsurf® uses the following steps:
– Import of 3D window menu option of Acrobat 3Dv.8.
– exportation with Acrobat in WRLM format
– Conversion of the wrml file into 3DS or OBJ format.

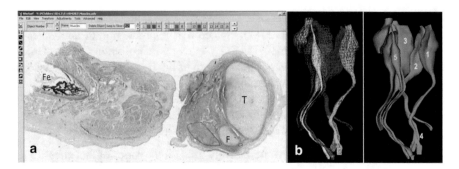

Fig. 4.4 Interface of the WinSurf™ software version 3.5 with segmentation of the muscles by drawing the boundaries on a series of slices. **a** The left window contouring the muscles on a slice, the software tools (pen, eraser, zoom…) are on the left side and the menu options on the top. T = tibia bone, F = fibula, Fe = femur. **b** The central window shows the 3D vectorial models in mesh form. Medial gastrocnemius muscle (1, green), lateral gastrocnemius muscle (2, blue), soleus muscle (3), Achille's tendon (4), hallux flexor longus (5, yellow), posterior tibial muscle (purple), and peroneus longus muscle (light blue)

Other softwares are available for 3D reconstruction: Detailed explanations about 3D reconstruction software and functions are downloadable on the Stanford website.

Finally, the 3D model could be handled interactively in any CAD software, like Autocad®, Maya®, CATIA®, Blender®. One can rotate the models around, move them, modify the color transparency of each object and even do things like run cutting planes into them.

In summary, the principle of 3D reconstruction by vectorial modeling from a series of transverse sections is the main technique. It is here associated with the identification of several anatomical structures (nerves) by a specific immunostaining, but it could be done alone.

4.3 Results

The results mainly relate to human embryos (carpal bones, mitral valve, urethra nerves, pelvic nerves, lower limbs veins) 3D reconstruction of embryos could be done without immunostaining.

We did several works using the Embryos of Rouvière collection (Anatomy institute of Paris), which was available in our University until 2010 (since then, the collection has been moved to Montpellier University):

a. Carpal bones (Durand et al. 2006)—The aim of this study was to provide a quantitative morphology analysis of the evolution of the carpal bones during embryogenesis. histological serial sections of 18 human embryos and early fetuses from the Rouvière collection (Paris). Results: According to inertia parameters, the geometry of carpal cartilaginous structure, initially plane, becomes curved

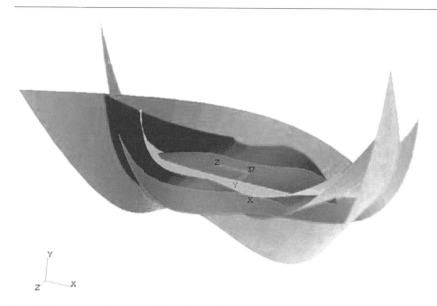

Fig. 4.5 Evolution of the carpal "envelope" during embryo development. Embryos size: Blue 17.5 mm purple 22 mm yellow 25 mm green 33.5 mm orange 44 mm fetus. The envelope passes from plane (blue) to curve geometry (green)

during embryogenesis (Fig. 4.5) Carpal bones growth follows non-homothetic transformation (Fig. 4.6)

b. Pelvic Region—The Pelvis of a 49 mm embryo including the hypogastric plexus was reconstructed (Hounnou et al. 2001) as shown in Fig. 4.7. The limitation of this work is the lack of use of immunomarkers, thus a less accurate reconstruction and the lack of differentiation between adrenergic and cholinergic nervous fibers.

c. Mitral Valve—The 3D reconstruction of the mitral valve has been achieved in 5 embryos from 12, 5 to 69 mm in the thesis of H.F Carton (Science thesis of H.F Carton Descartes University 1990). Descartes University 1990 (Fig. 4.8).

d. Human Embryo—An outstanding work has been done on the whole human embryo by the department of embryology of Amsterdam University (de Bakker et al. 2012). It uses all the digital images of serial sections of 34 human embryos of the *Carnegie Collection* between Carnegie stages 7 (15–17 days) and 23 (56–60 days). These 3D reconstructions of different organs and systems were done with the *Amira™* software package. The 3D models can be interactively viewed within an Acrobat™ 3D-pdf file (Fig. 4.9). This huge work, made from the Carnegie collection is the first complete digital 3D human embryology atlas of this size, containing all developing organ systems, at all stages.

Several studies using the CAAD technique itself with immunomarkers have been done since 2009 in collaboration with our research unit. This improves the technique by segmentation of specific structure, such as cholinergic and adrenergic nerves, that

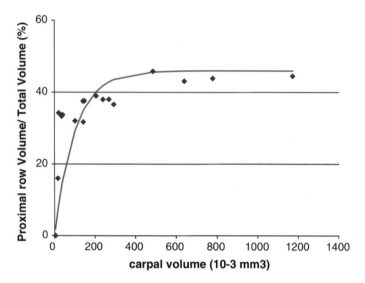

Fig. 4.6 Volumetric evolution of the embryo carpal bones. The red curve represents the correlation between the carpal volume and the proximal carpal bones (% of total carpal volume)

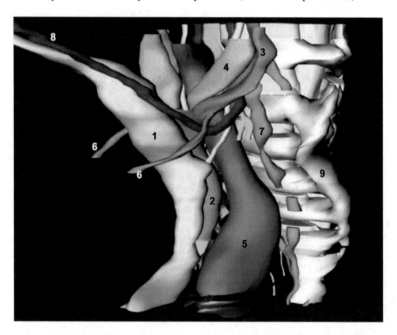

Fig. 4.7 Hypogastric plexus and pelvic 3D reconstruction of a 69 mm human embryo of the Rouvière collection (from former museum Delmas-Orphila-Rouvière (Paris University)). 1 = bladder 2 = uterus 3 = salpinx 4 = ovary 5 = rectum 6 = round ligaments 7 = hypogastric plexus 8 = umbilical artery. 9 = sacrum

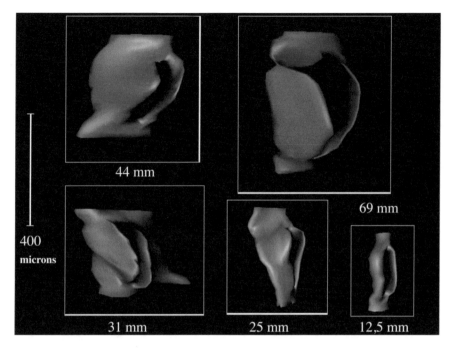

Fig. 4.8 3D Reconstruction of the mitral valve of 5 human embryos from 69 mm to 12, 5 mm, Rouvière collection (from former museum Delmas-Orphila-Rouvière (Paris University)

cannot be identified on the ordinary slices. They open the way of the new nerve sparing techniques during pelvic surgery:

a.   ***Innervation of male*** (Karam et al. 2005) ***and female urethra*** (Karam et al. 2005)

In a first study, Karam et al. showed the precise location of the nervous fibers of the male urethra: The majority of unmyelinated nerve fibers penetrates the male urethral smooth muscle layers at 5 O'clock and at 7 O'clock, while the majority of myelinated nerve fibers penetrates the striated muscles of the prostatic capsule and of the urethral sphincter at 9 O'clock and at 3 O'clock (Fig. 4.10).

The second study (female urethra) shows that in the proximal third of the urethral sphincter, myelinated fibers were identified running with unmyelinated fibers from the pelvic plexus. These fibers were closely related to the lateral and anterior aspects of the vagina. Unmyelinated fibers entered the smooth muscle part of the sphincter at 4 o'clock and at 8 o'clock. Most myelinated fibers entered the sphincter at 3 o'clock and at 9 o'clock. These two studies demonstrate that a better knowledge of neuroanatomy would allow preservation of all functional qualities of the urethra during pelvic surgery, in particular the preservation of urinary continence.

Prostatic neurovascular bundles (Alsaid et al. 2010) (Fig. 4.11): Bayan et al. showed in this paper the distribution of nerve fibers within the posterolateral prostatic neurovascular bundles and the existence of mixed innervation in the posterior and

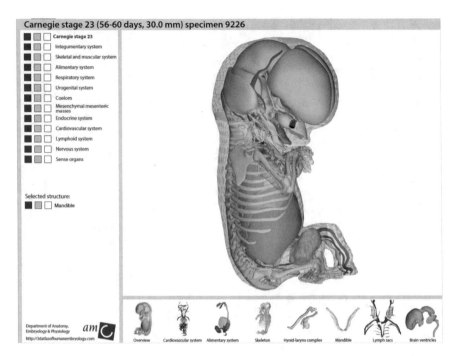

Fig. 4.9 Interactive 3D pdf interface of the 9226 embryo specimen (Carnegie stage 23) done by DeBaker and al., from the Amsterdam University (http://atlasofhumanembryology.com) The interactive 3D model could be displayed by system (horizontal icons below) or by organ (boxes on the left)

lateral fiber courses at the level of the prostate and seminal vesicles give us an insight into how to minimize effects on sexual function during prostatic surgery.

b. **CAAD for intra pelvic innervation** (Alsaid et al. 2012) (Fig. 4.12): This paper describes precisely the CAAD technique and its main interest: to improve the understanding of the complex anatomic regions such as the pelvis from both surgical and educational point of view.

c. ***Reconstruction of the IVC*** (Abid et al. 2013): The aim of this study of a 20 mm human embryo was to specify the path relative to the liver and initiate a series of computerized three-dimensional reconstruction that will follow the evolution of the retrohepatic segment of the inferior vena cava and this in a pedagogical and morphological research introducing the time as the fourth dimension.

d. ***Nerves of the rectum*** (Moszkowicz et al. 2011): This study was based on serial histological sections of the pelvic portion of five human female fetuses (18–31 weeks of gestation) showed that hypogastric nerves were located in the retrorectal multilaminar structure and joined the homolateral inferior hypogastric plexus at the lateral border of the recto-uterine pouch. The intersection of the ureter with the posterior wall of the uterine artery precisely located their junction. Antero-inferior branches supplying female sexual and continence

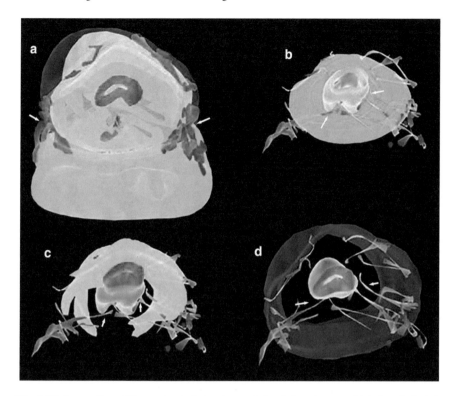

Fig. 4.10 Innervation of the anatomical structures of ejaculation: immuno-histochemical study and 3D reconstruction in a male fetus. **a** The cavernous nerves remain in posterolateral location (arrows). **b, d** Represents the unmyelinated nervous fibers giving branches into the prostatic gland, reach the external part of the smooth urethral muscle and the submucosa (arrows) which reach the external part of the smooth muscle along the anterolateral aspects of the prostatic gland (**c** and **d**: arrows). From Karam et al.

    organs originated from the anteroinferior angle of inferior hypogastric plexus and were bundled at the posterolateral vaginal wall. Their preservation prevents postoperative sexual and urinary dysfunction after rectal surgery.

e. ***Nerves of the uterus*** and preservation during radical hysterectomy for cancer (Balaya et al. 2017a, b; Li et al. 2019) (Fig. 4.13):

    Pelvic Splanchnic Nerves (PSN) are parasympathetic nerves that emerge from ventral rami of S2, S3 and S4. They run on the posterolateral side of the rectum under the middle rectal artery and the deep uterine vein until the posterior edge of the IHP. Sacral Splanchnic Nerves (SSN) are sympathetic nerves that come from sympathetic trunks of S2, S3 and S4 and have the same course as pelvic splanchnic nerves. IHP is constituted by these three previous nerves.

    In this study, authors explain the possible injuries of the pelvic nerves and how to prevent them: Superior hypogastric plexus could be injured during lombo-aortic lymphadenectomy and its preservation necessitates an approach on the right side

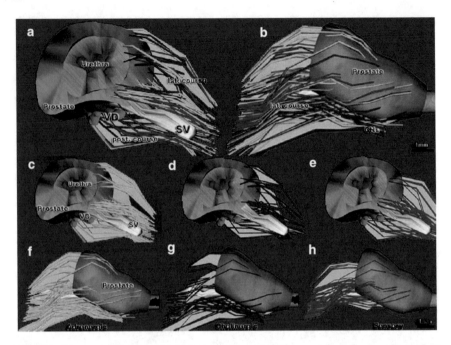

Fig. 4.11 Computer-assisted anatomic dissection, zoom-in of the prostatic region in a 17-week-old male fetus. **a** Posterior view of the prostatic base and the seminal vesicle (SV), **b** lateral view of the prostate, cavernous nerves (CNs) sectioned at the prostatic apex. Views illustrate the mixed innervation at prostate and seminal vesicle level; adrenergic fibers in green, cholinergic fibers in purple, and sensory fibers in blue. The nerve fibers follow two courses, posterior and lateral, innervating the seminal vesicles, the vas deferens (VD), the ejaculatory canal, and the prostatic gland. **c–e** Posterior views and **f–h** lateral views demonstrate the adrenergic, the cholinergic, and the sensory innervation (from Karam et al.)

of the aorta and a blunt dissection of the promontory. Injuries of hypogastric nerve occur during uterosacral ligament and/or rectovaginal resection, and therefore after dividing the uterosacral ligament only the medial fibrous part should be ablated. Pelvic splanchnic nerves and inferior hypogastric plexus can be injured during dissection and resection of paracervix and to preserve them, an attention should be paid to identify main anatomical landmarks as the middle rectal artery, the deep uterine vein and the ureter. Vesical branches can be preserved by blunt dissection of the posterior layer of the vesicouterine ligament after identifying the inferior vesical vein.

f. ***3D reconstruction of the Neurovascular and lymphatic vessels distribution in uterine ligaments*** (Li et al. 2019): This study shows a 3D reconstruction of the neurovascular and lymphatic vessels in uterine ligaments and depicts their precise location and distribution: The vessels were primarily located in the upper part of the ligaments model, while the pelvic autonomic nerves were primarily in the lower part.

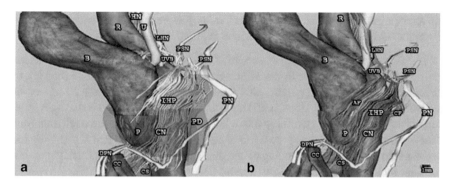

Fig. 4.12 Computer-assisted anatomic dissection of 17-week-old human fetal pelvis. **a** Lateral view of the inferior hypogastric plexus (IHP) without bones and with pelvic diaphragm (PD) transparency, **b** same view with transparency of plexus nerve fibers and without muscles, adrenergic fibers (AF) in green and cholinergic fibers (CF) in purple, hypogastric nerves (HN) and pelvic splanchnic nerves (PSN) contain both fibers' types, adrenergic fibers mostly situated in the superior portion of the plexus, cholinergic fibers intend to concentrate in inferior portion. (B Bladder, CC Corpus Cavernosum, CS Corpus Spongiosum, LHN Left Hypogastric Nerve, CN Cavernous Nerve, P Prostate, R Rectum, U Ureter, UVB Uretero-vesical Branches.) from Bayan et al.

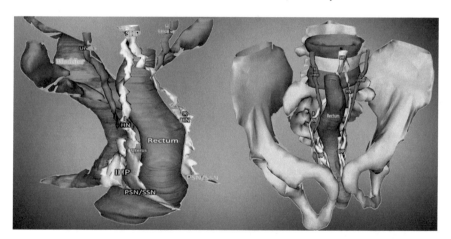

Fig. 4.13 CAAD3D reconstruction of hypogastric plexus. The Superior Hypogastric Plexus (SHP) is located at the level of the aortic bifurcation in front of the sacral promontory. Underneath the promontory, the SHP caudally and laterally divides into two filaments of variable width which are hypogastric nerves. (from Balaya et al. 29) Hypogastric Nerves (HN), that are sympathetic nerves, descend along the lateral side of the rectum through the mesorectum, then run posteromedially to the ureter and in the lateral part of the uterosacral ligament until the superior angle of the Inferior Hypogastric Plexus (IHP)

g. ***Embryogenesis of the lower limbs' veins*** has been studied with only three
embryos (Uhl and Gillot 2007; Kurobe et al. 2014) (Figs. 4.3, 4.4 and 4.14):

The aim of this research work is mainly for a better understanding of the venous
embryogenesis of the lower limbs, which remains a conundrum, due to the lack of
direct observations. It brings some data about the theory of the angioguiding nerves
of Claude Gillot (Uhl and Gillot 2007) and shows that the axial vein is the main thigh
venous axis at 10–14 weeks. It regresses later in the majority of the adult anatomy.
This explains the variations of the femoral vein (Uhl et al. 2010) and is a remarkable
database for a better understanding of the diagnosis, classification of the congenital
vascular malformations and their treatment (Belov St. 1993; Lee 2005).

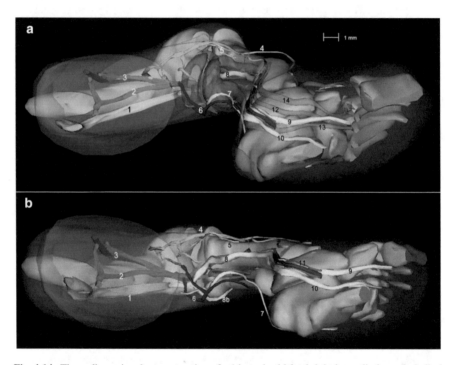

Fig. 4.14 Three-dimensional reconstruction of a 14-week-old fetal right lower limb. **a** whole limb
with muscles and **b** without muscles. 1 Ischiatic nerve, 2 axial vein, 3 femoral vein and artery, 4
great saphenous vein, 5 saphenous nerve, 6 small saphenous vein, 7 sural nerve, 8b fibular nerve,
8 tibial nerve, 9 medial plantar nerve, 10 lateral plantar nerve, 11 posterior tibial artery and two
veins, 12 tendon of the hallux flexor longus, 13 flexor digitorum longus, 14 posterior tibial tendon,
15 tendons of peroneus longus and brevis muscles

4.4   Discussion

4.4.1   *Limitations of the CAAD Technique*

The first issue of CAAD technique is the alignment of the slices. We saw that the FIJI software is the best cheap solution today (without landmarks), but it could be improved to automatize the alignment of a big number of slices, by the use of artificial intelligence and shape recognition. The addition of landmarks inside the blocks could simplify the alignment of long series.

The second issue is the tedious human–computer interface of Winsurf® software for manual segmentation: Even if a full automatization is impossible on such slices, an improvement of the interface could facilitate the manual outlining of each anatomical element. Here again, the use of artificial intelligence is the best way to simplify the segmentation process.

In summary, the user's interface of the tools to build the 3D models should be improved.

Fundamental human–computer interface contributions:

This human–computer interface is a true revolution in the field of morphology to investigate the 3D reconstructed micro anatomical structures, providing today:

– High resolution digitized images
– Accuracy of the segmentation tools with manual expertise
– High quality of the rendering providing realistic 3D objects
– Easy handling of the 3D models
– Availability of quantification tools
– Better understanding of the anatomical variability

The CAAD technique is of high interest to investigate, quantify and better understand complex micro anatomical structures of the human embryo, in particular the nerve distribution. The human–computer interface of the new cad software makes possible an accurate 3D reconstruction of human anatomy, leading to planning and simulation for pelvic surgery of the future. From the research point of view, it allows us a quantification of human morphology, opening the way of computational anatomy, which studies the biological variability of the anatomical structures and organs (see the chapter of X. Pennec)

4.5   Conclusion

Descriptive anatomy is still a dynamic science. The power of the new computer tools opens the world to new research techniques in the field of morphological sciences: CAAD is an original method in anatomic research which has progressively evolved during the last decade. This method allows perfect anatomical and functional descriptions of the intra-organic innervation, the nature of the nervous fibers and the distribution of neurotransmitters and receptors. The use of CAAD helps to improve our knowledge of the complex anatomical regions such as the pelvis in particular the inferior hypogastric plexus from both surgical and educational points of view. It will improve the nerve-sparing surgical techniques of the pelvis: a better nerve preservation should decrease postoperative sexual and urinary complications.

References

Abid B, Douard R, Hentati N, Ghorbel A, Delmas V, UhlJF, Chevallier JM (2013) Computerized three-dimensional reconstruction of the retrohepatic segment of inferior vena cava of a 20 mm human embryo. Morphologie 97(317):59–64

Alsaid B, Karam I, Bessede T, Abdlsamad I, Uhl J-F, Delmas V, Benoıt G, Droupy S (2010)Tridimensional computer-assisted anatomic dissection of posterolateral prostatic neurovascular bundles. Europ Urol 58:281–287

Alsaid B, Bessede T, Diallo D, Karam I, Uhl J-F, Delmas V, Droupy S, Benoıt G (2012) Computer-assisted anatomic dissection (CAAD): evolution, methodology and application in intra-pelvic innervation study. SurgRadiolAnat 34(8):721–729

Balaya V, Douard R, Uhl JF, Rossi L, Cornou C, Ngo C, Bensaid C, Guimiot F, Bat AS, Delmas V, Lecuru F (2017) Pelvic nerve injury during radical hysterectomy for cervical cancer: key anatomical zone. Eur J Gynaecol Oncol XXXVIII(5)

Balaya V, Douard R, Uhletal JF (2017b) Pelvic nerve injury during radical hysterectomy for cervical cancer: key anatomical zone. EJGO 38(5):657–666. https://doi.org/10.12892/ejgo4103.2017

Bardol T, Subsol G, Perez MJ et al (2018) Three-dimensional computer assisted dissection of pancreatic lymphatic anatomy on human fetuses: a step toward automatic image alignment SurgRadiolAnat 40:587. https://doi.org/10.1007/s00276-018-2008-2

Belov St. (1993) Anatomopathological classification of congenital vascular defects. VascSurg 6:219–224

Born G (1883) Die PlattenmodeliermethodeArchiv f. mikrosk. Anatomie 22: 584–599. https://doi.org/10.1007/BF02952679

de Bakker BS, de Jong BS, Hagoort J (2012) Towards a 3-dimensional atlas of the developing human embryo: the Amsterdam experience. Reproductive Toxicology 34(2): 225–236. https://doi.org/10.1016/j.reprotox.2012.05.087

Durand S, Delmas V, Ho Ba Tho MC, Batchbarova Z, UhlJ F, Oberlin C (2006) Morphometry by computerized 3D reconstruction of the human carpal bones during embryogenesis. Surg Radiol Anat 28:355–58

Eumenia Costa da Cunha Castro, CsabaGalambos (2011) Prox-1 and VEGFR3 Antibodies Are Superior to D2–40 in Identifying endothelial cells of lymphatic malformations -a Proposal of a New Immunohistochemical Panel to Differentiate Lymphatic from other Vascular Malformations. Pediatric Develop Pathol 12;3:169–176. https://doi.org/10.2350/08-05-0471.1

Hounnou GM, Uh lJF, Plaisant O, Delmas V (2001) Morphometry by computerized three-dimensional reconstruction of the hypogastric plexus of a human fetus. Surg Radiol Anat 25:21–31

Karam I, Moudouni S, Droupy S, Abd-Alsamad I, Uhl JF, Delmas V (2005a) The structure and innervation of the male urethra: histological and immunohistochemical studies with three-dimensional reconstruction. J Anat 206(4):395–403

Karam I, Droupy S, Abd-Alsamad I, UhlJ F, Benoit G, Delmas V (2005) Innervation of the female human urethral sphincter: 3D reconstruction of immunohistochemical studies in the fetus. EurUrol 2005 47(5):627–33

Kurobe N, Hakkakian L, Chahim M, Delmas V, Vekemans M, UhlJ F (2014) Three-dimensional reconstruction of the lower limb's venous system in human fetuses using the computer-assisted anatomical dissection (CAAD) technique. SRA 2014. https://doi.org/10.1007/s00276-014-1350-2

Lee BB (2005) New Approaches to the treatment of congenital vascular malformations (CVMs). Eur J VascEndovascSurg 30(2)184–97

Lewis DA, Campbell MJ, Foote SL et al (1987) The distribution of tyrosine hydroxylase-immunoreactive fibers in primate neocortex is widespread but regionally specific. J Neurosci 7(1):279–290

Li P, Duan H, Wang J, Gong S et al (2019) Neurovascular and lymphatic vessels distribution in uterine ligaments based on a 3D reconstruction of histological study: to determine the optimal plane for nerve-sparing radical hysterectomy. Arch Gynecol Obstetrics 299(5):1459–1465. https://doi.org/10.1007/s00404-019-05108-w

Lozanoff S, Diewert VM, Sinclair B, Wang K-Y (1988) Shape reconstruction from planar cross sections. Comput Vis Graph Image Proc 44:1–29

Moszkowicz D, Alsaid B, Bessede T, Penna C, Benoit G, Peschaud F (2011) Female pelvic autonomic neuroanatomy based on conventional macroscopic and computer-assisted anatomic dissections. SRA 33:397–404

Park JS, Chung MS, Hwang SB, Lee YS, Har DH, Park HS (2005) Technical report on semiautomatic segmentation by using the Adobe Photoshop. J Digit Imaging 18:333–343

Park JS, Shin DS, Chung MS, Hwang SB, Chung J (2007) Technique of semiautomatic surface reconstruction of the Visible Korean Human data using commercial software. ClinAnat 20:871–879

Pillet JC, Papon X, Fournier H-D, Sakka M, Pillet J (1995) Reconstruction of the aortic arches of a 28-day human embryo (stage 13) using the Born technique. SRA 17(2):129–132

Ren S, Abuel-Haija M, Khurana JS, Zhang X (2011) D2-40: an additional marker for myoepithelial cells of breast and the precaution in interpreting tumor lymphovascular invasion. Int J Clin Exp Pathol 4(2):175–182

Schaefer HJ (1957) A rapid trichrome stain of Masson type. Am J ClinPathol 28(6):646–647

Schneider CA, Rasband WS, Eliceiri KW (2012) NIH Image to ImageJ: 25 years of image analysis. Nature Methods 9(7):671–675, PMID 22930834

Science thesis of H.F Carton Descartes University (1990)

Stefansson K, Wollmann RL, Moore BW (1982) Distribution of S-100 protein outside the central nervous system. Brain Res 234(2):309–317

Uhl JF, Gillot C (2007) Embryology and three-dimensional anatomy of the superficial venous system of the lower limbs. Phlebology 22(5):194–206

Uhl JF, Gillot C, Chahim M (2010) The anatomical variations of the femoral vein. JVS 52:714-9

Usdin TB, Eiden LE, Bonner TI et al (1995) Molecular biology of the vesicular ACh transporter. Trends Neurosci 18(5):218–224

FIJI. https://imagej.net/Fiji/Downloads

RVSS. http://imagej.net/Register_Virtual_Stack_Slices

http://med.stanford.edu/biocomp/projects/3dreconstruction/software.html

Yucel S, Baskin LS (2004) An anatomical description of the male and female urethral sphincter complex. J Urol 171(5):1890–1897

Chapter 5
Volume Rendering Technique from DICOM® Data Applied to the Study of Virtual Anatomy

Juan Pablo Merino, José Alberto Ovelar, and Jorge Gustavo Cédola

Abstract Volume Rendering (VR) from Digital Imaging and Communications in Medicine (DICOM®) data is a powerful source for the analysis and virtual representation of the human anatomy. To achieve this, accurate virtual dissections must be performed by applying the tools and different reconstruction methods offered by Volume Rendering Techniques (VRT).

5.1 Introduction

New technologies applied to the field of Diagnostic Imaging have substantially expanded its implications in modern medicine.

As a result of the great anatomopathological detail achieved by the three-dimensional images acquired by volumetric biomedical scanners: Multidetector Computed Tomography (MDCT), Magnetic Resonance Imaging (MRI) and Ultrasonography (US); this discipline has expanded its horizons by venturing into a new challenge: to represent virtually the first of the medical sciences, anatomy.

When a patient is subjected to a volumetric medical imaging procedure, one or more regions of the body (Real Volume) are exposed for analysis.

This studied volume is digitized in "raw data" format by the scanner, converting it into an Acquired Volume.

By means of specialized software, the information is codified in DICOM® data, which allows the display of a Virtual Volume on the screen in different planes, which can only be used by specialists.

But when a series of Volume Rendering Techniques are applied, it is possible to express this volume as a three-dimensional, transferable and more easily interpreted Rendered Volume (RV).

Nowadays, due to technological advances such as submillimeter acquisitions and the use of contrast media, volumetric medical imaging procedures provide a faithful

J. P. Merino (✉) · J. A. Ovelar · J. G. Cédola
Digital Anatomy Laboratory CIMED-FUNDAMI, Partners of the Unesco Chair of Digital Anatomy, Calle 5 416, La Plata B1902, Buenos Aires, Argentina
e-mail: lad.3d.lp@gmail.com

© The Author(s), under exclusive license to Springer Nature Switzerland AG 2021
J.-F. Uhl et al. (eds.), *Digital Anatomy*, Human–Computer Interaction Series,
https://doi.org/10.1007/978-3-030-61905-3_5

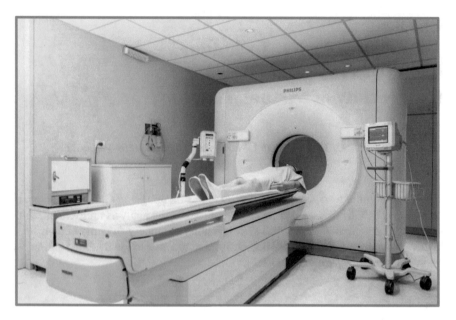

Fig. 5.1 MDCT room, CIMED, Argentina

correlation between the Real Volume and the final Rendered Volume, offering a considerable virtual alternative to the study and reproduction of human anatomy.

The purpose of this chapter is to describe the several methods and tools to virtually represent the human anatomy from a volume acquired by a biomedical scanner, in this case by a Philips Brilliance 64® multidetector tomograph or CT scanner.

In addition, technical concepts and definitions are introduced using MDCT venography as an example to illustrate the strengths and limitations of the technique.

All images were acquired at the bioimaging center CIMED, Argentina; and reconstructed by the Digital Anatomy Laboratory CIMED-FUNDAMI (Fig. 5.1).

5.2   What Is a DICOM® Data?

"DICOM® (Digital Imaging and Communications in Medicine)" is the international standard to transmit, store, retrieve, print, process, and display medical imaging information.

DICOM®:

– makes medical imaging information interoperable
– integrates image-acquisition devices, PACS, workstations, VNAs and printers from different manufacturers

– is actively developed and maintained to meet the evolving technologies and—
  needs of medical imaging
– is free to download and use (DICOM 2020).

In 1993 the American College of Radiology (ACR), in conjunction with the
National Electrical Manufacturers Association of the United States (NEMA), developed and applied these standards to all acquirable biomedical data beyond the differences in equipment and physical principles with the main objective of avoiding
technological conflicts between different digital image formats (Spanish Society of
Cardiac Imaging 2018).

The primary information acquired by the biomedical scanners is the "Raw Data",
a series of unprocessed numbers, which, through different advanced reconstruction
algorithms (filtered rear projection, iterative reconstruction, etc.), are encoded in
DICOM® data (Spanish Society of Cardiac Imaging 2018).

The core of DICOM® is the bioimage file (.dcm), which securely stores inseparable data about the patient and his study. The Information Object Definition (IOD),
provides it a unique identification available for transfer, even via the web (HL7 v3),
and to be reproducible beyond the manufacturer (Dahilys González López et al.
2014; ISO 1205).

5.3   What is a Volume Rendering Technique?

A Volume Rendering Technique is a set of techniques for the virtual three-
dimensional representation of a real prescanned volume. Applied to the field of
Diagnostic Imaging they become different methods of virtual reconstruction and
dissection in pursuit of the digital study of human anatomy (Fig. 5.2) (Espinosa
Pizarro 2012).

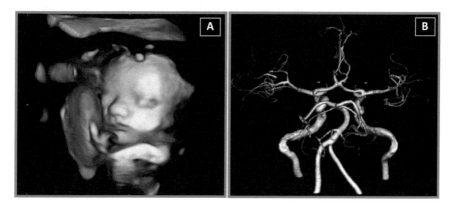

Fig. 5.2 Volume rendering from: **A** Ultrasonography; **B** Cerebral Angio MRI

5.3.1 MDCT Volume Rendering Technique

The Multidetector Computed Tomography ("tomo" = slice) procedure consists of a controlled emission of radiation from an X-ray tube as it rotates around a patient (Carro and Cecilia Lorena 2016).

The attenuated radiation that has not been absorbed by the body is received by multiple detector arrays, which allow the information to be digitized using advanced reconstruction algorithms.

The region scanned from the patient is called Acquired Volume and is formed by contiguous and submillimeter virtual "slices".

In turn, this volume is digitally composed of thousands of isometric voxels (cubes of exactly the same dimensions), which enable the Multiplanar (MPR) condition (Fig. 5.3).

This harmonious symmetry is the physical quality that allows a Rendered Volume to represent a Real Volume as accurately as possible (Perandini et al. 2010).

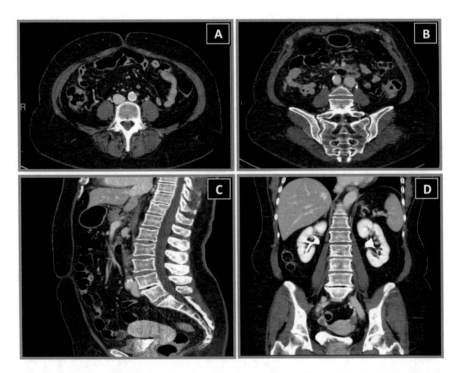

Fig. 5.3 Abdominal MDCT: **A** Axial plane; **B** Coronal plane oblique; **C** Sagittal plane; **D** Strict coronal plane

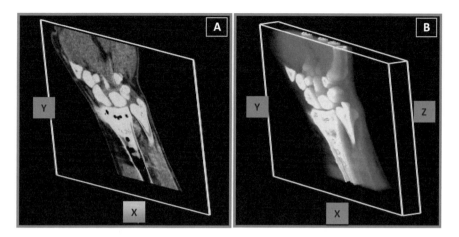

Fig. 5.4 Wrist MDCT: **A** MPR coronal oblique, Pixel, **B** Three-dimensional volume, Voxel

Once on the screen, voxels (x–y–z axis) that conform the Virtual Volume are synthesized into pixels (x–y axis). The set of these, one next to the other, constitutes a matrix (256, 512, 768, 1024) in which a representative, multiplanar and two-dimensional image of the Real Volume is generated, the typical CT image within a grayscale (Fig. 5.4A) (Romina Luciana Muñoz 2015).

On the other hand, when the virtual slices of the Acquired Volume are overlapped, that is, one voxel after another, a three-dimensional RV is formed on the screen (Fig. 5.4B)

Therefore, the VRT from a DICOM® data is the result of a long process in the search to represent as reliably as possible a scanned Real Volume (Acquired Volume) by means of a three-dimensional and comprehensible Rendered Volume on a screen (Fig. 5.5), 3D printing or in Augmented Reality.

5.4 Reconstruction Methods

The first step in VRT is done automatically by the biomedical scanner (MDCT) by showing on the screen the total surface of the Acquired Volume (Fig. 5.6B).

But for the analysis and representation of the human anatomy, it is necessary to model or "sculpt" the volume using several editing tools for its segmentation and the dissection of the different tissues.

This can be achieved by three methods of reconstruction: Automatic, Semi-automatic and Manual reconstruction method (Medina and Bellera 2017).

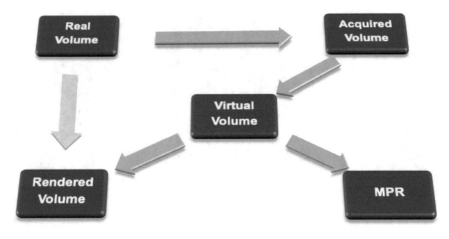

Fig. 5.5 VRT steps

5.4.1 Automatic Reconstruction Method

This method consists of applying a factory preset, which contains a series of parameters arranged so the Rendered Volume quickly exposes a specific structure such as bone tissue (Fig. 5.6D)

Although it obviously requires operator supervision, understandable RVs are achieved in a few seconds that help determine immediate behaviors (e.g., detection of complex fractures, Fig. 5.11); this is because machine learning improves its sensitivity and specificity year after year (Fig. 5.7).

5.4.2 Semi-automatic Reconstruction Method

In this method, the operator determines the structures to be highlighted by applying an automatic editable segmentation. Several tools allow to cut (virtual dissection), emphasize with different colors, illuminate and even superimpose different tissues previously separated, reaching high levels of discernment (Ignacio García Fenoll 2010).

An adequate knowledge of the anatomopathology is necessary, as well as of the limitations of the method in order to minimize errors. The sensitivity of the program is the same as that of the Automatic Method, but the details depend on the operator (Fig. 5.8).

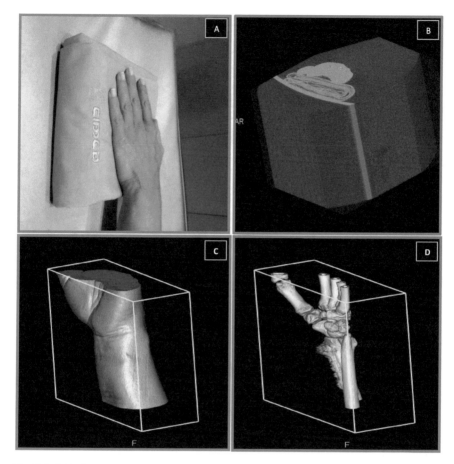

Fig. 5.6 A Real Volume; **B** Acquired Volume; **C** Rendered Volume; **D** Automatic Reconstruction
Method (bone tissue)

5.4.3 Manual Reconstruction Method

Finally, in this method, the operator must act as a Virtual Anatomist (VA). The
total surface of the Acquired Volume is presented on the screen without presets,
simulating a dissection table. Thanks to the available tools and the appropriate VA
training, accurate virtual dissections of organs and systems are possible.

At this point, knowledge of the anatomy and pathology is crucial because the
margin of error is much greater. In comparison, virtual dissection allows revealing
anatomical details that would not be achieved with automatic or semi-automatic
segmentation, achieving highly descriptive RV (Fig. 5.9).

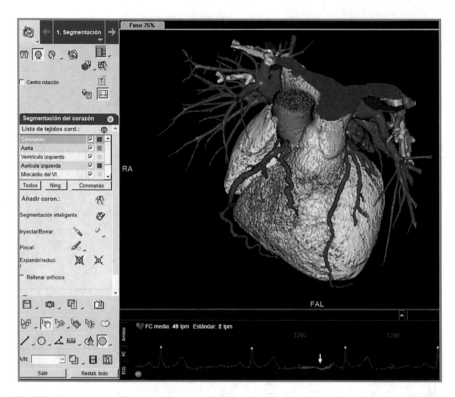

Fig. 5.7 Coronary MDCT, Comprehensive Cardiac Analysis®, Automatic Reconstruction Method: Three-dimensional Segmentation Automatic detection of cardiac structures

5.5 VRT Tools

For the Rendered Volume to achieve an effect similar to the Real Volume, the Virtual Anatomist has a wide range of manipulation provided by the virtual interaction between the light and the dissected object (shading process).

This includes changing the angle of illumination and its intensity, zooming in and out on the volume, modifying the density of certain segments, or assigning different colors, among other options (Tierny 2015).

The reason for this is that VRT uses many and varied engineering tools applied to 3D design, such as: gloss refraction; color mapping; protrusion mapping (exposes roughness); normal mapping (simulates texture); transparency and occlusion mapping; displacement mapping; etc (Fig. 5.10).

To apply colors and different opacities it is necessary to manipulate the Threshold tool, based on the Hounsfield scale and its units (Fig. 5.11).

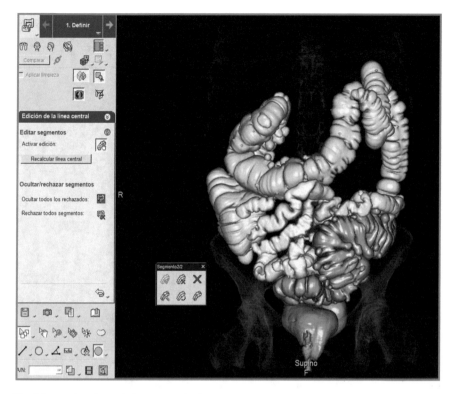

Fig. 5.8 MDCT Virtual Colonoscopy CT®, Semiautomatic Reconstruction Method for navigation and analysis of the colon. The acceptance and rejection of the automatically detected segments are operator dependent

5.5.1 Hounsfield Scale

Named after the engineer and Nobel Laureate in Physiology or Medicine Sir Godfrey Hounsfield, it is a quantitative scale of assigned radiation levels in Hounsfield Units (HU), which is used in the MDCT to describe its correspondence with human tissue.

Each pixel (Virtual Volume-MPR), expresses the attenuation coefficient of linear radiation corresponding to the exposed anatomical structure (Palacios Miras 2012).

For example, water corresponds to 0 HU; the liver, under normal conditions, has an attenuation of approximately 60 HU; while lung tissue is around -500 HU and compact bone 1000 HU (Fig. 5.12).

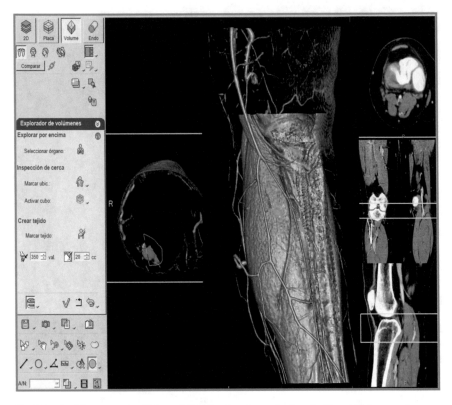

Fig. 5.9 MDCT venography Volume® Freehand sculptured using the Target Volume tool

5.5.2 Threshold

Due to the fact that each tissue has certain HU designated, a Threshold of visualization can be applied. For this, VA chooses one to highlight and assign as the Center of the Window (CW), within a Window Width (WW), which includes the secondary structures that are part of the same image (Fig. 5.13A–C).

The Threshold also allows the operator to assign colors (RGB) and manipulate the degree of opacity transparency of the structures that constitute the RV (Fig. 5.13D).

For greater accuracy, the threshold offers a dynamic histogram for editing that allows you to move the trapezes, modify their slopes or even add others to expose new tissue (Fig. 5.14) (Vicente Atienza Vanacloig 2011).

Even though the threshold is present in all three VRT Reconstruction Methods, it takes on greater importance in the semi-automatic and manual methods, since it is the basis for tissue segmentation by their HU.

By applying it to the appropriate WC is possible to remove complete structures easily.

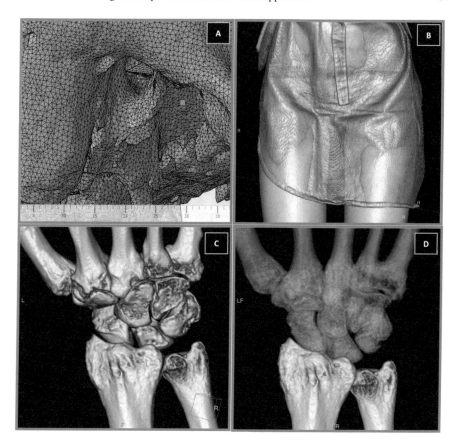

Fig. 5.10 **A** Mapping of normal on Temporal bone rendering; **B** Mapping of displacement in Pelvis; **C** Mapping of protuberances and **D** Mapping of Transparency and Opacity applied to wrist RV from MDCT

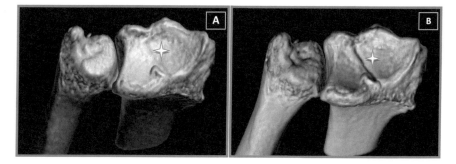

Fig. 5.11 **A** wrist RV from MDCT with automatic extraction of the carpal bones **B** Same volume with a change of the light angle showing a fracture (✦) of the radial facet of the joint

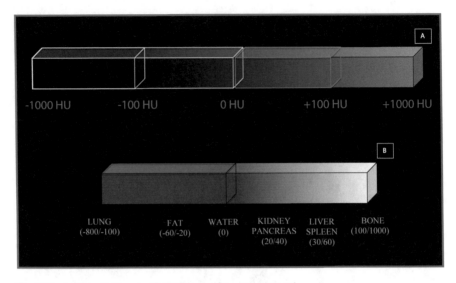

Fig. 5.12 **A** Hounsfield scale; **B** HU threshold for significant tissues

5.5.3 Automatic Tissue Removal

For the precise use of this tool, the technical parameters applied at the time of the volumetric acquisition of the MDCT are very important (Francisco Javier Olías Sánchez 2014).

This is because the Spatial Resolution (the ability to differentiate two tissues that are very close to each other) and the Contrast Resolution (the ability to differentiate two similar HU structures) drastically affect the sensitivity of the segmentation (Fig. 5.15).

To improve this, submillimeter virtual slices are performed and, in cases of vascular evaluation, the use of intravenous contrast media is added.

5.5.4 Smart Injection

By using intravenous contrast and adequate contrast resolution, it is possible to use this valuable tool. It allows to dye a vessel by injecting it virtually, making it easier to know its trajectory and dissection of other structures.

It is a semi-automatic tool that allows editing and is widely used in the evaluation of tortuous vessels as in MDCT venography (Fig. 5.16).

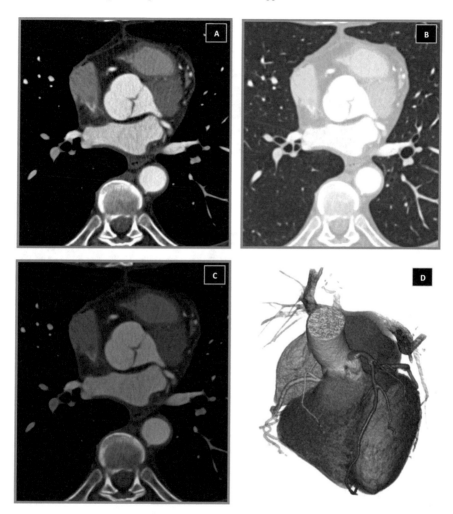

Fig. 5.13 Coronary MDCT, axial plane MPR displayed by different windows: **A** Soft tissue window (60 CW/360 WW); **B** Lung window (-600 CW/1600 WW); **C** And bone window (800 CW/2000 WW) **D** RV, Manual Reconstruction Method, designation of colors to the Coronary arteries for pedagogical purposes. Right Coronary = Green; Left and Left Anterior Descendent Coronary = Purple; Circumflex Coronary = Blue

5.5.5 Target Volume

When the surface of the Virtual Volume is irregular, the VA has this tool. It consists of applying blocks of a certain thickness to visualize and dissect freehand, virtually sculpt, deep structures without damaging the surface planes that remain unaltered for the time necessary to be replaced. As the block is thinner, more anatomical details can be achieved, being proportional to the effort required (Fig. 5.17).

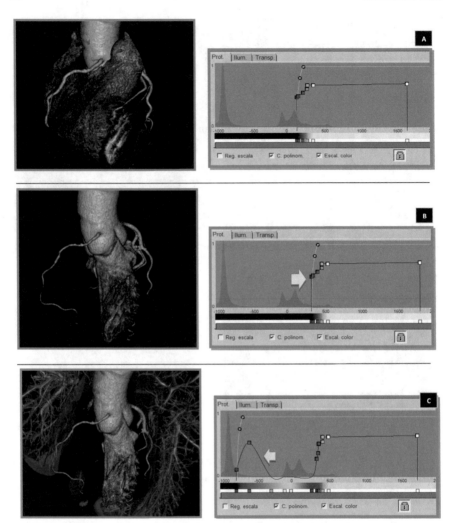

Fig. 5.14 **A** Coronary RV from MDCT with cardiac threshold and its respective histogram with the assigned colors; **B** The same volume with a coronary threshold achieved from the displacement of the trapezium correcting the WC; **C** Addition of a trapezium for the addition of a pulmonary threshold with certain color and opacity. New width of the WW

5.5.6 Tissue Management

Tissue management allows overlapping of different previously dissected tissues. The segmentation can be set aside and then superimposed on other structures to form—reconstruct—an *Enriched Rendered Volume* (ERV), so called because it allows to emphasize each dissected tissue by applying different opacities and colors, obtaining highly descriptive RV (Fig. 5.18) (Ovelar et al. 2015).

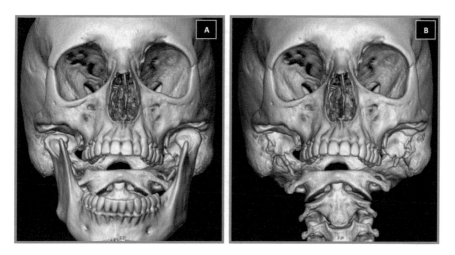

Fig. 5.15 A Craniofacial RV from MDCT; **B** Automatic removal of the jaw bone by choosing the threshold corresponding to the bone tissue (approximately 700 HU) Virtual acquisition slices of 0.6 mm prevented the removal of the rest of the skull

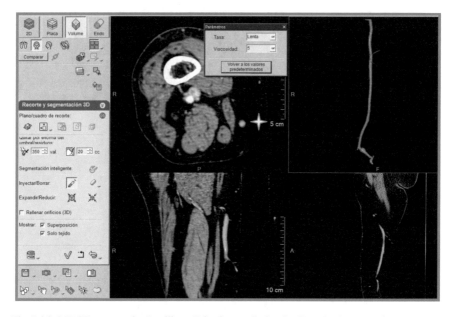

Fig. 5.16 MDCT venography. Intelligent Injection applied to the Great Saphenous vein discovering its disposition (star glyph ⛄)

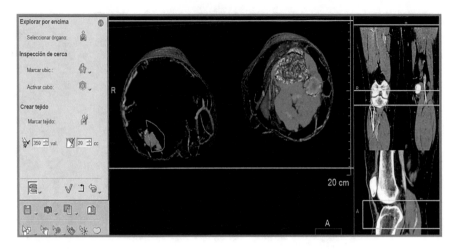

Fig. 5.17 MDCT Venography. Target volume in axial plane allowing specific "sculpting" without damaging other structures

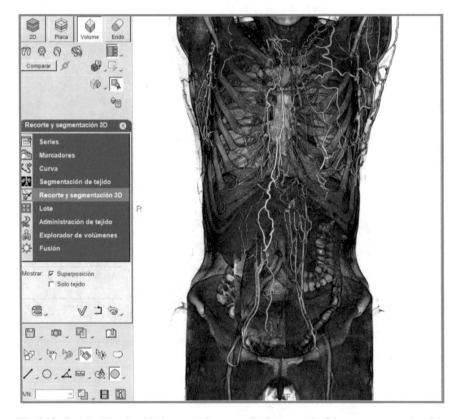

Fig. 5.18 Enriched Rendered Volume. Assignment of colors to each of the venous networks of the anterior thoracic-abdominal wall superimposed with other tissues

5.6 VRT Applied in MDCT Venography

The angio-MDCT of the venous system or Phlebotomography (PMDCT) or MDCT-venography (Uhl et al. 2002, 2003, 2003, 2008, 2012; Uhl and Caggiati 2005; Uhl and Gillot 2007; Uhl 2009, 2001; Ovelar et al. 2014) has acquired in the last years great importance in the field of Phlebology because it offers morphological and structural information complementing the functional information provided by the Doppler ultrasound, called *hemodynamical venous mapping* (Uhl et al. 2018), mandatory before any surgical or endovenous treatment.

Mostly applied to the study of the veins of the lower limbs and the thoraco-abdominal region, it is a very useful tool for the analysis of acute and/or chronic venous diseases since it provides a wide venous mapping of the region.

This virtual map provides information that allows the specialists to know which are the affected veins, the extension of their condition, their relations and superficial and/or deep communications; with the purpose of an accurate treatment.

5.6.1 PMDCT Protocols

Two protocols are practiced in the bioimaging center CIMED, Argentina: Indirect and Direct Protocol. Each of them has a precise medical indication to reach the diagnosis (Fig. 5.19) (Uhl et al. 2003, 2018).

For both cases, deciding the position of the patient, the amount and flow of the intravenous contrast medium injection and other technical parameters, is critical to accomplish a proper venous mapping (Fig. 5.20).

5.6.2 PMDCT Enriched Rendered Volume (ERV)

Once the MDCT venography of lower limbs has been acquired, the Digital Anatomy Laboratory CIMED-FUNDAMI complements the diagnosis with an ERV of the venous mapping that allows distinguishing the Hypodermic Reticular Venous System and the affected veins.

In these cases, the Virtual Anatomists of the laboratory opt for the Manual Reconstruction Method to achieve an adequate virtual dissection of the Dermal tissue, of the Subcutaneous Cellular tissue preserving, in this way, the Communicating and Perforator veins.

Steps for an ERV of PMDCT

1. The Acquired Volume is presented through an automatic reconstruction, visualizing the skin of the lower limbs (Fig. 5.21).

Protocol Technique

-Patient in supine position. -Venipuncture in the affected arm or lower limb. -High pitch, short rotation time, modulated radiation. -Slice of 1 mm. overlapping. -Non-ionic iodine contrast 370mg/350mg -Flow: 4 ml/sec; 1-2 ml/kg. -Foot-head acquisition, delay 180 seconds or more.	**Direct Protocol:** It consists of injecting contrast medium into the tributary vein most distal to the limb. In this way, an excellent appreciation of the venous system of the affected limb is obtained [31]. **Indirect protocol:** In this case, the location of the venipuncture is not crucial because venous enhancement is caused physiologically. It allows both extremities to be visualized and also the arterial system previously [32].

Fig. 19 When the disease is in the deep venous system, for example in syndrome of May-Thurner, an Indirect protocol is applied. On the other hand, when it is necessary to study venous insufficiency and ulcers in a certain lower limb, it is convenient to employ the Direct protocol

2. Using Threshold, select the CW and WW needed to distinguish the tissues (Fig. 5.22).
3. In coronal plane, the thickness of the Target Volume block and where to start is decided. With the volume block positioned axially, unnecessary structures are dissected by freehand sculpting to avoid damaging other tissues (Fig. 5.23).
4. The same meticulous procedure is performed on the entire limb taking care not to discard important structures or affect its continuity (Fig. 5.24A).
5. Once the epithelial, subcutaneous tissue and its respective venous structures have been dissected, they are assigned color and opacity (Fig. 5.24B).
6. With the Smart Injection, certain vessels are selected by assigning them a different color and opacity (Fig. 5.24C)
7. By means of tissue management, all dissected structures are superimposed obtaining an ERV, highlighting the Hypodermic Reticular Venous System (shades of blue), the skeletal muscle volume (shades of red) and some perforating veins (shades of pink) (Fig. 5.24D)
8. As a result of the spatial manipulation of the volume, the change of the angle of the light and the realization of a virtual anatomical cut, it is possible to achieve understandable and didactic ERV (Fig. 5.25) (Willemink and Noël 2018).

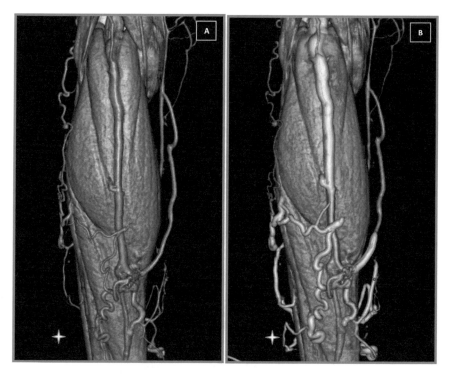

Fig. 5.20 **A** PMDCT RV, Indirect protocol, rear view of the left leg; **B** PMDCT Direct protocol of the same patient. An enhanced communicating vein (star glyph ⟨glyph⟩) is observed, not visualized in the other protocol

5.7 Discussion

It is important to note that in most cases, the analysis and representation of the human anatomy are immediate consequences of the main objective of medical imaging exams, i.e., to reach a diagnosis. Due to this, high-quality VRs are not always obtained so this demands a constant development of new and more effective acquisition protocols.

We believe that VR, besides collaborating with the diagnosis of a disease and facilitating surgical planning, offers an innovative morphological and structural analysis, but not functional, of the organs and tissues of living bodies (vivisection) that are sometimes difficult to dissect corpses.

While though numerous examples are lacking, ERV of the superficial venous system acquired from PMDCT is a good starting point.

Even so, we must recognize the weaknesses of each medical imaging method. In the specific case of MDCT, we must always consider that it uses harmful ionizing radiation and its choice must be justified. On the other hand, its strength resides in

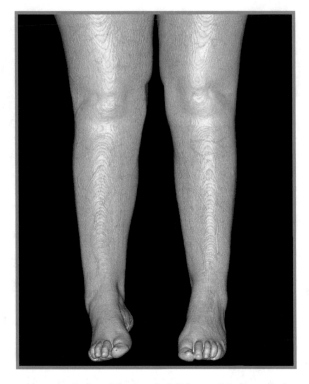

Fig. 5.21 Automatic reconstruction of the Acquired Volume of the lower limbs

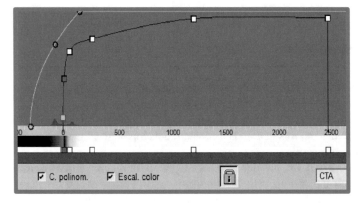

Fig. 5.22 Adaptation of the center and width of the window at the Threshold to distinguish selected tissues

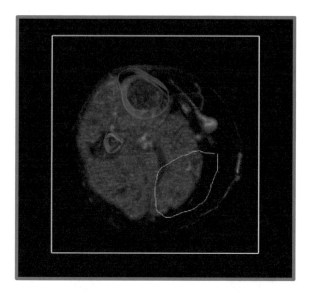

Fig. 5.23 Using the Target Volume block, the selected tissues are sculpted freehand, excluding, for example, the muscles of the limb

its wide availability all over the world, its fast acquisition process and the possibility of evaluating multiple tissues at the same time.

Although MRI allows better representation of certain tissues such as the brain and without harming the patient, the limited availability of scanners and the slowness of their examinations are a major difficulty (Zorzal et al. 2019).

In contrast, the Ultrasonography has the highest availability of equipment, but only allows for few tissues to be rendered [35].

This is why we must know and consider all the advantages and disadvantages of each biomedical scanner at the moment of studying and representing virtually the human anatomy.

Finally, from the Digital Anatomy Laboratory CIMED-FUNDAMI, we consider it essential to establish the complementary place of the study and the virtual representation of the human anatomy from the DICOM® data in relation to the cadaveric study:

1. This information is added to the knowledge based on the classical study of the cadaveric anatomy.
2. It is trying to be considered as another tool for modern education.
3. It could be a valid alternative for all those who do not have access to corpse studies.

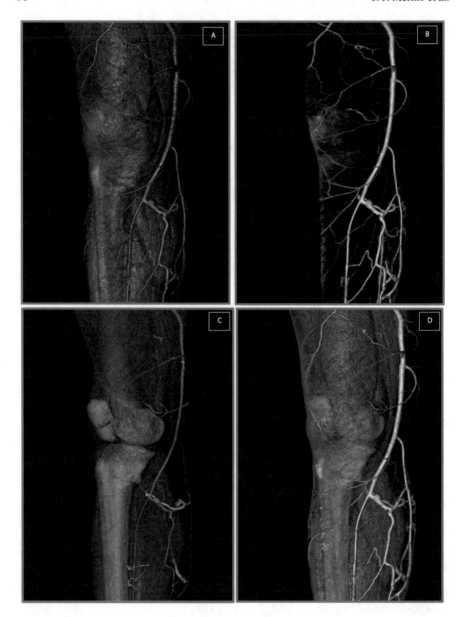

Fig. 5.24 Final steps to achieve an Enriched Rendered Volume: **A** Rendered Volume of the affected lower limb; **B** Dissected Volume of the Hypodermic Reticular Venous System; **C** Superposition of the highlighted vessels over the Musculoskeletal Volume; **D** Final superposition of all previously dissected volumes Enriching the Rendered Volume

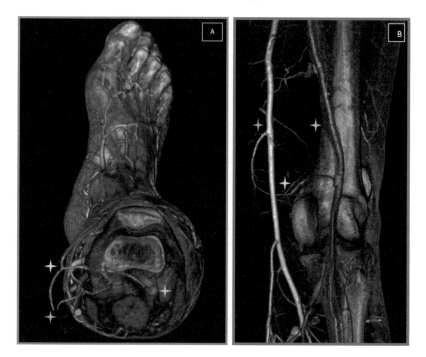

Fig. 5.25 Virtual Anatomists are able to illustrate, for example, how the Supracondylar Perforating Vein (⋆) communicates the Deep Venous System (⋆) with the Great Saphenous Vein (⋆) of the right leg. **A** Axial section; **B** Coronal section, posterior view

5.8 Conclusion

Rendered Volumes performed from DICOM® data prove to be, day after day, a very useful tool to achieve a proper analysis and dynamic virtual representations of human anatomy.

Due to the different disciplines it covers, we consider it is crucial to conform multi-disciplinary teams with anatomists, doctors in Diagnostic Imaging, radiology technicians, biomedical engineers and specialists from different areas such as surgeons, cardiologists, traumatologists, among others.

Also, it is essential to maintain a continuous education, to know in detail all the tools provided by the Volume Rendering Techniques and to train skilled and creative Virtual Anatomists. This is demanded by the constant technological development applied to medicine in general and specifically to the field of Diagnostic Imaging such as Artificial Intelligence, Machine Learning, Deep Learning, 3D Printing and Augmented Reality (Willemink and Noël 2018; Zorzal et al. 2019).

We also point out that all this is possible thanks to the characteristics of DICOM® data: it can be anonymized, portable and replicated by any digital anatomy laboratory.

For this reason, we recommend increasing the intercommunication to improve the performance of the discipline and we propose to develop a DICOM® database for the study and virtual representation of human anatomy.

Acknowledgements Authors acknowledge Celia Ferrari, MD for collaborating in the revision process.

Glossary

Digitize Convert or encode into digit numbers, data or information such as an image or a document.
Voxel Cubic unit making up a three-dimensional object, mm^3.
Pixel Homogeneous unit forming the matrix of a digital image, mm^2.
Virtual Image Representation that produces a similar effect to the real object.

References

Carro, Cecilia Lorena (2016) Generación de imágenes tridimensionales en tomografía helicoidal, [On line] http://www.unsam.edu.ar. Accessed 23 Feb 2019
Claudio Silva F-A et al (2014) Análisis del rol de la venografía de extremidades inferiores por tomografía computada en pacientes con sospecha clínica de tromboembolismo pulmonar. Revista Chilena de Radiología. 20(2):51–55
Ing. Dahilys González López, Ing. Liset M. Álvarez Barreras, Ing. Adrián Fernández Orozco (2014) Implementación de estándares DICOM sr y HL7 CDA para la creación y edición de informes de estudios imagenológicos (1):71–86, [On line] http://scielo.sld.cu
A. Espinosa Pizarro (2012) Técnicas de postprocesado de las imágenes (TC y RM): qué, como, cuando y porque s-1551, SERAM, [On line]. https://posterng.netkey.at
Dra. Eugenia Lucía Saldarriaga Cardeño (2016) May-Thurner syndrome as a differential diagnosis in recurrent thrombosis. Acta Med Colombia 41: 67–70
Francisco Javier Olías Sánchez (2014) Segmentación en 3d de huesos en imágenes TAC. Teoría de la señal y comunicaciones, escuela técnica superior de ingeniería, Universidad de Sevilla. [On line]. http://bibing.us.es
Gillot C, Uhl J-F, Ovelar J, Merino J (2019) Anatomy of the bony perforators veins of the knee. Annals Med 51: 60. https://doi.org/10.1080/07853890.2018.1561943
Huang Q, Zeng Z (2017) A review on real-time 3D ultrasound imaging technology. BioMed Res Int 1–20. https://doi.org/10.1155/2017/6027029
Ignacio García Fenoll (2010) Aportaciones a la segmentación y caracterización de imágenes médicas 3d. [On line]. http://bibing.us.es/proyectos/abreproy/11854. Accessed 1 May 2021
ISO 12052: 2017, [On line]. https://www.iso.org/standard/72941.html
Medina R, Bellera J (2017) Bases del procesamiento de imágenes médicas. Universidad de los Andes, facultad de ingeniería, grupo de ingeniería biomédica de la ULA, Venezuela, [On line]. http://www.saber.ula.ve. Accessed 23 Feb 2019
Ovelar JA, Dr. Cédola J, Tr. Merino JP (2014) Importancia de la fascia safénica en el desarrollo de la patología venosa de safena magna y de su menor porcentual patológico en su sector infrapatelar,

en relación a la indicación de tratamiento láser endoluminal. Ediciones de la Universidad del salvador, Bs. As.[On line]. https://issuu.com/revistasocflebologia

Ovelar JA, Dr. Cédola J, Tr. Merino JP (2015) Importancia de los afluentes proximales en la integración venosa torácica abdominal. Manual para el diagnóstico y tratamiento de las flebopatías[On line]. https://issuu.com/revistasocflebologia/docs/flebo_n___3_web_2015

Palacios Miras C (2012) Fundamentos de imagen digital aplicados a radiología. SERAM [On line]. http://pdf.posterng.netkey.at/download/index.php

Perandini S, Faccioli N, Zaccarella A, Re T, Mucelli RP (2010) The diagnostic contribution of CT volumetric rendering techniques in routine practice. Indian J Radiol Imaging 20(2):92–97. https://pubmed.ncbi.nlm.nih.gov/20607017/. Accessed 1 May 2021

Romina Luciana Muñoz (2015) La matemática en las imágenes médicas: tomografía computarizada. Universidad Nacional de La Pampa, Facultad de ciencias exactas y naturales [On line]1. http://redi.exactas.unlpam.edu.ar

Spanish Society of Cardiac Imaging. Los dogmas de fe: retroproyección filtrada, reconstrucción iterativa y filtros de Kernel, [On line]. https://ecocardio.com. Accessed 20 Aug 2018

Tierny J (2015) Introduction to volume rendering [On line]. https://www-apr.lip6.fr/~tierny/stuff/teaching/tierny_intro_vol_rend09.pdf. Accessed 20 Aug 2018

Uhl JF (2001) The Progress in imaging of the leg veins: Multislice CT venography Mayo Clinic International Vascular Symposium 2011. Editions Minerva Medica, pp 393–400

Uhl JF (2009) 3D investigation of the venous system by MSCT venography. In: Innovative treatment of venous disorders (7):61–73. Ed Cees Wittens. Editioni Minerva Medica 2009

Uhl JF, Caggiati A (2005) 3D evaluation of the venous system in varicoselimbs by multidetector spiral CT Multidetector row CT angiography. Springer Catalano C, Passariello R (Eds), 199–206

Uhl JF, Gillot C (2007) Embryology and three-dimensional anatomy of the superficial venous system of the lower limbs. Phlebology 22(5):194–206

Uhl JF, Gillot C, Verdeille S, Martin-Bouyer Y, Mugel T (2002) Three dimensional CT-Venography: a promising tool to investigate the venous system. Phlebolymphology 38:74–80

Uhl JF, Verdeille S, Martin-Bouyer Y (2003a) Three-dimensional spiral CT venography for the pre-operative assessment of varicose patients. VASA 32(2):91–94

Uhl JF, Verdeille S, Martin-Bouyer Y (2003) Springer Verlag Ed pavone, Debatin Pre-operative assessment of varicose patients by veno-CTwith 3D reconstruction. 3rd International workshop on multisliceCT 3D imaging, pp 51–53

Uhl JF, Ordureau S, Delmas V (2008) Les nouveaux outils de dissection anatomique virtuelle. e-mémoires de l'Académie Nationale de Chirurgie 7(2):39–42

Uhl JF, Chahim M, Verdeille S, Martin-bouyer Y (2012) The 3D modeling of the venous system by MSCT venography (CTV): technique, indications and results Phlebology 27:270–288

Uhl JF, Prat G, Costi D, Ovelar JA, Scarpelli F, Ruiz C, Lorea B (2018) Modelado 3d del sistema vascular. Flebología 44:17–27, [On line]. http://www.sociedadflebologia.com

Vicente Atienza Vanacloig (2011) El histograma de una imagen digital.[On line]. https://riunet.upv.es

Wang W, Lin J, Knosp E, Zhao Y, Xiu D, Guo Y (2015) Application of MSCTA combined with VRT in the operation of cervical dumbbell tumors. Int J Clin Exp Med 8(8):14140–14149

Willemink MJ, Noël PB (2018).The evolution of image reconstruction for CT—from filtered back Projection to artificial intelligence. [On line]. https://www.researchgate.net

DICOM digital imaging and communications in medicine, history, [On line]. https://www.dicomstandard.org. Accessed 7 Sep 20

UHL JF, Chahim M, Cros F, Ouchene A.3D modeling of the vascular system. J Theoret Appl Vascular Res JTAVR 1(1):28

Zorzal ER, Sousa M, Mendes D, dos Anjos RF, Medeiros D, Paulo SF, Rodrigues P, Mendes JJ, Delmas V, Uhl J-F, Mogorrón J, Jorge JA, Lopes DS (2019) Anatomy studio: a tool for virtual dissection through augmented 3D reconstruction. Comput Graph 85:74–84. https://doi.org/10.1016/j.cag.2019.09.006

Chapter 6
The Virtual Dissection Table: A 3D Atlas of the Human Body Using Vectorial Modeling from Anatomical Slices

Jean François Uhl, José Mogorron, and Maxime Chahim

Abstract Teaching morphological sciences' suffers from the lack of human corpses for dissection due to ethical or religious issues, worsened by increasing students' demand for educational anatomy. Fortunately, the technological revolution now put at our disposal new virtual reality tools to teach and learn anatomy. These multimedia tools are changing the way students engage and interact with learning material: they can engage in meaningful experiences and gain knowledge. This evolution is particularly true for the virtual dissection table, based on 3D vectorial atlases of the human body. This chapter describes the manual segmentation methodology from the anatomical slices of the Korean visible human project with Winsurf® software. Although using the same slices, our segmentation technique and refinement of the 3D meshes are quite different from the Korean team (Ajou University, Séoul, Korea). The resulting 3D vectorial models of the whole body of men and women include 1300 anatomical objects. After improvement with a modeler (Blender® version 4.79), we export 3D atlas into a ".u3d" format to take advantage of the powerful interface of the 3Dpdf Acrobat® file working in four different languages. The user interface is simplified by a touch screen to manipulate and dissect the virtual body with three fingers easily. Anatomical regions, systems, and structures are selected and controlled by javascript buttons. We developed this educational project under the Auspices of the Unesco chair of digital anatomy (www.anatomieunesco.org).

6.1 Introduction

Cadaver dissection remains the gold standard of knowledge in anatomy. Still, it has significant drawbacks: The lack of cadavers, which cannot cover the demand of the medical schools, and the limited location of the activity (anatomical theater).

J. F. Uhl (✉) · J. Mogorron · M. Chahim
Unesco Chair of Digital Anatomy, Descartes University, Paris, France
e-mail: jeanfrancois.uhl@gmail.com

M. Chahim
Service de médecine vasculaire HEGP, 75015 Paris, France

103

Moreover, we can only perform each dissection once because it relies on a destructive, irreversible process on the human tissues.

For all these reasons, the 3D reconstruction of anatomical structures promotes new educational methods widely used worldwide, most successful for their novel realistic and interactive interfaces.

This visualization is a splendid tool for the students willing to learn the human body and the teachers in anatomy and for interactive clinical simulation for practitioners (Spitzer and Whitlock 1998; Ackerman 1999; Spitzer et al. 2006). Finally, it is a revolution for surgeons to help preoperative planning, simulation, and augmented reality during surgical achievement.

3D digital visualizations open a new way to *teach anatomy* due to the data's digital nature, enabling a quantitative morphological analysis in the frame of computational anatomy (Miller 2004).

It also opens a new way for the young to *learn anatomy*: by drawing the boundaries of the anatomical structures on the slices manually, they progressively build the whole 3D models. These manipulations lead to the sustainable learning of anatomy by a better comprehension of spatial relationships between anatomical structures.

6.2 History of the Visible Human Projects

The US Visible Human Project (VHP; the male and then female) conducted by John Ackerman in 1999, was the first in the world (Ackerman 1999; Ackerman 1998; Spitzer et al. 1996).

Since 2005, several similar projects have seen the light. First, was the visible Korean project. Its researchers digitized and processed the whole male body, then the male head, and finally, the whole female body (Park et al. 2005a, 2006, 2015). More recently, the Chinese Visible Human, developed digital representations of male and female bodies (Zhang et al. 2006).

Finally, the Virtual Chinese Human project, aiming to digitize male and female anatomies, started in 2008 (Tang et al. 2010).

Figure 6.1 shows the different teams worldwide who produced different datasets of thin anatomical slices (0.2–1 mm in thickness) of the human body. Researchers created these parts using a cryomacrotome slicing of the frozen bodies.

The educational use of these 3D anatomical models of the whole body provides high added value thanks to the Korean team's various computer and imaging tools:

– Software tools for Browsing the anatomical and segmented images (Shin et al. 2011). These programs could be used by the students to identify different anatomical structures from the slice data.
– Virtual dissection software working with Acrobat® 3D PDF interface (Shin et al. 2012, 2013; Chung et al. 2015) is straightforward and displays the vectorial models of about 1000 anatomical structures together with the horizontal slices.

Fig. 6.1 The teams included in the Visible human projects over the world. From left to right: Mr. Tang (virtual Chinese human), Pr Chung (Séoul, Korean visible human), Pr Ackerman (visible human project, USA), Pr Zhang (Chinese visible human), and Pr Liu (new visible Chinese)

Most importantly, private companies created dedicated virtual dissection tables, using the same anatomical data. The most famous, named Anatomage® (https://www.anatomage.com/), is probably the most sophisticated virtual dissection system (but not the cheapest) commercially available worldwide.

6.3 Virtual Reality Techniques: A New Human–Computer Interface for Education

The virtual dissection table is today the main educational tool in Anatomy Courses. But other techniques have been developed, using similar 3D vector models:

Computer Science researchers used Virtual reality head-mounted displays with a dedicated software working, in particular, with Samsung Oculus GearVR® (http://

3Dorganon.com, https://www.roadtovr.com). The unique advantage of learning in Virtual Reality is that immersive virtual environments keep distractors away due to a high degree of absorption and responsiveness by kicking the boredom factor out of the classroom.

The 3D VR tool interface has many unique advantages. Stereoscopic presentation of 3D visuals makes for enticing content; immediate response and interactivity foster user immersion; elimination of distractors, gamification of learning, multi-sensory experience, and high interactivity may improve the learning experience and retention. Indeed, virtual reality techniques have a high emotional impact on the learning process.

6.3.1 Objectives

This chapter aims to describe our methodology and results regarding the 3D vectorial modeling of the human body from anatomical slices and compare them with the *Korean visible project team of Seoul.*

6.3.2 Materials and Methods

1. Material

Our database was a series of anatomical slices of the *Korean visible project* (man and woman). Our department has signed a contract of partnership with the Korean team of Seoul in 2012, and we collaborate together for 14 years.

The *KVH male* cadaver (Park et al. 2005a) was a young adult (33 years old) without obesity (height, 1.65 m; weight, 55 kg).

A cryomacrotome (Fig. 6.2) was used to obtain 8506 sectioned images of the whole body.

For the *KVH female*, the subject was aged 26 years of standard body size: length, 1690 mm; weight, 52 kg, frozen at $-70\,°C$. The serial-sectioning was achieved from the vertex to the perineum at 0.2 mm interval and from the perineum to the toes at 1 mm interval. That produced 5960 anatomical slices of the whole body (Park et al. 2015). For photographing the sectioned surfaces of both bodies, a Canon™ EOS-1Ds Mark III™ was used, equipped with a Canon™ EF 50 mm f/1.2L USM lens. The resolution of the resulting images was 5616×3744 pixels, saved in the TIFF format.

The Korean team achieved since 2010, the 3D reconstruction by vectorial modeling from these data with the following steps:

- The first step was the colored segmentation slice by slice by using Photoshop. Each anatomical element was outlined with the magic wand (Park et al. 2005b).

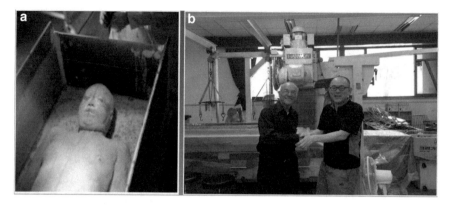

Fig. 6.2 a The frozen body of the KV woman. **b** Cryomacrotome (Pr Chung, Ajou University, Seoul, Korea)

– The second step was to create a 3D vector model with Mimics® software from the series of colored segmented slices.
– The next step was the embedding of the slices into the 3D model with Maya®…

Finally, the resulting 3D model was converted into a ".u3d" file format in order to take advantage of the 3Dpdf Acrobat interface, for an interactive display of the 3D model together with the anatomical slices.

This huge work of segmentation was achieved in 8 years by the Korean team. We collaborate with them since 2005 and wrote a paper together showing a reconstruction of the urogenital tract in the man (Uhl et al. 2006).

2 **Our own Methodology** uses the same slices with the five following steps: (Fig. 6.3)

- Segmentation and 3D vectorial modeling of anatomical elements with Winsurf® software version 3.5 (Moody and Lozanoff 1997) from the anatomical slices.
- Exportation of the Winsurf® mesh into a cad format
- Mesh refinement, cleaning, and arrangement with Blender®
- Ranking of 1300 anatomical elements with Acrobat 3D toolkit®
- Building the final user interface with Acrobat pro®.

2.1 *Segmentation and 3D reconstruction of mesh models*

Instead of firstly performing a segmentation with Photoshop like Park et al., we directly achieved the segmentation on the anatomical slices of the KVW dataset by using the Winsurf® software (Moody and Lozanoff 1997). The main steps and functions of the software were the following, previously explained in our papers to reconstruct the urogenital tract (Uhl et al. 2006) and the heart (Uhl et al. 2017) of the Korean visible human.

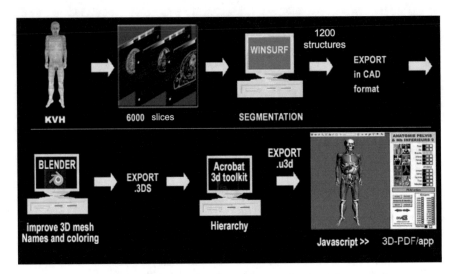

Fig. 6.3 Methodology from the anatomical slices to the final body Atlas. It follows 5 steps: 1—Starting with 6000 anatomical slices done with the cryomacrotome. 2—Segmentation by outlining each anatomical structure over the anatomical slices, which creates 1200 3D vector objects. 3—Exportation of the meshes in cad format. 4—Refinement and improvement of the meshes with Blender version 2.79b. This includes the use of the "Skin" modifier to obtain a more realistic display of the nerves and of the vessel trees. 5—Ranking the list of anatomical objects with Acrobat toolkit®. 6—Building a 3Dpdf interface with JavaScript language (Acrobat®)

Creation of new 3D objects: Each 3D object representing an anatomical structure was built separately, identified by placing manually chains of points around the object edges, by this way each distinct anatomical structure was segmented by mapping. Figure 6.2a shows the software interface, with the available tools on the left upper side of the window to edit or modify the contours. Here the boundaries of the veins of the left upper limb were drawn.

Scale measurement of the 3D model. The scale parameters were measured on a slice as well as the slice thickness by a special function of the software. The resulting value was 7000/2/x for the woman, where x represents the jump between slices during display. For the man, the parameters were 5650/2/x. The same scale parameters are to be used for all anatomical objects.

Creation of 3D anatomical objects is achieved by outlining them slice by slice: it is a manual thorough segmentation of the objects. This step can seem long and tedious, but it is an essential part of the process of learning anatomy.

Once the contours had been assigned to each serial slice containing the object, vertices were connected using a surfacing routine developed especially for Winsurf® software (Uhl et al. 2006). This routine computes and refines a best-fit solution for assembling the vertices into a volumetrically optimized 3D object. Figure 6.4b shows how to use several color channels to build the arborescence of upper limb's veins. This option is very useful for the user to localize a mistake or missing information on one of the colored channels.

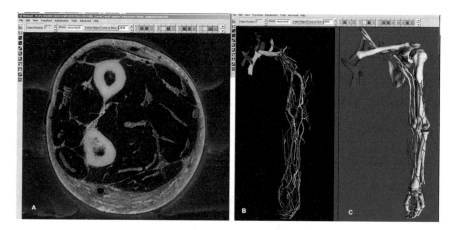

Fig. 6.4 Creation of 3D mesh of the upper limb's veins with Winsurf software. **a** Outline of the 2D slice of veins of the upper limb by using several colored channels. **b** Resulting 3D model with the colored 3D mesh in "wireframe" mode. **c** Resulting modeling of the bones, nerves, arteries, and veins displayed inside the 3D window

When finished, all objects could be gathered by using the "add surface file" option of the software, in order to obtain the final 3D model. In addition, textures could be applied on the objects to obtain a more realistic 3D reconstruction (Fig. 6.4c).

2.2   **Exportation** *of the 3D mesh model Mesh produced by Winsurf*® software version 3.5

The second step was the export of the whole 3D vectorial model built by Winsurf® into 3D PDF format in order to improve the anatomical model. In fact, Winsurf® is an old and cheap software working well with the old windows® xp Operating system, and easy to use.

Before this exportation, it is important to ensure that every anatomical object has a different RGB color, in order to import it as a separate anatomical element under the Acrobat format. The format conversion was achieved by using a simple function of Adobe Acrobat® 3D version 9, in the menu option of the software: "create a pdf file by 3D capture". It could be then exported in .wrl file format and exported to Blender software.

2.3   **Mesh** *refinement with a powerful free modeler: Blender*® *v 2.79b* (http://ble nder.com)

Improvement of the mesh models and correction of some anatomical issues were done with Blender®, as well as the embedding of a series of 100 anatomical slices into the 3D model. This step is very important to improve the quality of the models and the anatomical accuracy, as well as to improve the student's understanding of anatomical relationships in the 3D space.

Regarding the vessels, the nerves, and the tendons, the ugly 3D mesh of Winsurf®
were replaced by "skins" (with the skin modifier of Blender®), to produce more
realistic anatomical structures (Fig. 6.5).

2.4 *Exportation of the 3D model into Acrobat3D Toolkit*®

After exportation of the whole Blender 3D model and slices into .obj or .3Ds format,
Acrobat 3D toolkit® software was used to produce a .u3d compatible file and to set
up the hierarchical list of the anatomical elements, displaying them in the model tree
window located on the left side of the Acrobat window.

Fig. 6.5 Reconstruction of the great vessels of the thorax and abdomen with the anatomical slice
corresponding to the 5th thoracic vertebra. (Blender® v 2.79 window interface). 1 = ascending
aorta. 2 = pulmonary artery trunk. 3 = trachea. 4 = right bronchia. 5 = right subclavian artery. 6
= common carotid artery. 7 = vertebral artery. 8 = splenic artery. 9 = common hepatic artery. 10
= right coronary artery

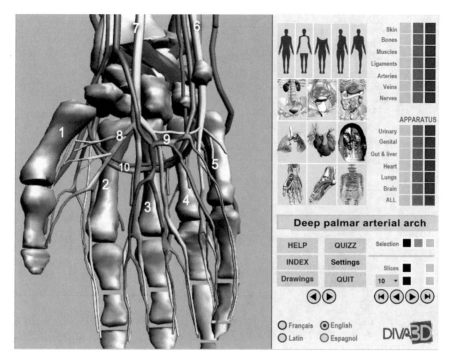

Fig. 6.6 **Interface screen of the virtual dissection table** showing the 3D model of the right hand (palm view). On the left, the 3D window. On the right, the function buttons to choose the display mode to select by area, apparatus, system, organ, and/or slices

2.5   *3D interactive display with Acrobat*® *3D PDF interface*

The interface is very versatile with the command panel on the right, giving access to the different anatomical regions of interest, here the hand (Fig. 6.6).

A selection of the 3D interactive model is possible:

– by area (full body, trunk, head & neck, upper limb, lower limb - icons on the top)
– by system (integumentary, urinary, genital, alimentary, respiratory, circulatory)
– by region of interest (pelvis, neck, brain, hand, foot …)

For any of these options, by clicking the color boxes of this list of anatomical structures (on the right) one can make any element visible, hidden or transparent. You can also select and display the horizontal slices together with the 3D models (Fig. 6.7)

Other functions of the Acrobat® interface are also available by using the 3D toolbar located at the top of the display window (Fig. 6.8).

Menu of interactive handling functions of the 3D model, choice of lighting effects, background color, 3D rendering options (solid, wireframe, shaded wireframe, transparent wireframe, illustration…), 3D labels, cross section in any plane, and "menu"

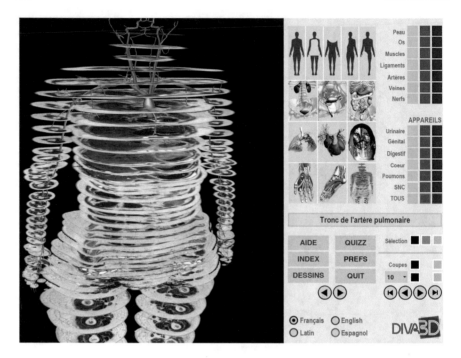

Fig. 6.7 Interface screen of the virtual dissection table showing the 3D model with the anatomical slices (in French). The 3D window on the left shows the display of the original anatomical slices together with the vectorial models (here the arteries)

memorizing the display view together with the whole parameters of the 3D anatomical model. We finally created a user friendly interface with JavaScript® programming with the help of our colleagues of Amsterdam University who created an outstanding 3D tool dedicated to the human embryo development (de Bakker et al. 2016; de Boer et al. 2011) available on: www.3Dembryoatlas.com/publications.

By clicking on a structure, the 3D object is highlighted and its name is displayed in 4 languages inside the blue window.

As a result, all these functionalities are available through the simplified interface of the Diva3D® virtual dissection table represented by a big touch screen working with only 3 fingers. (Fig. 6.9) A demonstration of this tool is available on the website www.diva3D.net, extended to the whole female body and the male.

Results

We have segmented 1131 anatomical structures to build the whole 3D model: (Table 6.1) Skin (1), muscles & tendons (370), ligaments (82), arteries (91), veins (150), nerves (66), and Organs (40).

The user interface includes 3D handling, zoom, navigation, and slicing in any plane. Students can arrange anatomical elements into systems and areas of interest

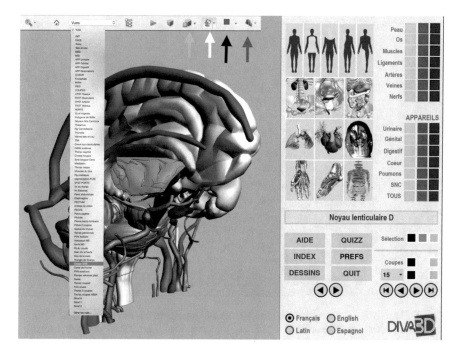

Fig. 6.8 Interface screen of the virtual dissection table showing the 3D model of the brain and vessels. The toolbar options show access to the list of the anatomical regions available directly to give courses of anatomy. On the top right, interface settings could be changed: 3D slicing in any plan (red arrow)Background color (black arrow), Lighting (white arrow), and 3D rendering options (green arrow)

by using menus provided by Acrobat pro®, directly available by buttons on the touch screen.

The terminology of all anatomical structures is available in 4 languages: French, English, Latin, and Spanish by clicking on each 3D object.

In practice, we have used this table for two years in our university and several schools for courses and practical work in anatomy. The student satisfaction index reaches 85% according to our post-course evaluations.

Discussion

The Interest in 3D models and digital teaching tools in anatomy is increasing.

From the data set of the KVH, the Korean team has also built a set of 4 outstanding learning tools free of charge, downloadable on their website www.vkh. ajou.ac.kr/#vk.

As we saw previously, the authors used a different methodology to build the vectorial models. They first made a colored segmentation on the slices with Photoshop® tools (Park et al. 2005b) and then use Mimics® to build the 3D vectorial models,

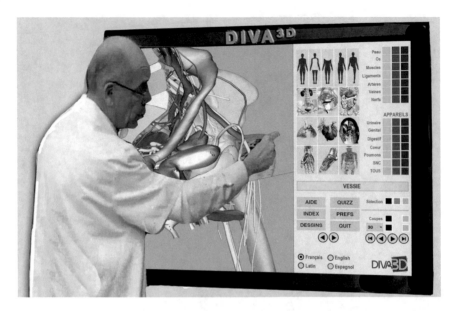

Fig. 6.9 The virtual dissection Table Diva3D interactively handled with 3 fingers with a 65 inches touch screen (demonstration available on the website www.diva3D.net). The touch screen interface allows the user to easily manipulate the 3D model using only 3 fingers and to access all the functions and menus of the program. Here is shown the right view of the pelvis, using a sagittal cutting plane to give access to the organs

lastly embedding the original slices with Maya®. This nontrivial work took about 8 years.

They have provided a different kind of computer–human interface with the 4 following learning tools: (Chung and Chung 2019)

- *The first tools* are browsers to identify the structures on the sectioned and color-filled images.
- *The second tool* interactively shows surface models of individual structures. Automatic annotation of the segmented structures is possible, so it can be used similarly to our file as a virtual dissection table.
- *The third tool handles a volume model* that was continuously peeled and could be 3D printed out.
- *The fourth tool* is the most powerful, with the possibility of handling interactively a volume model that is freely sectioned in any plane. It works with a special software: MRIcroGL® from Professor Chris Rorden available on www.mccaus landcenter.sc.edu/mricrogl/home.

The user can display the horizontal, coronal, and sagittal planes simultaneously as if a block part of the volume model is removed (Fig. 6.10).

Table 6.1 1131 structures in the female body segmented and reconstructed as anatomical 3D objects

Systems	N	Polygons (thousands)	Names of anatomical elements reconstructed
Integumentary	1	20	Skin
Skeleton	230	443	Head & spine, ribs pelvis and limbs
Muscles & tendons	370	410	Head & neck, thorax, abdomen pelvis and 4 limbs
Ligaments	82	16	Spine pelvis and 4 limbs
Respiratory	18	190	Right & left lungs with segments, trachea, bronchi and vessels
Heart	6		Left & right atrium & ventricles–valves
Arteries	91	131	Head & neck, thorax, abdomen pelvis and 4 limbs
Veins	150	354	Vena Cava, Azygos and Portal systems
Nerves	55	77	Thorax, abdomen pelvis and 4 limbs
Brain	27	57	7 cortex lobes, central grey nuclei, ventricles, pituitary gland
Alimentary	25	51	Oesophagus, Gaster, duodenum, small intestine, ascending colon, transverse colon, descending colon, sigmoid, rectum, Liver with VIII segments, Gallbladder, biliary tract
Urinary	6	22	Bladder, Right & left kydneys & Ureters
Genital	10	15	Uterus, 2 salpinx and ovaries, vagina, round and uterosacral ligaments
Glands	10	9	Thyroid, Surrenals, Parotids, submaxillar, sublingual
Anat. slices	50	0	Selected original anatomical slices
Total	1131	1795	

In addition, several 3Dpdf files can simulate simple surgical procedures. They are freely available on the Korean website: dorsal foot flap (Shin et al. 2015), anterior rib flap, deep circumflex artery, virtual colonoscopy.

Other Educational Tools

Another interesting educational tool using the same data allows *3D printing* of the anatomical elements (AbouHashem et al. 2015).

We also previously mentioned the outstanding *3D embryo atlas* built by the team of Amsterdam University (de Boer et al. 2011; Chung and Chung 2019).

In the *field of neuroanatomy* with a similar interface, splendid tools are proposed by Prats-Galino et al. (2015): 3D interactive model of the lumbar spine shown in Fig. 6.11 and simulation of endoscopic endonasal surgical approaches (Mavar-Haramija et al. 2015).

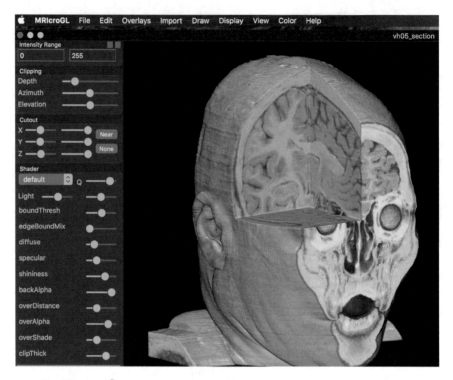

Fig. 6.10 MRIcroGL® software using the dataset of the Korean visible human's head. Software from Chris Rorden. One can handle interactively the Korean head model that is freely sectioned in any plane. The control panel (on the left) modifies the axis, direction, and depth of the clipping planes demonstrated in the figure

Moreover, commercial software programs are available on the web for virtual anatomy, running on tablets or smartphones. Visible body® is probably the main one, but they suffer from the lack of accuracy. Big issue: they are built by graphic artists and not experts in morphology working on real human bodies.

A powerful simulator "Simanato project" is currently developed by Dr. René Cejas Bolecek et al. (https://www.youtube.com/watch?v=h51C8LIxBZE&t=3s (simanato software)) from Centro Atómico Bariloche (San Carlos de Bariloche, Rio Negro, Argentina) with several other institutions.

Analysis of traditional versus 3D augmented curriculum on anatomical learning has been studied in several papers:

Peterson and Mlynarczyk (2016) showed that the addition of 3D learning tools can influence long-term retention of gross anatomy material and should be considered as a beneficial supplement for anatomy courses.
Moro et al. (2017), Erolin (2019), Birt et al. (2018) showed that Both virtual reality (VR) and augmented reality (AR) techniques are as valuable for teaching anatomy as tablet devices, but also promote intrinsic benefits such as increased

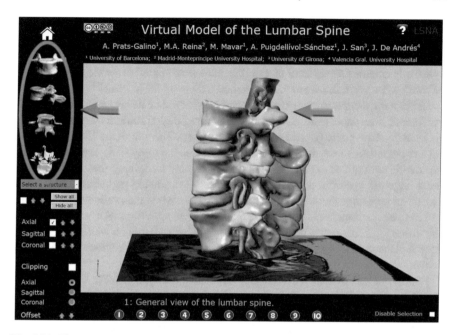

Fig. 6.11 Virtual 3D model of the lumbar spine by Prats-Galino et al. (Prats-Galino et al. 2015). The CT slices could be displayed in the axial, sagittal, and coronal planes combined with the 3D interactive vector model using the same interface of an Acrobat® 3Dpdf file

learner immersion and engagement. These outcomes show great promise for the effective use of virtual and augmented reality as a means to supplement lesson content in anatomical education. The interest for the students is also to use a mobile application (Kurniawana et al. 2018).

In summary, the main interest of our offline educational 3Dpdf tool is the simplicity and the power of the human–computer interface using a body-size touch screen.

The command by only 3 fingers gives access to the whole functions of the device: 2D handling and rotating, zooming, and cutting in any plane. Selection of anatomical elements (with their names available in 4 languages) and modify their transparency. One can also display the horizontal anatomical slices, and cut in any direction of the 3D model.

There is *another educational interest of our work* related to the user interface. The segmentation process done by outlining the anatomical slices to build the vectorial model is one of the most powerful training to memorize the 3D anatomical volumes inside the student's brain. The progressive building of the 3D model from slice to slice is the best way for accurate and sustainable learning of the relationships between anatomical structures.

However, there are some *limitations of our work*:

We did not achieve the reconstruction of the cranial and thoracic nerves, as well as the peritoneum.

For some structures as vertebra, heart, and cerebrum, it is really difficult to achieve a perfectly accurate segmentation. For this reason, we started to develop a web-based software to improve the segmentation process done by the Winsurf® software.

In the near future, these new computer tools should be developed in the frame of a partnership available through the Unesco Chair of digital anatomy, created recently in Descartes University (www.anatomieunesco.org). The aim of the UNESCO Chair is to promote all these new educational tools for anatomy. It is also to set up a worldwide network of experts in morphology willing to develop these digital tools, and a huge 3D vectorial database in anatomy. A worldwide partnership is the best way to set up a huge databank of accurate 3D vectorial models validated by experts.

These educational tools could be used in different situations: Self-directed study, classroom teaching, workshops. They also could be available on web servers. Specimen replacement is also necessary to study the anatomical variations of the human body.

Finally, in addition to research and educational anatomy, these tools could be adapted and used in other fields, in particular surgical training, simulation, and planning.

This opens the way for image-guided operations, especially in abdominal surgery: mini-invasive, more controlled, more accurate, safer because avoiding the main complications.

An outstanding example of computer and imaging assisted surgery is given by Marescaux and Soler from IRCAD (www.ircad.fr), who created a web server and an application named visible patient (Fig. 6.12). The system provides the vascular

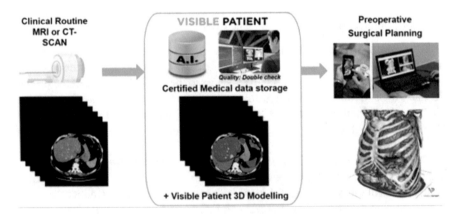

Fig. 6.12 Visible patient software™ is a company resulting from 15 years of research of the IRCAD R&D department in computer assisted surgery. Visible Patient proposes a connected solution providing a 3D model of a patient from his/her medical image sent through a secured internet connection. The surgeon can make a simulation and preoperative planning on a tablet for each patient by using the interactive 3d model (from IRCAD website)

anatomy with the segmentation of the liver, the kidney, and the lung (Soler et al. 2017). More than 5700 surgeons over the world participate in training organized by IRCAD each year.

A number of other applications of 3D modeling are available today including the possibility of *3D printing anatomical objects* in biocompatible materials for orthopedics, dental implantology, and neurosurgery.

6.4 Conclusion

By producing these interactive and accurate 3D anatomical models, our goal is to provide anatomical data and processing systems that contribute to the student's education, but also for basic, clinical medicine and surgical simulation.

The Acrobat® 3Dpdf interface provides a powerful and easy, offline educational tool for students. It could also be used to simulate surgical operations and training. A large network of partners through the UNESCO Chair of Digital anatomy (www.anatomieunesco.com) should contribute to develop these promising tools in a near future.

References

AbouHashem Y, Dayal M, Savanah S, Štrkalj G (2015) The application of 3D printing in anatomy education. Med Educ Online 20:29847

Ackerman MJ (1998) The visible human project. Proc IEEE 86:504–511

Ackerman MJ (1999) The visible human project. A resource for education. Acad Med 74:667–670

Birt J, Stromberga Z, Cowling M, Moro C (2018) Mobile mixed reality for experiential learning and simulation in medical and health sciences education. Information 9:1–14. https://doi.org/10.3390/info9020031

Chung BS, Chung MS (2019) Four learning tools of the Visible Korean contributing to virtual anatomy. SRA (in press)

Chung BS, Shin DS, Brown P, Choi J, Chung MS (2015) Virtual dissection table including the visible Korean images, complemented by free software of the same data. Int J Morphol 33(2):440–445

de Bakker BS, de Jong KH, Hagoort J, de Bree K et al (2016) An interactive three-dimensional digital atlas and quantitative database of human development. Science 354(6315)

de Boer BA, Soufan AT, Hagoort J et al (2011) The interactive presentation of 3D information obtained from reconstructed datasets and 3D placement of single histological sections with the 3D portable document format. Development 138:159–167. https://doi.org/10.1242/dev.051086

Erolin C (2019) Interactive 3D digital models for anatomy and medical education. In: Rea P (eds) Biomedical visualisation. Advances in experimental medicine and biology, vol 1138. Springer, Cham

http://3Dorganon.com

http://blender.com

https://www.anatomage.com/

https://www.roadtovr.com

https://www.youtube.com/watch?v=h51C8LIxBZE&t=3s (simanato software)

Kurniawana MH, Suharjitoa, Dianab, Witjaksonoa G (2018) Human anatomy learning systems using augmented reality on mobile application. Procedia Comput Sci 135:80–88

Mavar-Haramija M, Prats-Galino A, Juanes Méndez JA, Puigdelívoll-Sánchez A, de Notaris M (2015) Interactive 3D-PDF presentations for the simulation and quantification of extended endoscopic endonasal surgical approaches. J Med Syst 39:127 https://doi.org/10.1007/s10916-015-0282-7

Miller M (2004) Computational anatomy: shape, growth, and atrophy comparison via diffeomorphisms. NeuroImage 23(Suppl 1):S19–S33

Moody D, Lozanoff S (1997). SURFdriver: a practical computer program or generating 3D models of anatomical structures. In: 14th Annual Meeting of the American Association of Clinical Anatomists, Honolulu, Hawaii

Moro C, Štromberga Z, Raikos A, Stirling A (2017) The effectiveness of virtual and augmented reality in health science and medical anatomy. Ana Sci Educ. https://doi.org/10.1002/ase.1696

Park JS, Chung MS, Hwang SB et al (2005a) Visible Korean human: improved serially sectioned images of the entire body. IEEE Trans Med Imag 24:352–360

Park JS, Chung MS, Hwang SB, Lee YS, Har DH, Park HS (2005b) Technical report on semiautomatic segmentation by using the Adobe Photoshop. J Digit Imag 18(4):333–343

Park JS, Chung MS, Hwang SB et al (2006) Visible Korean human: its techniques and applications. Clin Anat 19:216–224

Park HS, Choi DH, Park JS (2015) Improved sectioned images and surface models of the whole female body. Int J Morphol 33(4):1323–1332

Peterson DC, Mlynarczyk GS (2016) Analysis of traditional versus three-dimensional augmented curriculum on anatomical learning outcome measures. Anat Sci Educ 9(6):529–536. ISSN: 1935-9780

Prats-Galino A, Reina MA, Haramija MM, Puigdellivol-sanchez A, Juanes Mendez JA, De Andrés JA (2015) 3D interactive model of lumbar spinal structures of anesthetic interest. Clin Anat 28:205–212 (2015)

Shin DS, Kim HJ, Kim BC (2015) Sectioned images and surface models of a cadaver for understanding the dorsalis pedis flap. J Craniofacial Surg 26(5):1656–1659

Shin DS, Chung MS, Park HS, Park JS, Hwang SB (2011) Browsing software of the Visible Korean data used for teaching sectional anatomy. Anat Sci Educ 4(6):327–332

Shin DS, Chung MS, Park JS et al (2012) Portable document format file showing the surface models of cadaver whole body. J Korean Med Sci 27(8):849–856

Shin DS, Jang HG, Hwang SB, Har DH, Moon YL, Chung MS (2013) Two-dimensional sectioned images and three-dimensional surface models for learning the anatomy of the female pelvis. Anat Sci Educ 6(5):316–323

Soler L, Nicolau S, Pessaux P, Mutter D, Marescaux J (2017) Augmented reality in minimally invasive digestive surgery. In: Lima M (ed) Pediatric digestive surgery. Springer, pp 421–432

Spitzer VM, Whitlock DG (1998) The visible human dataset. The anatomical platform for human simulation. Anat Rec 253: 49–57

Spitzer VM, Ackerman MJ, Scherzinger AL (2006) Virtual anatomy: an anatomist's playground. Clin Anat 19:192–203

Spitzer VM, Ackerman MJ, Scherzinger AL et al (1996) The visible human male: a technical report. J Am Med Inform Assoc 3:118–130

Tang L, Chung MS, Liu Q et al (2010) Advanced features of whole body sectioned images: virtual Chinese human. Clin Anat 23:523–529

Uhl JF, Park JS, Chung MS et al (2006) Three-dimensional reconstruction of urogenital tract from visible Korean human. Anat Rec A Discov Mol Cell Evol Biol 288:893–899

Uhl JF, Hautin R, Park JS, Chung BS, Latremouille C, Delmas V (2017) Tridimensional vectorial modeling of the heart and coronary vessels from the anatomical slices of the Korean Visible human. J Hum Anat 1(3): 1–9

www.ircad.fr

Zhang SX, Heng PA, Liu ZJ (2006) Chinese visible human project. Clin Anat 19:204–215

Chapter 7
Segmentation and 3D Printing of Anatomical Models from CT Angiograms

Guillermo Prat, Bárbara Constanza Lorea, Camilo Ruiz, and Franco Saúl Scarpelli

Abstract Many fields have adopted 3D technologies, and medicine is no exception. Their use ranges from educational purposes to skill training and clinical applications. This chapter proposes a possible protocol related to obtaining 3D anatomical models from Computed Tomography Angiogram (CTA) data and its subsequent 3D printing. We describe relevant features of free software available for this process as an introductory guide to those who want to make their first steps. We briefly discuss some of the benefits and drawbacks of applying 3D anatomy in pedagogical and surgical areas.

7.1 Introduction

Gross anatomy is one of the pillars of medical instruction as it is in other health sciences. The courses aim to create a three-dimensional brain image of the body's architecture, with clinical relevance (Arráez-Aybar et al. 2010) during a physician practice and other health sciences. The classical study of gross anatomy is generally based on a cadaver's dissection or learning from a pre-dissected piece. This pedagogical strategy was questioned from the late 1990s to the early 2000s, and alternate methods were proposed (Gregory and Cole 2002). Some medical schools have gone even further and proposed to teach anatomy without using cadavers but alternate methods (McLachlan and Regan De Bere 2010). Advances in technology defined a new paradigm in which three-dimensional (3D) objects can be easily created from reliable data such as the visible human project (Ackerman 1998) or anatomical slices from imaging methods (developed later in this chapter).

3D models can be created either from anatomical slices or using graphics and animation software. The first will have a trustworthy origin, whereas the second relies on the knowledge of the operator. The data from imaging methods allow automatic reconstruction algorithms, including volume and surface rendering (Chiorean

G. Prat (✉) · B. C. Lorea · C. Ruiz · F. S. Scarpelli
Laboratorio de Investigaciones Morfológicas Aplicadas "Dr. Mario H. Niveiro", Facultad de Ciencias Médicas, Universidad Nacional de La Plata, La Plata, Argentina
e-mail: gdprat@gmail.com

et al. 2011) (described later in this chapter). Though software generates many of the features automatically, there is an important work to be done manually. Obtaining high-quality organ models is not a trivial process, Human–Computer Interface (HCI) being the key.

After reconstructing the raw 3D model from anatomical slices, various software programs are used to refine and edit it. The final 3D model merges descriptive anatomy (i.e., the morphology of an organ) with topographic anatomy (i.e., relationship with surrounding organs). They can be displayed on screens, from a computer or other devices, using different methods (e.g., pdf files, upload to a web page, virtual reality, among other methods) or made into physical objects by 3D printing. This technique allows interacting with models by looking at them and giving back tactile and spatial feedback.

Based on patents from 1984, the first enterprise selling 3D printers incorporated in 1992 as 3D Systems (Rock Hill, South Carolina) (Horvath 2014). In 2004, Reprap software was created (Jones et al. 2011). Reprap is an open-source code project that aims to build a 3D printer to print most of its components. From this point onwards, 3D printing started expanding, being today broadly available. Since its foundation, much research has focused on the potential application of 3D printing in medicine. Wake Forest's Regenerative Institute is one of the main referents in the field, working with organ regeneration implants based on 3D scaffolds (Prince 2014).

3D models can be used both for pedagogical and clinical matters. 3D prints can complement cadaveric dissection to teach gross anatomy courses (Hashem et al. 2015). Various areas can benefit from 3D reconstruction. Surgery is undoubtedly one of the fields where these techniques have greater use (Rengier et al. 2010). We discuss some of the most relevant uses of 3D modeling in the following sections.

7.2 Objectives

This chapter aims to describe one of the protocols to create 3D models from Computed Tomography Angiograms (CTA) and 3D print. There is a wide range of software that provides the tools that are necessary for this process. We will focus on those that are free and provide a more accessible first approach to the matter.

7.3 Materials and Methods

Creating 3D objects from bi-dimensional (2D) images is based on following a surface's perimeter through a stack of anatomical slices, which must follow a regular space interval and be aligned. The software will compute the perimeters marked and create a mesh of triangles of the selected surface representing the 3D model

(Fig. 7.1). When selecting a perimeter corresponding to a particular anatomical structure, marking it while ignoring other structures is called segmentation. An example of the step-by-step process will be developed in the following sections.

Thus, the model can be edited using compatible software to improve the surfaces, clean artifacts, make 3D segmentations, select a certain part of the segmentation, or scale the model, among other actions. Many software programs can be used to this end. In the following sections, we will detail some of them.

There are several ways to 3D printing a mesh based on additive manufacturing. This means creating an object layer by layer, unlike subtractive manufacturing (creating an object by eliminating parts of a solid object). Furthermore, there are different kinds of additive manufacturing, such as fused deposition modeling, stereolithography, digital light processing, among others (Jasveer and Jianbin 2018). In

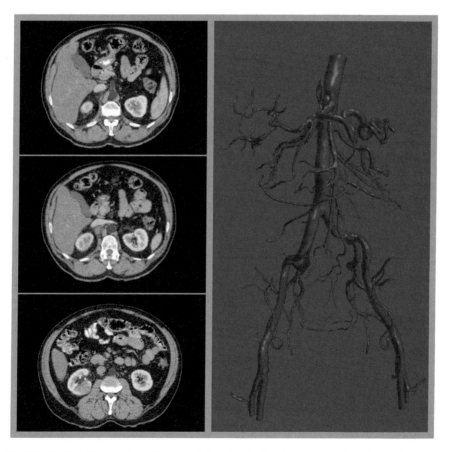

Fig. 7.1 On the left panel three slices out of three hundred and ninety-seven have been selected as representative of an abdomen CTA, the abdominal aorta perimeter, as well as the ones corresponding to its branches, has been highlighted in red. On the right panel the 3D reconstruction of the aorta

this chapter, we will describe Fused Deposition Material (FDM). FDM is an easy-to-learn technique. Compared to the other materials, thermoplastics are not expensive. While the method is accurate, it is limited by the thermodynamic characteristics of the material. The end product has a standard finish on its surface, and there are limitations to its flexibility depending on the plastic filament.

7.3.1 3D Model Creation

Computed Tomography (CT) is a radiological imaging method. A source emits X-rays that penetrate the different tissues. When they reach the detector, a computer can process the information to create cross-sectional images (Kalender 2011). A constant interval separates slices from one another, which makes the reconstruction process more manageable. The final image of a CT is a slice that will represent the body according to the penetration of the X-rays measured in Hounsfield Units (HU). The threshold goes from −1000 HU for the air to over 1000 HU for the cortical bones and metals. The remaining structures will have different values, −100 to 60 HU for soft tissue, including fat, 0 HU for water, and different values for other tissues (Hsieh 2003).

Angiotomography (CTA) in use since the late 1970s (Weinstein et al. 1977) combines a conventional CT scan with the injection of a radiopaque contrast to the vascular system. This process allows visualizing the vessels, and the blood within them, with an enhancement of HU. The CTA study requires careful planning; it is mandatory to keep in mind the patient preparation, the adequate scan protocol, and the post-processing techniques to achieve high-quality images.

Both CT and CTA and other digital imaging methods need an articulation between the data acquisition hardware and the final image displayed. The imaging process standard language is known as Digital Imaging and COmmunication in Medicine (DICOM) (Pianykh 2009). More than a file format, DICOM is a way to transmit, store, and retrieve (among other uses) data. The DICOM format encodes relevant study attributes such as patient name, study protocol, acquisition equipment, ID, among other data, and one particular attribute containing the image pixel data. The image pixel data corresponds either to a single image, or it may contain multiple frames, allowing the storage of cine loops or other multi-frame datasets (DICOM, Digital imaging and communications in medicine. Key concepts, Data format, https://www.dicomstandard.org).

For the 3D modeling process described in this chapter, the DICOM standard represents the 2D data (anatomical slices) from where the 3D model is created. Special software is required to read DICOM files. While many software programs can read DICOM, not many of them can create 3D models. Several open-source programs can perform volume or surface rendering (e.g., Osirix, Horos, Slicer, InVesalius, among others). This chapter will focus on performing surface rendering and anatomical segmentation from DICOM models.

Anatomical segmentation of data obtained from CT scans is based on the fact that every anatomical structure (skin, fat, muscle, bone, injected vessels) has a specific HU density. Several programs can automatically segment bones (cortical bone has a high HU value) from soft tissue and other structures using surface reconstruction to create a mesh (Fig. 7.2). Another way of creating a 3D model is to set a HU threshold to be included inside the perimeter and create a surface rendering of those values. Structures can be defined manually by drawing a contour in each of the images to get a mesh. Some programs include other features, making it possible to segment different structures semi-automatically.

The arterial contrast phase in CTA studies makes it possible to create a semiautomatic segmentation of the arterial system due to the contrast enhancement inside the blood vessels. As previously mentioned, the cortical bone also has a high HU value, so the surface reconstruction can also segment bone structures. In some anatomical regions, the vascular system is intimately enmeshed with the skeletal structures. This may lead to artifacts, for example, when segmenting the internal carotid inside the carotid canal in the temporal bone. These artifacts may be rectified manually by correcting the contours created automatically by the software to get a better final representation.

After creating a 3D model using DICOM compatible software, we need to export it to specialized programs for editing or 3D printing. The edition process is not mandatory, but the authors strongly recommend it because editing the mesh leads to better results in the 3D printing process. There are standard file formats supported by both software program classes (both exporting and importing) such as.stl (3D Systems),.obj (Wavefront), to ease this task.

During editing, the modifications done to the model will depend on its intended use. Many programs support this, including free software such as Meshmixer (Autodesk), MeshLab (STI-CNR), and Blender. Models created for pedagogical use may be changed and divided according to the lesson's purpose. However, when working with models with clinical use, the modifications should be minimal to give the physician who will use them the most accurate version. Editing triangle meshes acquires relevance at this stage. A mesh refers to the connected triangle structure that describes the geometry of 3D models. Each triangle defines vertices, lines, and faces as well as neighbors (Fig. 7.3).

Many actions can be useful in 3D modeling software, a few of which will be described in this paragraph. Many of these support cleaning up the model geometry by identifying islands that are not in contact with the main mesh. We can also print a large model in stages by cutting it into smaller pieces. This is important when the mesh is larger than the maximum size supported by a 3D printer. Finally, smoothing can be applied to get a better-looking final model and improve the 3D printing (reduce the overhang angles and imperfections, developed later in this chapter). Some of the modifications that can be made are shown in Fig. 7.4.

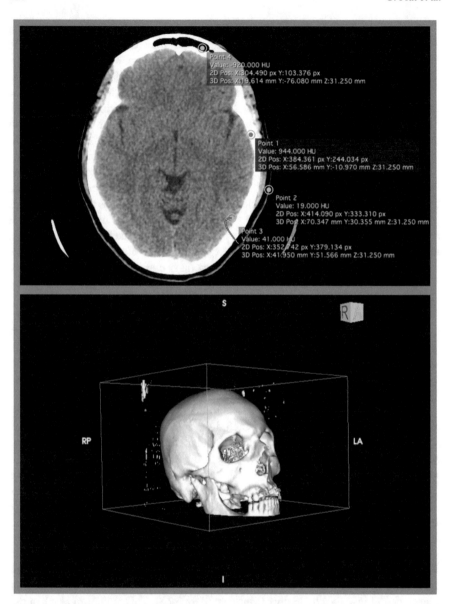

Fig. 7.2 A screenshot taken from Horos, a Free and Open-Source code Software (FOSS) program that is distributed free of charge under the LGPL license at Horosproject.org and sponsored by Nimble Co LLC d/b/a Purview in Annapolis, MD USA. The CT slice on the top image shows the HU values of different tissues surrounding the bone (Point 1, highlighted in red). The 3D model created from this CT using surface rendering shows some artifacts in the mesh (e.g., the head holder) which can be corrected using edition software

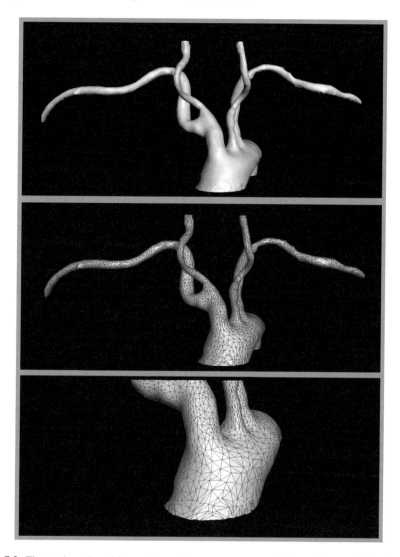

Fig. 7.3 The aortic arch and its main branches were reconstructed by surface rendering from a CTA of head and neck. The mesh of the 3D model is based on a wireframe of triangles as shown in the image

7.3.2 3D Printing

FDM is based on the extrusion of melted filament over a build plate while moving it to create the computerized model's physical prototype. This technique has gained popularity in 3D printing because it is relatively easy to learn; it has low cost and

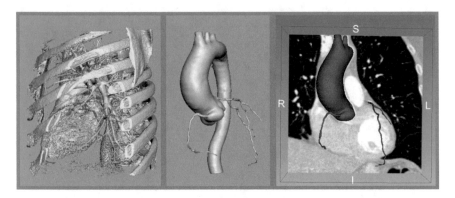

Fig. 7.4 From left to right the figure shows the edition process in which the surface render is segmented to isolate the aorta and the coronary arteries to finally over impose the edited surface rendering to the original CTA data

allows fast manufacturing. Although there are a lot of computerized automatic procedures, the printing process needs patience. Starting with simple objects allows getting proper knowledge of the operation of the machine. Once the technology is mastered, printing larger and more complex objects becomes more straightforward.

FDM 3D printers operate by pushing the thermoplastic filament (using an electrical motor) through the extruder, which has two main parts: the cold and the hot end. The hot end has a nozzle heated to a specific temperature (depending on the material used), bringing the plastic into a viscous state, allowing it to be deposited on the printer's bed following a defined shape. Over the hot end is the cold end that prevents the material from melting upward and clogging the system. Three motors are set on rails following space coordinates (x, y, and z) to move the extruder and the build plate (Fig. 7.5).

The 3D printer needs instructions to deposit the fused filament following a particular shape. This information is communicated via gcode (Kramer et al. 2000), a numerical control language composed of a letter code and numbers that carry information to drive the 3D printer. The printer can read the code and translate it into movements and other actions (heating the extruder, setting the fan speed, for example) necessary for the printing process.

Slic3r and **Cura** are examples of free software that can convert 3D model file formats into gcode. They are not only useful for the creation of the code but also offer customization of the printing characteristics. This last fact is important because, according to the kind of model to be printed, modifying specific characteristics leads to better results. For example, printing the outer wall of a model at a lower speed than the rest of the model improves its final look; modifying the first layers' print temperature can improve bed adhesion, among other features. An exciting feature is creating supports for the overhang parts. This concept refers to those layers created without a layer below when printing the model from the base upwards (Fig. 7.6).

Fig. 7.5 Front view (left) and rear view (right) of the extruder of a 3D printer operating on FDM. 1 Build Plate; 2 Extruder; 3 Rails; 4 Stepper motor; 5 Gear that pulls PLA; 6 cooler for material deposition; 7 Hotend; 8 cooler for hot end/cold end limit

Supports give the overhang layer a surface upon which it can be built. There are distinct characteristics that supports can have according to different needs.

7.3.3 Abdominal Aorta 3D Modeling and Printing

This section describes the 3D reconstruction of the abdominal aorta from a CTA using free software and subsequent 3D printing of the mesh. The model was created from an angio-CT of the thorax with vascular pathology, using a slice thickness of 1.25 mm. Horos DICOM viewer has a 3D surface reconstruction feature, under the 3D viewer menu, which automatically creates a mesh and exports it. The resulting mesh can be edited in Meshmixer. In this example, we separated the aorta from the other elements and simplified it to isolate the part to be printed. We accomplished this via different features of Meshmixer. For example, the first parts of the mesh to be discarded are those not connected to the vascular structures, as previously discussed. We could complete this process without severe modifications of the original mesh, as can be checked while overlaying the mesh with the original DICOM data. We imported the model into Slic3r and set appropriate printing parameters, such as relocating the model to reduce overhang parts and defining appropriate support structures. After printing the model, we gave it to the surgeon performing the endovascular procedure. Figure 7.7 shows a reduced version of the protocol followed.

Fig. 7.6 Shows in sequence a 3D model of the Willis polygon imported in Slic3r panel, how the software automatically places support material for overhang angles (on yellow the outer walls, green for the support material) and the 3D printed model placed in its location on a dissected skull base

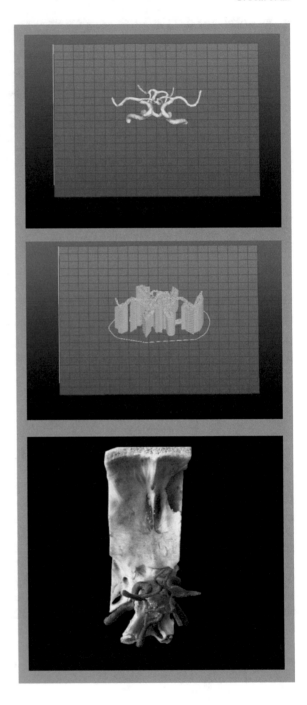

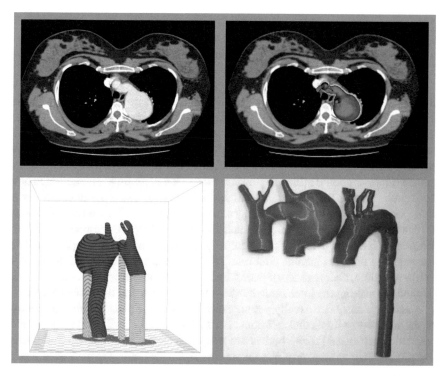

Fig. 7.7 The sequence shows images representing the process of creating a 3D model from DICOM data of a CTA using Slic3r (www.slic3r.org), the preview of the mesh in the printer bed using Cura (from Ultimaker), and the comparison between a normal aortic arch and a pathological one

7.4   Discussion

One of the challenges that education faces in this millennium is how to adapt new technologies to the teaching and learning processes. The medical field is not the exception to this fact: applying these new tools to the anatomy, clinical and surgical fields should continue to grow (McMenamin 2014; Abou Hashem 2015).

Through this chapter, certain aspects of HCI have been developed separately, considering that it is crucial for the whole process. The next few paragraphs will focus on it. Obtaining the patient's data using a tomograph needs the interaction between a human with technical knowledge and the workstation to make a correct acquisition. However, this is beyond the scope of this chapter.

The protocol shown above is heavily dependent on mastering different user interfaces. While the software used can become easy to use, operators need to master a learning curve to become proficient. Indeed, the time spent creating models and editing them is proportional to the control of the programs. The segmentation process, which resembles dissection, needs strong anatomical knowledge to recognize and adequately segment different structures. Many steps can be done automatically using appropriate computer programs. However, automatic processing, if not properly corrected by an anatomist, can cause mistakes. These can lead to a learning misinterpretation or an imperfect correlation between the 3D model and the anatomy during clinical procedures.

Besides being printed as previously discussed, the final models can also be useful in a digital format. Depending on their intended use, the visualizations can vary. For educational purposes, a friendly interface could aid learners in exploring anatomical features. These include making groups of similar structures that can either be shown or hidden to study specific anatomical regions—cutting planes along the three axes to study the topography of a region. For clinical use of the digital models, 3D free camera rotations should allow the physician a quick and intuitive interaction to have a complete view of the structures from different angles. Using 3D anatomy inside the Operating Room has its advantages, as discussed in what follows. However, the interaction between the surgeon and the three-dimensional model should not interfere with the surgical procedure properly. Some promising developments being researched in this area include controlling the movement of the 3D digital model as well as real-time modifications of juxtaposed images between laparoscopic video and 3D models, making the interaction between the physician and the software easier.

Anatomy lectures are usually accompanied by cadaveric pieces, which are controversial in many cultures and religions. Furthermore, making dissections is a time-consuming effort, and many bodies are needed to show the structures at different levels. This leads to another problem: preserving the anatomical structures is expensive. Usually, they are immersed in irritating chemists putting at risk the health of educators and students. There are also legal concerns regarding body donation in many countries (Lim et al. 2016).

3D modeling and printing seem promising tools to complement the traditional teaching of anatomy using dissected cadavers. They have the advantage of showing structures in the way they appear inside a living human being, the relationships they have. Furthermore, they offer the possibility of scaling specific structures as needed. There are also many anatomical variations and pathological structures that can be reconstructed and 3D printed as another tool for the learning process. However, not every anatomical structure can be segmented and reconstructed from anatomical slices. Indeed, textures and other physical features of organs are not the same. Students and pedagogs can benefit from the wide range of possibilities resulting from combining 3D anatomy (in both digital and physical formats) with cadaveric anatomy. In this way, the advantages of both can be leveraged (Chien et al. 2010).

The surgical field is one of the clinical areas where 3D anatomy has been broadly used. 3D printing has mainly been used to create and shape implants, selecting patients, surgical planning, molds for prosthetics, surgical guides, and postgraduate pedagogical applications. However, more research is needed to assess the advantages and disadvantages of 3D technologies in this area (Tack et al. 2016). These models need to be highly precise, lest they interfere with the proper development of the procedure. Digital models can potentially be a powerful tool in surgeries. An example of this is using augmented and virtual reality in the operating room to provide information regarding the region's anatomy (Quero et al. 2019).

7.5   Conclusion

3D modeling and technologies have proven to be very powerful and are finding many uses in medical sciences. Though their cost could be high, and in-depth technical knowledge is involved, creating 3D models can be done using free/open-source software with flexible and relatively simple user interfaces. We have found that FDM is a cheap and easy way to print these models. This chapter has explored the actual and potential use of 3D anatomy to encourage the widespread use of these resources in medical education and clinical matters.

References

Abou Hashem Y (2015) The application of 3D printing in anatomy education, 20:29847. https://doi.org/10.3402/meo.v20.29847

Abouhashem YA, Dayal M, Savanah S, Štrkalj G (2015) The application of 3D printing in anatomy education. Med Educ Online 20:1. https://doi.org/10.3402/meo.v20.29847

Ackerman MJ (1998) The visible human project. Proc IEEE 86(3):504–511

Arráez-Aybar, Luis-Alfonso et al (2010) Relevance of human anatomy in daily clinical practice. Ann Anat-Anatomischer Anzeiger 192(6):341–348. https://doi.org/10.1016/j.aanat.2010.05.002

Chien C-H, Chen C-H, Jeng T-S (2010) An interactive augmented reality system for learning anatomy structure. In: Proceedings of the international multiconference of engineers and computer scientists. International Association of Engineers, Hong Kong, China, pp 17–19

Chiorean L-D et al (2011) 3D reconstruction and volume computing in medical imaging. Acta Technica Napocensis 52(3):18–24

Gregory SR, Cole TR (2002) The changing role of dissection in medical education. JAMA 287(9):1180–1181. https://doi.org/10.1001/jama.287.9.1180-JMS0306-4-1

DICOM, Digital imaging and communications in medicine. Key concepts, Data format, https://www.dicomstandard.org

Horvath J (2014) A brief history of 3D printing. In: Mastering 3D printing. Apress, Berkeley, CA, pp 3–10

Hsieh J (2003) Preliminaries. In: Computed tomography: principles, design, artifacts, and recent advances, vol 114. SPIE Press, pp 19–36

Jasveer S, Jianbin X (2018) Comparison of different types of 3D printing technologies. Int J Sci Res Publ (IJSRP) 8(4):1–9. http://dx.doi.org/10.29322/IJSRP.8.4.2018.p7602

Jones R et al (2011) RepRap–the replicating rapid prototyper. Robotica 29(1):177–191

Kalender WA (2011) Principles of computed tomography. In: Computed tomography: fundamentals, system technology, image quality, applications. Wiley, pp 18–31

Kramer TR et al (2000) The NIST RS274NGC interpreter: version 3. Commerce Department, National Institute of Standards and Technology (NIST). https://www.nist.gov/publications/nist-rs274ngc-interpreter-version-3. Accessed 1 Aug 2000

Lim KHA et al (2016) Use of 3D printed models in medical education: a randomized control trial comparing 3D prints versus cadaveric materials for learning external cardiac anatomy. Anat Sci Educ 9(3):213–221. https://doi.org/10.1002/ase.1573

McLachlan JC, Regan De Bere S (2010) How we teach anatomy without cadavers. Clin Teach 1:49–52. https://doi.org/10.1111/j.1743-498x.2004.00038.x

McMenamin PG (2014) The production of anatomical teaching resources using three-dimensional (3D) printing technology, 7(6):479–86. https://doi.org/10.1002/ase.1475

Pianykh OS (2009) What is DICOM? In: Digital imaging and communications in medicine (DICOM): a practical introduction and survival guide. Springer Science & Business Media, pp 3–6

Prince JD (2014) 3D printing: an industrial revolution. J Electron Resourc Med Libr 11(1):39–45. https://doi.org/10.1080/15424065.2014.877247

Quero G et al (2019) Virtual and augmented reality in oncologic liver surgery. Surg Oncol Clinics 28(1):31–44

Rengier F, Mehndiratta A, von Tengg-Kobligk H et al (2010) 3D printing based on imaging data: review of medical applications. Int J CARS 5:335–341. https://doi.org/10.1007/s11548-010-0476-x

Tack P, Victor J, Gemmel P, Annemans L (2016) 3D-printing techniques in a medical setting: a systematic literature review. BioMed Eng OnLine 15:115

Weinstein MA, Duchesneau PM, Weinstein CE (1977) Computed angiotomography. Am J Roentgenol 129(4):699–701. https://doi.org/10.2214/ajr.129.4.699

Chapter 8
3D Reconstruction from CT Images Using Free Software Tools

Soraia Figueiredo Paulo⊙, **Daniel Simões Lopes**⊙, **and Joaquim Jorge**⊙

Abstract Computed Tomography (CT) is a commonly used imaging modality across a wide variety of diagnostic procedures (World Health Organisation 2017). By generating contiguous cross-sectional images of a body region, CT has the ability to represent valuable 3D data that enables professionals to easily identify, locate, and accurately describe anatomical landmarks. Based on 3D modeling techniques developed by the field of Computer Graphics, the Region of Interest (ROI) can be extracted from the 2D anatomical slices and used to reconstruct subject-specific 3D models. This chapter describes a 3D reconstruction pipeline that can be used to generate 3D models from CT images and also volume renderings for medical visualization purposes (Ribeiro et al. 2009). We will provide several examples on how to segment 3D anatomical structures with high-contrast detail, namely skull, mandible, trachea, and colon, relying solely on the following set of free and open-source tools: ITK-SNAP (Yushkevich et al. 2006) and ParaView (Ahrens et al. 2005).

8.1 3D Reconstruction Pipeline

The process of 3D reconstruction extends from 2D medical image visualization to 3D anatomical model visualization, incorporating several image processing and mesh processing blocks assembled in a sequential manner, resulting in a pipeline. The input of the 3D reconstruction pipeline is an ordered stack of 2D medical images which is submitted to a cascade of computational geometry operations. As a result, this pipeline outputs accurate, patient-specific 3D geometric models composed of

S. F. Paulo (✉) · D. S. Lopes · J. Jorge
INESC-ID Lisboa, Instituto Superior Técnico, Universidade de Lisboa, Lisbon, Portugal
e-mail: soraiafpaulo@inesc-id.pt

D. S. Lopes
e-mail: daniel.lopes@inesc-id.pt

J. Jorge
e-mail: jorgej@acm.org

© The Author(s), under exclusive license to Springer Nature Switzerland AG 2021
J.-F. Uhl et al. (eds.), *Digital Anatomy*, Human–Computer Interaction Series,
https://doi.org/10.1007/978-3-030-61905-3_8

135

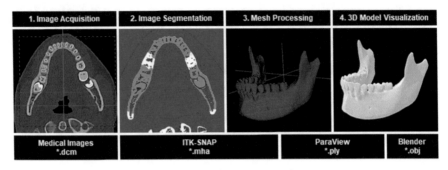

Fig. 8.1 3D Reconstruction pipeline for extracting anatomical structures from CT images: each stage (top row) is associated with a specific software and file extension (bottom row)

thousands of vertices and triangular facets, which constitute a digital equivalent to the human organ being modeled (Fig. 8.1).

Following medical image acquisition and 2D visualization, the onset of the geometric modeling pipeline consists of the identification of tissues and their boundaries by segmentation of a 3D image dataset. Given several contiguous tomographic images, the modeler can extract the geometric information that is necessary to 3D reconstruct the anatomical structures of interest with high accuracy and voxel resolution.

To perform this task, we chose ITK-SNAP (Yushkevich et al. 2006), a free and open-source software offering several medical image segmentation tools, in particular semi-automatic active contours based on region competition or edge-based algorithms, which are well-suited to segment high-contrast CT images. Region competition methods estimate the probability of a voxel belonging to the foreground (i.e., the region of interest) and to the background, given that voxel's intensity value in the input image (Yushkevich et al. 2006). This starts with a global thresholding (Fig. 8.2) which partitions the original grayscale image into a binary image with a background and a foreground intensity.

This separation relies on the user's specifications of the range of intensities to be considered, as pixel values within this range are considered foreground, while the pixel values outside this range are dismissed and considered background. Thus, global thresholding enables the separation of the ROI from the background, in the sense that thresholding can be defined to make the ROI correspond to the non-null pixel values of the foreground. Although in high-contrast medical images the thresholding technique can be very quick, it generally does not account for spatial information, which leads to rough results. Therefore, it requires a complementary technique to be applied, such as the active contours method.

Unlike thresholding, edge-based techniques do not depend on the intensity levels of the original image, but rather on the differences between neighboring pixel values, i.e., on the image gradient. These differences are translated by gradient magnitude and normalized to a range from 0 to 1, where larger differences correspond to values closer to zero (Fig. 8.3).

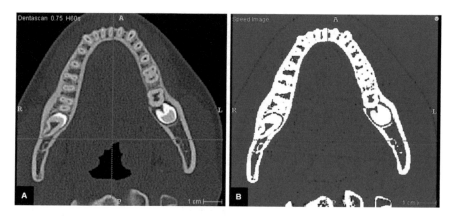

Fig. 8.2 Global thresholding: **a** original grayscale image; **b** binary image in which the pixel intensities of bony structures of the mandible are above the threshold value and are considered non-null (white), while soft tissue intensities rest below the threshold and are dismissed (blue)

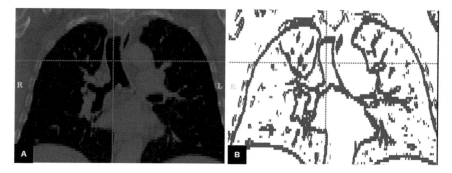

Fig. 8.3 Edge-based technique: the large differences between the intensities of the air inside the trachea and the neighboring pixels of **a** the original grayscale image produce **b** a clear contour of the tracheal wall (blue)

Both region competition and edge-based techniques are used to produce a feature image, which guides an automatic active contour progression. In ITK-SNAP, the active contours model is called snake evolution and aims at complementing the segmentation results obtained by the previous step. Snakes consist of closed curves (2D) or surfaces (3D) which adapt their motion to the image patches created by the feature images. The user initializes the snakes as spherical bubbles that should be strategically placed along the ROI, evolving from rough estimates to a close approximation of the anatomical structure to be segmented (Fig. 8.4).

The output of the segmentation stage is represented as a collection of voxels that define the geometric *locus* surrounded by a non-belonging voxel background.

The next step in the pipeline consists of converting the segmented data into a triangular surface using a mesh-based technique denominated as marching cubes (Lorensen and Cline 1987). Considering the voxels of the 3D image generated by

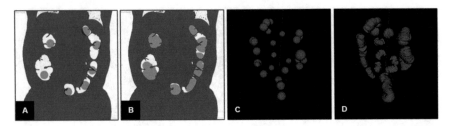

Fig. 8.4 Snake evolution in 2D (**a** and **b**) and 3D (**c** and **d**): the user placed the initial snakes along the colon's feature images, which allowed the snakes to grow inside the non-null pixel values and adapt to the shape of the colon wall

the previous stage, the marching cubes algorithm creates a triangular surface mesh from the binary or quantized 3D image. A 3D mesh is a discrete entity formed by geometrical elements such as vertices, edges, polygons, or polyhedral finite elements (Viceconti and Taddei 2003). It can be interpreted as a 3D model tiled with polygons or polyhedrons over a 3D space with geometrical elements (e.g., triangles, quadrangles, tetrahedrons, or prisms) arranged to intersect along a face, an edge, or a vertex. The algorithm marches through each parallelepiped in the segmented outlines and generates a surface boundary of the anatomical structure composed of triangular elements. By connecting all the triangular facets, an isosurface will be produced, i.e., a 3D surface representation of points with equal values in a 3D distribution.

The model created by the marching cubes algorithm presents two major features that require further processing: (i) due to the digital nature of the images, the surface mesh initially presents a ladder-like aspect, which does not correspond to the natural surface curvature; and (ii) an excess of vertices and faces that express irrelevant information and hamper further computational processes. To deal with such processing requirements, we chose ParaView (Ahrens et al. 2005), an open-source application that enables surface adjustments and surface visualization. After importing the output of the marching cubes algorithm into ParaView, the isosurface is put through a process of smoothing and decimation, in order to attenuate the unwanted geometric features and to reduce the number of nodes in the 3D model. While smoothing works as a spatial filter to improve the overall mesh appearance, decimation is used to simplify the surface mesh and reduce its number of nodes and triangles. The latter involves a trade-off between the level of detail and the computational power required to process it, which should guarantee that lowering the number of nodes to reduce computational costs does not hamper the representation of the resulting 3D content. Finally, this mesh can be exported to Blender (2020), which serves as a 3D modeling tool to further edit and improve the surface mesh.

The output produced by this pipeline has two major applications that depend on the mesh quality: visualization and 3D printing. To visualize a 3D model, a complex mesh with a large number of nodes is the best option to enhance highly detailed features. Besides desktop computers, such 3D models can be visualized in Virtual and Augmented Reality displays, which provide better depth perception and enable true 3D object interaction.

8.2   3D Reconstruction of Subject-Specific Anatomical Structures

In this section, we illustrate the reconstruction of subject-specific 3D models of the skull, mandible, trachea, and the colon from CT images, which display the high-contrast detail required for the use of this pipeline. Each example is based on the same geometric modeling pipeline, relying on free and open-source software (ITK-SNAP and ParaView), which makes it readily accessible to practitioners, researchers, and educators.

Skull

For the purpose of this example, we used a CT image dataset available at the OsiriX DICOM Image Library (Rosset et al. 2004), alias name PHENIX. To import it to ITK-SNAP, the user must access the *File* menu and select the *Open Main Image* option (Fig. 8.5a). Then, the user must press the *Browse* button on the *Open Image* dialog box (Fig. 8.5b), select one of the images from the folder where the dataset is stored, and proceed to press *Next* and *Finish*.

To start the segmentation process, we define the ROI, which is represented by the red dashed rectangles (Fig. 8.6). To do so, the user must click on the *Active Contour Segmentation Mode* icon, in the *Main Toolbar* panel, which enables the *Snake Inspector* and displays an initial ROI. By default, ITK-SNAP considers the whole image dataset as the ROI. To adjust it, the user can edit the (x, y, z) values in the *Snake Inspector*, where *Position* corresponds to the center of the 3D ROI and *Size* corresponds to its dimensions. Another option is to manipulate the ROI by left-clicking inside the red dashed rectangle and dragging it to the target position, in order to adjust its position, and clicking and dragging each limit of the ROI horizontally (left and right) or vertically (top and bottom), to set its size.

Fig. 8.5 Opening an image dataset using ITK-SNAP: **a** File menu; and **b** Open Image dialog box

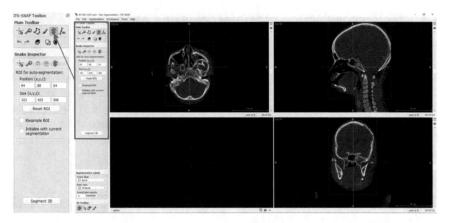

Fig. 8.6 In ITK-SNAP, the ROI corresponds to the area limited by the red dashed lines. By dragging the red lines or the whole rectangle, the user can adjust the ROI in any of the windows corresponding to the orthogonal planes: axial (top-left), sagittal (top-right), and coronal (bottom-right)

Head CT images present good contrast between the hard tissues (bony structures: light grey to white) and soft tissues (e.g., muscles, glands: grey), which enables image segmentation by global thresholding followed by semi-automatic active contours based on region competition. To divide the original grayscale dataset into binary images where the non-null pixel values correspond to the intensities of the skull, the user must define a *Lower threshold* value during the *Presegmentation* step. This can be done either by dragging the corresponding slider to adjust the value or by inserting the threshold value in the appropriate text box (Fig. 8.7). While the user must choose the *Lower threshold mode*, it is necessary to select the highest *Upper threshold* value beforehand, to consider all pixel intensities belonging to the skull.

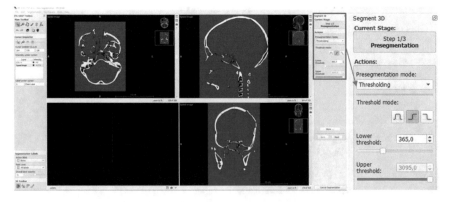

Fig. 8.7 After defining the ROI, the user is able to select *Thresholding* as the *presegmentation mode*, which enables the selection of the lower and upper thresholds (in Hounsfield Units) to partition the binary image into bony structures (white) and soft tissues and air (blue)

After pressing *Next*, the following step is to initialize the set of snakes for the progression of the active contours. In order to place a bubble, the user must click on the 2D image in the orthographic window of choice to define the 3D position of the bubble, and click on the *Add Bubble at Cursor* button in the *Actions* panel (Fig. 8.8).

Then, the bubble radius can be set using the corresponding text box or slider. To edit a bubble that was previously added, the user must select it from the list of *Active bubbles* and proceed to adjust it. To enable these snakes to evolve, the user must advance to the next step and press the play button. This will allow the user to visualize the active contours progression both in 2D (Fig. 8.9) and 3D (Fig. 8.10). However, while 2D views are automatically updated, the user is required to press the *Update* button, under the 3D window, to visualize the evolution in 3D.

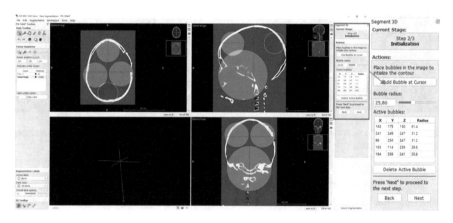

Fig. 8.8 Snake initialization: snakes start as spherical bubbles (yellow) which are placed using the cursor. The *Actions* panel is used to confirm bubble placement, adjusting each bubble's size, and deleting it

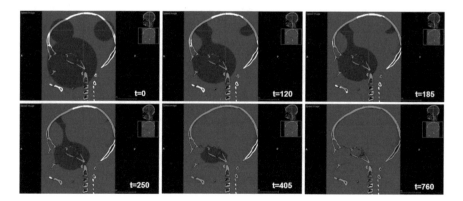

Fig. 8.9 Active contours progression (red) at different time steps: sagittal view of the skull

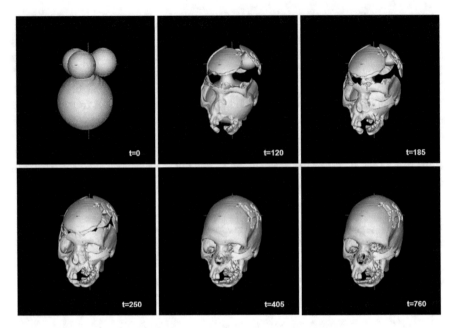

Fig. 8.10 Active contours model progression at different time steps: 3D view of the evolving contours of the skull

The next stage of the geometric modeling pipeline requires the user to export the segmentation output to ParaView. To do so, the user should access the *Segmentation* menu and select *Save Segmentation* Image (Fig. 8.11a). This will open a dialog box (Fig. 8.11b) where the user should choose *MetaImage* from the *File Format* dropdown menu, fill in the filename for the segmentation image, and press *Finish*.

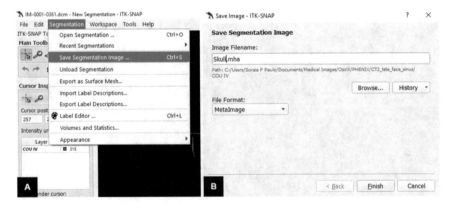

Fig. 8.11 Saving a segmentation image on ITK-SNAP: **a** Segmentation dropdown menu; **b** Save Image dialog

In ParaView, the user should click on the *File menu* and *Open* the segmentation image. Then, it is necessary to press the *Contour* button, above the *Pipeline Browser*, followed by *Apply* on the *Properties* panel (Fig. 8.12).

The following step is the process of smoothing and decimation, which involves accessing the navigation bar, opening the *Filters* dropdown menu, and choosing the *Smooth* filter, under the *Alphabetical* list (Fig. 8.13). Again, the user needs to define the *Number of Iterations* on the *Properties* panel and *Apply* this filter (Fig. 8.14a). Similarly, the *Decimation* filter should be applied next by defining the *Target reduction* (Fig. 8.14b), which will reduce the number of mesh triangles, followed by a new cycle of smoothing. In this example, smoothing was carried on for 750 iterations, decimation had a target reduction of 75% and the final smoothing was applied for 500

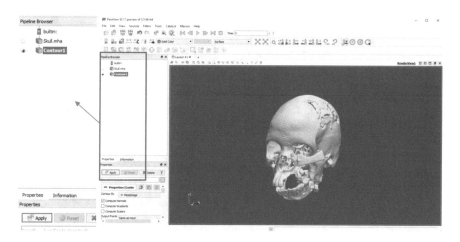

Fig. 8.12 Contour display on ParaView

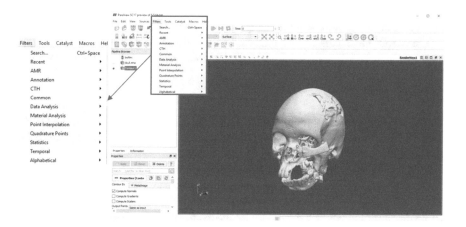

Fig. 8.13 Smooth and decimate filters are available under the Filters → Alphabetical menus

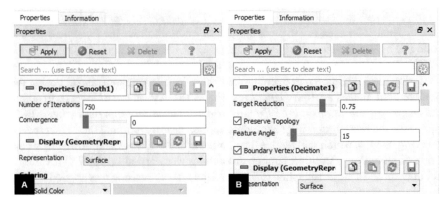

Fig. 8.14 Properties used to define **a** Smooth and **b** Decimation filters

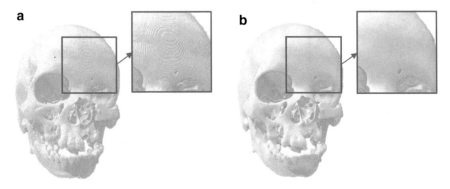

Fig. 8.15 3D mesh of the skull **a** before and **b** after Smoothing → Decimation → Smoothing

iterations to produce the final surface mesh (Fig. 8.15). Such values assure enough level of detail while reducing the computational cost.

The output mesh can then be exported as a *.ply (ASCII) file, through the *Save Data* option, on the *File* dropdown menu, and converted to an *.obj file via Blender, if necessary.

<u>Mandible</u>

For the purpose of this example, we used a CT image dataset available at the OsiriX DICOM Image Library (Rosset et al. 2004), alias name INCISIX. Firstly, we define the mandible as the ROI (Fig. 8.16).

Similarly to a head CT, dental scan images display good contrast between soft and hard anatomical tissues, namely teeth and bones. Thus, image segmentation can be performed via global thresholding and semi-automatic active contours techniques. While thresholding enables the binary image to consider bony structures as non-null pixels (Fig. 8.17), the active contours evolve through a set of snakes placed along the mandible (Figs. 8.18 and 8.19).

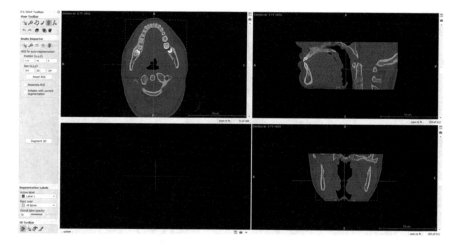

Fig. 8.16 Defining the mandible as the ROI in the CT image dataset

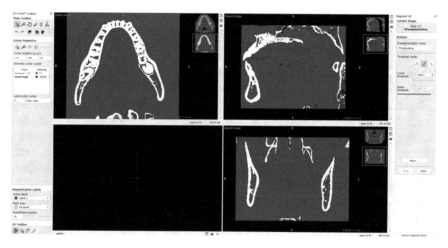

Fig. 8.17 Thresholding the original grayscale dental scan to produce a binary image where bone and teeth pixels are considered non-null, while adjusting the lower threshold value to dismiss soft tissues

Given the points of contact between teeth from the lower and upper jaw, which share identical intensity values, the 3D model of the mandible produced by the active contours may also include teeth from the upper jaw. This is a case of over-segmentation which requires the user to correct the contours manually. To do so, the user must access the *Segmentation Labels* panel and select the *Clear Label* from the *Active label* options, to *Paint over* the label used to segment the ROI. Then, the user can choose from two editing tools: the *Paintbrush Inspector* (Fig. 8.20) or the

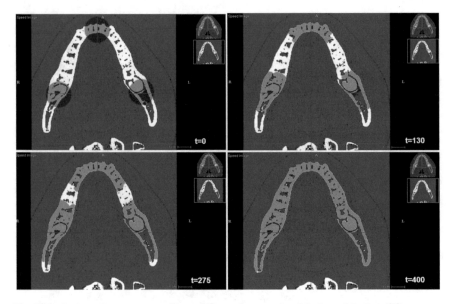

Fig. 8.18 Active contours progression at different time steps: axial view of the mandible

Polygon Inspector (Fig. 8.21). The *Paintbrush Inspector* erases the segmentation data by clicking and dragging the cursor over the unwanted features.

Another possibility is to perform this correction through the *Polygon Inspector*, where the user draws a polyline point-by-point to define the area to be erased, pressing *Accept* to apply such changes. Since this is a 2D process, it must be carried out throughout the 2D slices containing segmentation data to be erased, in order to produce the final 3D model (Fig. 8.22). Although this is an example of over-segmentation, where the manual editing consisted of eliminating information, the contrary can also occur, i.e., under-segmentation. In that case, the same editing tools can also be used to fill in the missing information (Fig. 8.23).

In ParaView, a smoothing filter is carried out for 500 iterations, followed by a decimation aimed at a 75% reduction of the mesh triangles, and finally the last smoothing filter is applied for 300 iterations to produce the final model (Fig. 8.24).

Trachea

For the purpose of this example, we used a Chest CT image dataset available in The Cancer Imaging Archive (National Cancer Institute Clinical Proteomic Tumor Analysis Consortium (CPTAC) 2018; Clark et al. 2013) (Collection ID: CPTAC-PDA, subject ID: C3N-00249, Description: CHEST 3.0 B40f). Given the contrast between the tracheal air column (air: black) and the tracheal surface (wall: light grey), image segmentation can also be performed by edge-based techniques followed by the semi-automatic active contours. This mode should be selected during the *Presegmentation* step, in the *Presegmentation mode* under the *Edge Attraction* option (Fig. 8.25). In this case, the edge-based feature assigns near-zero values to the pixels close to intensity edges in the original grayscale image, which drive the snakes that

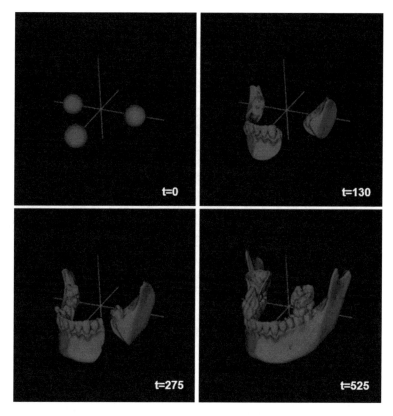

Fig. 8.19 Active contours model progression at different time steps: 3D view of the evolving contours of the mandible

are placed inside the trachea to evolve into the shape of the tracheal wall (Figs. 8.26 and 8.27).

In ParaView, a smoothing filter is applied for 50 iterations, followed by a decimation with a 60% target reduction, and the last smoothing filter is applied for 50 more iterations to create the final 3D model (Fig. 8.28).

Colon

In this example, we used a single CT Colonography (CTC) dataset available in The Cancer Imaging Archive (Clark et al. 2013; Smith et al. 2015; Johnson et al. 2008) (subject ID: CTC-3105759107), that was acquired in a supine position, had almost no liquid and presented large (>10 mm) and quite visible polyps along with several diverticula. CTC images present high contrast between the luminal space (air: black) and luminal surface (wall: light grey), which also facilitates image segmentation by global thresholding followed by semi-automatic active contours based on region competition. As a result, the original grayscale image is partitioned into a binary image where the resulting non-null pixel values correspond to the intensities of the

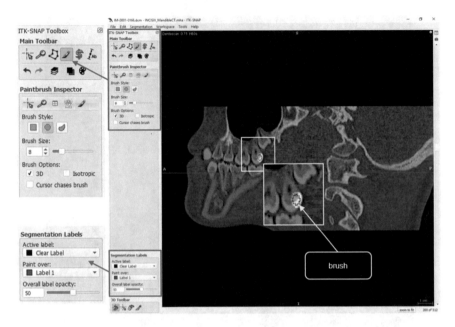

Fig. 8.20 Editing the result of the active contours progression (red) allows the user to eliminate the maxillary teeth from the final 3D model by erasing the contours corresponding to undesired structures

luminal space. Thus, this allows the active contours to iteratively evolve from a very rough estimate obtained by global thresholding to a very close approximation of the colon (Figs. 8.29 and 8.30).

After segmentation, the 3D surface mesh goes through smoothing and decimation, in order to produce the final mesh. The smoothing process consists of a low-pass filter carried on for 500 iterations, followed by decimation, to remove at least 75% of the mesh triangles, and a new cycle of 500 iterations of smoothing (Fig. 8.31).

8.3   Conclusions

Given the ambiguity of certain tissue boundaries, a fully automatic segmentation process is not viable to obtain high-quality 3D reconstructions based on 2D medical images. Considering the lack of contrast between different anatomical structures, this will induce the segmentation procedure to either consider voxels that are outside the ROI, i.e., over-segmentation, or the opposite, as several voxels belonging to the ROI may not be included, i.e., under-segmentation. For these reasons, manual and semi-automatic techniques must be combined to obtain accurate results.

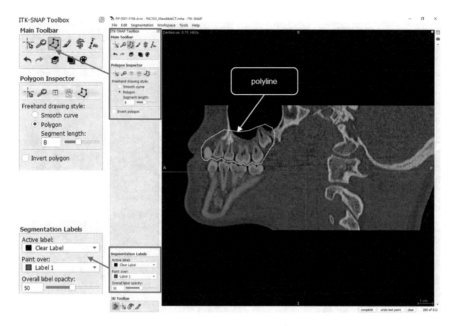

Fig. 8.21   Editing the result of the active contours progression (red) using a polyline. To determine the area to be erased, the user defines points, which are connected by straight line segments, until a closed polygon is defined

User interaction plays an important role in correcting the segmentation errors produced by semi-automatic segmentation. In this sense, it is important to emphasize that despite the usage of various computational methods, modeling human structures requires keen visual capabilities and a profound anatomical knowledge. If so, manual segmentation alone guarantees a highly accurate result, but the time, effort, and training involved are impractical for large-population studies. Combining semi-automatic with manual segmentation provides a powerful and reliable instrument for 3D image segmentation, but it strongly depends on users being anatomically savvy and highly experienced in Radiographic Anatomy.

Still, more effective segmentation interfaces are sorely needed, as conventional Windows, Icons, Menus, and Pointing interfaces hamper image segmentation by relying on flat 2D media which promote timely slice-by-slice segmentation and are unsuited to deal with anatomical complexity, miss direct spatial input, and afford limited 3D navigation control (Meyer-Spradow et al. 2009; Olsen et al. 2009; Mühler et al. 2010; Kroes et al. 2012; Paulo et al. 2018; Lopes et al. 2018a).

In the last decades, sketch-based interfaces have addressed semi-automatic segmentation issues that typically arise during anatomical modeling, allowing clinicians to rapidly delineate ROIs and explore medical images (Olabarriaga and Smeulders 2001; Heckel et al. 2013; Peng et al. 2014). Although delineation can be guided by simple edge-seeking algorithms or adjustable intensity thresholds, these often fail

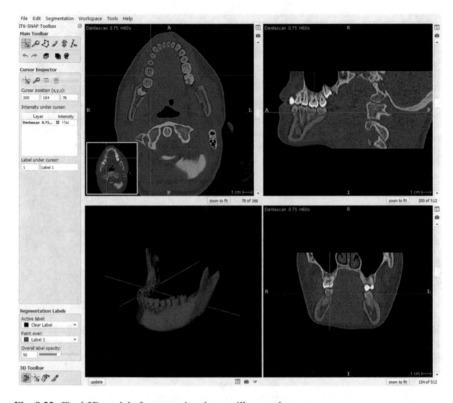

Fig. 8.22 Final 3D model after removing the maxillary teeth

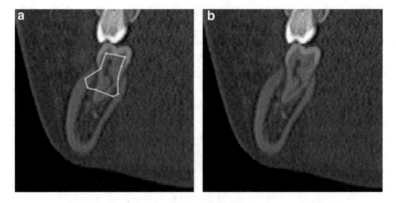

Fig. 8.23 Example of under-segmentation: **a** Manual correction; **b** Segmentation fault corrected

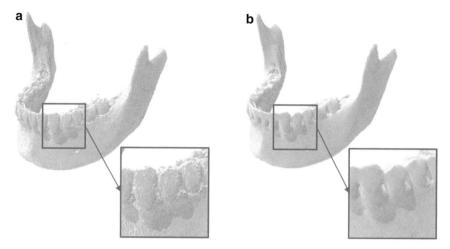

Fig. 8.24   3D mesh of the mandible **a** before and **b** after Smoothing → Decimation → Smoothing

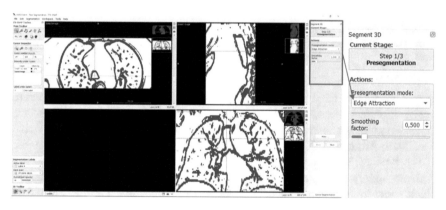

Fig. 8.25   Selecting the edge attraction presegmentation mode

to produce sufficiently accurate results (Shepherd et al. 2012; Van Heeswijk et al. 2016; Lopes et al. 2018b).

Besides sketch-based interfaces, head-mounted displays have proven to be suitable devices for analyzing medical volume datasets (Coffey et al. 2012; Sousa et al. 2017) and specifically adopted for rendering medical images to aid surgery (Robison et al. 2011; Vosburgh et al. 2013), virtual endoscopy (John and McCloy 2004), and interventional radiology (Duratti et al. 2008).

Only recently have Virtual Reality approaches been applied to the segmentation process, but resulting models continue to be rough representations of subject-specific anatomy, which require sub-millimetric precision (Johnson et al. 2016; Jackson and Keefe 2016). A potential option would be to consider multi-scale interfaces to enable more accurate, precise, and direct interaction with 3D content. Other spatial input

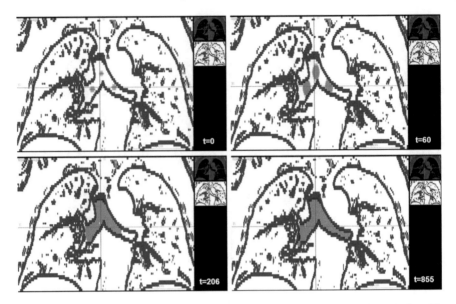

Fig. 8.26 Active contours progression at different time steps: coronal view of the trachea. The edges (blue) guide the growth of the snakes, which must be placed within the area limited by such edges

approaches such as hand tracking or 3D pens have not been properly developed for precise 3D input, whereas hybrid techniques exploring the use of physical surfaces and visual guidance seem to be the most accurate approach for VR settings (Bohari et al. 2018; Arora et al. 2017).

The type of 3D reconstruction pipeline described in this chapter opens way to create high-quality, patient-specific 3D meshes from a medical image dataset, which show great accuracy and, in some cases, remarkable mesh fidelity. Such strategies may provide the medical community an important 3D model to visualize and teach, not only normal, but also pathological anatomy. This would potentially enable medical education through a virtual reality environment (Lopes et al. 2018a; Figueiredo Paulo et al. 2018a, b, using interactive 3D tables (Mendes et al. 2014) and gamification techniques (Barata et al. 2013), or in Augmented Reality Settings (Zorzal et al. 2020) reducing the cost and access problems of human cadavers.

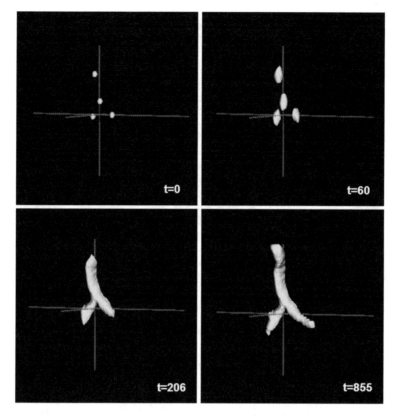

Fig. 8.27   Active contours progression at different time steps: 3D view of the evolving contours of the trachea

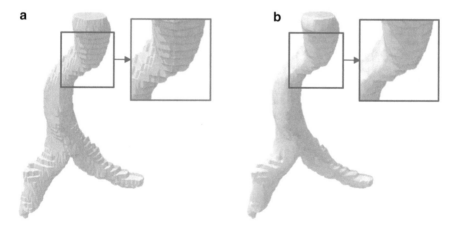

Fig. 8.28   3D mesh of the trachea **a** before and **b** after Smoothing → Decimation → Smoothing

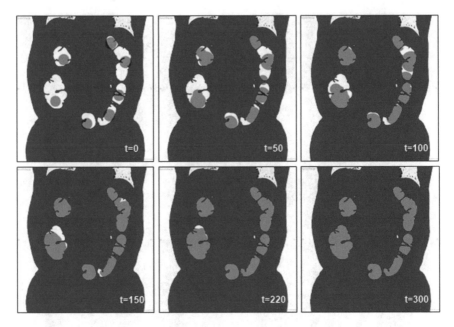

Fig. 8.29 Active contours progression at different time steps: coronal view of the colon

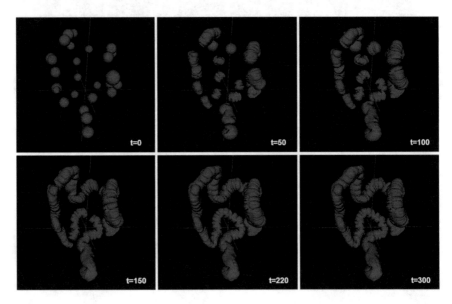

Fig. 8.30 Active contours progression at different time steps: 3D view of the evolving contours of the colon

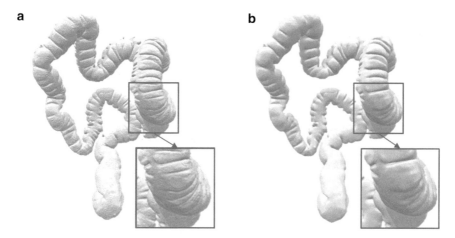

Fig. 8.31 3D mesh of the colon **a** before and **b** after Smoothing → Decimation → Smoothing

Acknowledgments All authors are thankful for the financial support given by Portuguese Foundation for Science and Technology (FCT). In particular, the first author thanks for the doctoral grant SFRH/BD/136212/2018. This work was also partially supported by national funds through FCT with reference UID/CEC/50021/2019 and IT-MEDEX PTDC/EEI-SII/6038/2014.

References

Ahrens J, Geveci B, Law C (2005) ParaView: an end-user tool for large-data visualization. Vis Handb 717–731. https://doi.org/10.1016/b978-012387582-2/50038-1

Arora R, Kazi RH, Anderson F, Grossman T, Singh K, Fitzmaurice G (2017) Experimental evaluation of sketching on surfaces in VR. In: Conference on human factors in computing systems—proceedings, vol 2017, May 2017, pp 5643–5654. https://doi.org/10.1145/3025453.3025474

Barata G, Gama S, Jorge J, Gonçalves D (2013) So fun it hurts—Gamifying an engineering course. In: Schmorrow DD, Fidopiastis CM (eds) Foundations of augmented cognition. AC 2013. Lecture notes in computer science, vol 8027. Springer, Berlin, Heidelberg. https://doi.org/10.1007/978-3-642-39454-6_68

Blender Online Community, Blender project—Free and Open 3D Creation Software. https://www.blender.org/. Accessed 17 Jun 2020

Bohari U, Chen TJ, Vinayak (2018) To draw or not to draw: recognizing stroke-hover intent in non-instrumented gesture-free mid-air sketching. In: International conference on intelligent user interfaces, proceedings IUI, March 2018, pp 177–188. https://doi.org/10.1145/3172944.3172985

Clark K et al (2013) The Cancer Imaging Archive (TCIA): maintaining and operating a public information repository. J Digit Imaging 26(6):1045–1057. https://doi.org/10.1007/s10278-013-9622-7

Coffey D et al (2012) Interactive slice WIM: Navigating and interrogating volume data sets using a multisurface, multitouch VR interface. IEEE Trans Vis Comput Graph 18(10):1614–1626. https://doi.org/10.1109/TVCG.2011.283

Duratti L, Wang F, Samur E, Bleuler H (2008) A real-time simulator for interventional radiology. In: Proceedings of the ACM symposium on virtual reality software and technology, VRST, 2008, pp 105–108. https://doi.org/10.1145/1450579.1450602

Figueiredo Paulo S, Figueiredo N, Armando Jorge J, Simões Lopes D (2018a) 3D reconstruction of CT colonography models for VR/AR applications using free software tools, Granada. https://miccai-sb.github.io/materials.html#mec2018. Accessed 17 June 2020

Figueiredo Paulo S, Belo M, Kuffner dos Anjos R, Armando Jorge J, Simões Lopes D (2018b) Volume and surface rendering of 3D medical datasets in unity®, Granada. https://miccai-sb.github.io/materials.html#mec2018. Accessed 17 June 2020

Heckel F, Moltz JH, Tietjen C, Hahn HK (2013) Sketch-based editing tools for tumour segmentation in 3d medical images. Comput Graph Forum 32(8):144–157. https://doi.org/10.1111/cgf.12193

Jackson B, Keefe DF (2016) Lift-off: using reference imagery and freehand sketching to create 3D models in VR. IEEE Trans Vis Comput Graph 22(4):1442–1451. https://doi.org/10.1109/TVCG.2016.2518099

John NW, McCloy RF (2004) Navigating and visualizing three-dimensional data sets. Br J Radiol 77(suppl2):S108–S113. https://doi.org/10.1259/bjr/45222871

Johnson CD et al (2008) Accuracy of CT colonography for detection of large adenomas and cancers. N Engl J Med 359(12):1207–1217. https://doi.org/10.1056/NEJMoa0800996

Johnson S, Erdman AG, Jackson B, Keefe DF, Tourek B, Molina M (2016) Immersive analytics for medicine: hybrid 2D/3D sketch-based interfaces for annotating medical data and designing medical devices. In: Companion proceedings of the 2016 ACM international conference on interactive surfaces and spaces: nature meets interactive surfaces, ISS 2016, November 2016, pp 107–113. https://doi.org/10.1145/3009939.3009956

Kroes T, Post FH, Botha CP (2012) Exposure render: an interactive photo-realistic volume rendering framework. PLoS One 7(7):e38586. https://doi.org/10.1371/journal.pone.0038586

Lopes DS et al (2018a) Interaction techniques for immersive CT colonography: a professional assessment. In: Lecture notes in computer science (including subseries Lecture notes in artificial intelligence and lecture notes in bioinformatics), vol 11071, September 2018, pp 629–637. https://doi.org/10.1007/978-3-030-00934-2_70

Lopes DS et al (2018b) Explicit design of transfer functions for volume-rendered images by combining histograms, thumbnails, and sketch-based interaction. Vis Comput 34(12):1713–1723. https://doi.org/10.1007/s00371-017-1448-8

Lorensen WE, Cline HE (1987) Marching cubes: a high resolution 3D surface construction algorithm. ACM SIGGRAPH Comput Graph 21(4):163–169. https://doi.org/10.1145/37402.37422

Mendes D, Fonseca F, Araùjo B, Ferreira A, Jorge J (2014) Mid-air interactions above stereoscopic interactive tables. In: 2014 IEEE symposium on 3D User Interfaces (3DUI), Minneapolis, MN, pp 3–10. https://doi.org/10.1109/3DUI.2014.6798833

Meyer-Spradow J, Ropinski T, Mensmann J, Hinrichs K (2009) Voreen: a rapid-prototyping environment for ray-casting-based volume visualizations. IEEE Comput Graph Appl 29(6):6–13. https://doi.org/10.1109/MCG.2009.130

Mühler K, Tietjen C, Ritter F, Preim B (2010) The mduedical exploration toolkit: an efficient support for visual computing in surgical planning and training. IEEE Trans Vis Comput Graph 16(1):133–146. https://doi.org/10.1109/TVCG.2009.58

National Cancer Institute Clinical Proteomic Tumor Analysis Consortium (CPTAC), Radiology data from the clinical proteomic tumor analysis consortium pancreatic ductal adenocarcinoma [CPTAC-PDA] Collection [Data set]. Cancer Imag Arch. https://doi.org/10.7937/k9/tcia.2018.sc20fo18

Olabarriaga SD, Smeulders AWM (2001) Interaction in the segmentation of medical images: a survey. Med Image Anal 5(2):127–142. https://doi.org/10.1016/S1361-8415(00)00041-4

Olsen L, Samavati FF, Sousa MC, Jorge JA (2009) Sketch-based modeling: a survey. Comput Graph 33(1):85–103. https://doi.org/10.1016/j.cag.2008.09.013

Paulo SF et al (2018) The underrated dimension: How 3D interactive mammography can improve breast visualization. Lect Notes Comput Vis Biomech 27:329–337. https://doi.org/10.1007/978-3-319-68195-5_36

Peng H et al (2014) Virtual finger boosts three-dimensional imaging and microsurgery as well as terabyte volume image visualization and analysis. Nat Commun 5(1):1–13. https://doi.org/10.1038/ncomms5342

Ribeiro NS et al (2009) 3-D solid and finite element modeling of biomechanical structures-a software pipeline. In: Proceedings of the 7th EUROMECH Solid Mechanics Conference (ESMC2009), September 2009. http://citeseerx.ist.psu.edu/viewdoc/summary?doi=10.1.1.544.2385. Accessed 17 June 2020

Robison RA, Liu CY, Apuzzo MLJ (2011) Man, mind, and machine: the past and future of virtual reality simulation in neurologic surgery. World Neurosurg 76(5):419–430. https://doi.org/10.1016/j.wneu.2011.07.008

Rosset A, Spadola L, Ratib O (2004) OsiriX: an open-source software for navigating in multidimensional DICOM images. J Digit Imaging 17(3):205–216. https://doi.org/10.1007/s10278-004-1014-6

Shepherd T, Prince SJD, Alexander DC (2012) Interactive lesion segmentation with shape priors from offline and online learning. IEEE Trans Med Imaging 31(9):1698–1712. https://doi.org/10.1109/TMI.2012.2196285

Smith K et al., Data from CT colonography. Cancer Imag Arch https://doi.org/10.7937/K9/TCIA.2015.NWTESAY1

Sousa M, Mendes D, Paulo S, Matela N, Jorge J, Lopes DS (2017) VRRRRoom: virtual reality for radiologists in the reading room. In: Conference on human factors in computing systems—proceedings, vol 2017, May 2017, pp 4057–4062. https://doi.org/10.1145/3025453.3025566

Van Heeswijk MM et al (2016) Automated and semiautomated segmentation of rectal tumor volumes on diffusion-weighted MRI: can it replace manual volumetry? Int J Radiat Oncol Biol Phys 94(4):824–831. https://doi.org/10.1016/j.ijrobp.2015.12.017

Viceconti M, Taddei F (2003) Automatic generation of finite element meshes from computed tomography data. Crit Rev Biomed Eng 31(1–2):27–72. https://doi.org/10.1615/CritRevBiomedEng.v31.i12.20

Vosburgh KG, Golby A, Pieper SD (2013) Surgery, virtual reality, and the future. Stud Health Technol Inform 184:vii–xiii. http://www.ncbi.nlm.nih.gov/pubmed/23653952. Accessed 22 June 2020

World Health Organisation, Global atlas of medical devices, no. Licence: CC BY-NC-SA 3.0 IGO. 2017

Yushkevich PA et al (2006) User-guided 3D active contour segmentation of anatomical structures: significantly improved efficiency and reliability. Neuroimage 31(3):1116–1128. https://doi.org/10.1016/j.neuroimage.2006.01.015

Zorzal E, Gomes J, Sousa M, Belchior P, Silva P, Figueiredo N, Lopes D, Jorge J (2020) Laparoscopy with augmented reality adaptations. J Biomed Inform 107:103463. ISSN 1532-0464, https://doi.org/10.1016/j.jbi.2020.103463

Chapter 9
Statistical Analysis of Organs' Shapes and Deformations: The Riemannian and the Affine Settings in Computational Anatomy

Xavier Pennec

Abstract Computational anatomy is an emerging discipline at the interface of geometry, statistics, and medicine that aims at analyzing and modeling the biological variability of organs' shapes at the population level. Shapes are equivalence classes of images, surfaces, or deformations of a template under rigid body (or more general) transformations. Thus, they belong to non-linear manifolds. In order to deal with multiple samples in non-linear spaces, a consistent statistical framework on Riemannian manifolds has been designed over the last decade. We detail in this chapter the extension of this framework to Lie groups endowed with the affine symmetric connection, a more invariant (and thus more consistent) but non-metric structure on transformation groups. This theory provides strong theoretical bases for the use of one-parameter subgroups and diffeomorphisms parametrized by stationary velocity fields (SVF), for which efficient image registration methods like log-Demons have been developed with a great success from the practical point of view. One can further reduce the complexity with locally affine transformations, leading to parametric diffeomorphisms of low dimension encoding the major shape variability. We illustrate the methodology with the modeling of the evolution of the brain with Alzheimer's disease and the analysis of the cardiac motion from MRI sequences of images.

9.1 Introduction

At the interface of geometry, statistics, image analysis, and medicine, computational anatomy aims at analyzing and modeling the biological variability of the organs' shapes and their dynamics at the population level. The goal is to model the mean anatomy, its normal variation, its motion/evolution, and to discover morphological differences between normal and pathological groups. For instance, the analysis of population-wise structural brain changes with aging in Alzheimer's disease requires first the analysis of longitudinal morphological changes for a specific subject, which can be done using non-linear registration-based regression, followed by a longi-

X. Pennec (✉)
Université Côte d'Azur and Inria, Nice, France
e-mail: xavier.pennec@inria.fr

tudinal group-wise analysis where the subject-specific longitudinal trajectories are transported in a common reference (Lorenzi et al. 2011; Hadj-Hamou et al. 2016). In both steps, it is desirable that the longitudinal and the inter-subject transformations smoothly preserve the spatial organization of the anatomical tissues by avoiding intersections, foldings, or tearing. Simply encoding deformations with a vector space of displacement fields is not sufficient to preserve the topology: one needs to require diffeomorphic transformations (differentiable one-to-one transformations with differentiable inverse). Space of diffeomorphisms are examples of infinite-dimensional manifolds. Informally, manifolds are spaces that locally (but not globally) resemble a given Euclidean space. The simplest example is the sphere or the earth surface which looks locally flat at a scale which is far below the curvature radius but exhibit curvature and a non-linear behavior at larger scales.

Likewise, shape analysis most often relies on the identification of features describing locally the anatomy such as landmarks, curves, surfaces, intensity patches, and full images. Modeling their statistical distribution in the population requires to first identify point-to-point anatomical correspondences between these geometric features across subjects. This may be feasible for landmark points, but not for curves or surfaces. Thus, one generally considers relabelled point-sets or reparametrized curve/surface/image as equivalent objects. With this geometric formulation, shapes spaces are the quotient the original space of features by their reparametrization group. One also often wants to remove a global rigid or affine transformation. One considers in this case the equivalence classes of images, surfaces, or deformations under the action of this space transformation group, and shape spaces are once again quotient spaces. Unfortunately, even if we start from features belonging to a nice Euclidean space, taking the quotient generally endows the shape space with a non-linear manifold structure. For instance, equivalence classes of k-tuples of points under rigid or similarity transformations result in non-linear Kendall's shape spaces (see, e.g., Dryden and Mardia (2016) for a recent account on that subject). The quotient of curves, surfaces, and higher dimensional objects by their reparametrizations (diffeomorphisms of their domains) produces in general even more complex infinite-dimensional shape spaces, see Bauer et al. (2014).

Thus, shapes and deformations belong in general to non-linear manifolds, while statistics were essentially developed for linear and Euclidean spaces. For instance, adding or subtracting two curves does not really make sense. It is thus not easy to average several shapes. Likewise, averaging unit vectors (resp. rotation matrices) do not lead to a unit vector (resp. a rotation matrix). It is thus necessary to define a consistent statistical framework on manifolds and Lie groups. This has motivated the development of Geometric Statistics during the last decade, see Pennec et al. (2020). We summarize below the main features of the theory of statistics on manifolds, before generalizing it in the next section to more general affine connection spaces.

9.1.1 Riemannian Manifolds

While being non-linear, manifolds are locally Euclidean, and an infinitesimal measure of the distance (a metric) allows to endow them with a Riemannian manifold structure. More formally, a Riemannian metric on a manifold $\mathcal{M}$ is a continuous collection of scalar products on the tangent space $T_x\mathcal{M}$ at each point x of the manifold. The metric measures the dot product of two infinitesimal vectors at a point of our space; this allows to measure directions and angles in the tangent space. One can also measure the length of a curve on our manifold by integrating the norm of its tangent vector. The minimal length among all the curves joining two given points defines the intrinsic distance between these two points. The curves realizing these shortest paths are called geodesics, generalizing the geometry of our usual flat 3D space to curved spaces among which the flat torus, the sphere, and the hyperbolic space are the simplest examples.

The calculus of variations shows that geodesics are the solutions of a system of second-order differential equations depending on the Riemannian metric. Thus, the geodesic curve $\gamma_{(x,v)}(t)$ starting at a given point x with a given tangent vector $v \in T_x\mathcal{M}$ always exists for some short time. When the time-domain of all geodesics can be extended to infinity, the manifold is said to be geodesically complete. This means that the manifold has no boundary nor any singular point that we can reach in a finite time. As an important consequence, the Hopf-Rinow-De Rham theorem states that there always exists at least one minimizing geodesic between any two points of the manifold (i.e., whose length is the distance between the two points), see do Carmo (1992). Henceforth, we implicitly assume that all Riemannian manifolds are geodesically complete.

The function $\exp_x(v) = \gamma_{(x,v)}(1)$ mapping the tangent space $T_x\mathcal{M}$ at x to the manifold $\mathcal{M}$ is called the exponential map at the point x. It is defined on the whole tangent space but it is diffeomorphic only locally. Its inverse $\log_x(y)$ is a vector rooted at x. It maps each point y of a neighborhood of x to the shortest tangent vector that allows to join x to y geodesically. The maximal definition domain of the log is called the injectivity domain. It covers all the manifold except a set of null measure called the cut-locus of the point. For statistical purposes, we can thus safely neglect this set in many cases. The Exp and Log maps $\exp_x$ and $\log_x$ are defined at any point x of the manifold (x is called the foot-point in differential geometry). They realize a continuous family of very convenient charts of the manifold where geodesics starting from the foot-point are straight lines, and along which the distance to the foot-point is conserved. These charts are somehow the "most linear" chart of the manifold with respect to their foot-point (Fig. 9.1)

In practice, we can identify a tangent vector $v \in T_x\mathcal{M}$ within the injectivity domain to the end-points of the geodesic segment $[x, y = \exp_x(v)]$ thanks to the exponential maps. Conversely, almost any bi-point (x, y) on the manifold where y is not in the cut-locus of x can be mapped to the vector $\overrightarrow{xy} = \log_x(y) \in T_x\mathcal{M}$ by the log map. In a Euclidean space, we would write $\exp_x(v) = x + v$ and $\log_x(y) = y - x$. This reinterpretation of addition and subtraction using logarithmic and exponential

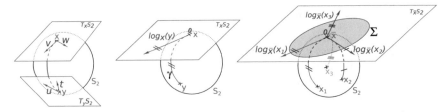

Fig. 9.1 Riemannian geometry and statistics on the sphere. Left: The tangent planes at points x and y of the sphere S_2 are different: the tangent vectors v and w at the point x cannot be compared to the vectors t and u that tangent at the point y. Thus, it is natural to define the scalar product on each tangent plane. Middle: Geodesics starting at x are straight lines in a normal coordinate system at x and the distance is conserved up to the cut-locus. Right: the Fréchet mean $\bar{x}$ is the point minimizing the mean squared Riemannian distance to the data points. It corresponds to the point for which the development of the geodesics to the data points on the tangent space is optimally centered (the mean $\sum_i \log_{\bar{x}}(x_i) = 0$ in that tangent space is zero). The covariance matrix is then defined in that tangent space. Figure adapted from Pennec (2006)

maps is very powerful to generalize algorithms working on vector spaces to algorithms on Riemannian manifolds. It is also very powerful in terms of implementation since we can express many of the geometric operations in these terms: the implementation of the exp and log maps at each point is thus the basis of programming on Riemannian manifolds.

9.1.2 Statistics on Riemannian Manifolds

The Riemannian metric induces an infinitesimal volume element on each tangent space, denoted $d\mathcal{M}$, that can be used to measure random events on the manifold and to define intrinsic probability density functions (pdf). It is worth noticing that the measure $d\mathcal{M}$ represents the notion of uniformity according to the chosen Riemannian metric. With the probability measure of a random element, we can integrate functions from the manifold to any vector space, thus defining the expected value of this function. However, we generally cannot integrate manifold-valued functions since an integral is a linear operator. Thus, one cannot define the mean or expected "value" of a random manifold element using a weighted sum or an integral as usual.

The main solution to this problem is to redefine the mean as the minimizer of an intrinsic quantity: the Fréchet (resp. Karcher) mean minimizes globally (resp. locally) the sum of squared Riemannian distance to our samples. As the mean is now defined through a minimization procedure, its existence and uniqueness may be questioned. In practice, one mean value almost always exists, and it is unique as soon as the distribution is sufficiently peaked. The properties of the mean are very similar to those of the modes of a distribution in the Euclidean case. The Fréchet mean was used since the 1990s in medical image analysis for redefining simple statistical methods on Riemannian manifolds (Pennec 1996; Pennec and Ayache 1998; Pennec 1999, 2006; Fletcher et al. 2004).

To compute the Fréchet mean, one can follow the Riemannian gradient of the variance with an iteration of the type:

$$\bar{x}_{t+1} = \exp_{\bar{x}_t} \left(\alpha \frac{1}{n} \sum_i \log_{\bar{x}_t}(x_i) \right).$$

The algorithm essentially alternates the computation of the tangent mean in the tangent space at the current estimation of the mean, and a geodesic marching step toward the computed tangent mean. The value $\alpha = 1$ corresponding to a Gauss–Newton scheme is usually working very well, although there are examples where it should be reduced due to the curvature of the space. An adaptive time-step in the spirit of Levenberg–Marquardt is easily solving this problem.

When the Fréchet mean is determined, one can pull back our distribution of data points on the tangent space at the mean to define higher order moments like the covariance matrix $\Sigma = \frac{1}{n} \sum_{i=1}^{n} \log_{\bar{x}}(x_i) \log_{\bar{x}}(x_i)^{\mathsf{T}}$. Seen for the most central point (the Fréchet mean), we have somehow corrected the non-linearity of our Riemannian manifold. Based on this mean $\bar{x}$ and this covariance matrix Σ, we can define the Mahalanobis distance in the tangent space by

$$\mu^2_{(\bar{x}, \Sigma)}(y) = \log_{\bar{x}}(y)^{\mathsf{T}} \Sigma^{(-1)} \log_{\bar{x}}(y).$$

It is worth noticing that the expected Mahalanobis distance of a random point is independent of the distribution and is equal to the dimension of the manifold when its mean and covariance are known, as in the vector case (Pennec 1996, 2006). A very simple extension of Principle Component Analysis (PCA) consists in diagonalizing the covariance matrix Σ and defining the modes using the eigenvectors of decreasing eigenvalues in the tangent space at the mean. This method usually works very well for sufficiently concentrated data. More complex methods like Principal Geodesic Analysis (PGA), geodesic PCA, or Barycentric Subspace Analysis (BSA) may be investigated for data distributions with a larger support as in Pennec (2018).

A notion of Gaussian may also be defined on a manifold by choosing the distribution that minimizes the entropy knowing the mean and the covariance. It was shown in (Pennec 1996, 2006) that this amounts to consider a truncated Gaussian distribution on the tangent space at the mean point which only covers the injectivity domain (i.e., truncated at the tangential cut locus): the pdf (with respect to the Riemannian measure) is $N_{(\bar{x}, \Sigma)}(y) = Z(\bar{x}, \Sigma) \exp(-\frac{1}{2} \log_{\bar{x}}(y)^{\mathsf{T}} \Gamma \log_{\bar{x}}(y))$. However, we should be careful that the relation between the concentration matrix Γ and the covariance matrix Σ is more complex than the simple inversion of the Euclidean case since it has to be corrected for the curvature of the manifold.

Based on this truncated Gaussian distribution, one can generalize the multivariate Hotelling T-squared test using the Mahalanobis distance. When the distribution is Gaussian with a known mean and covariance matrix, the law generalizes the χ^2 law and Pennec (2006) showed that it has the same density as in the vector case up to order 3. This opens the way to the generalization of many other statistical tests, as we should obtain similarly simple approximations for sufficiently centered distributions.

Notice that the reformulation of the (weighted) mean as an intrinsic minimization problem allows to extend quite a number of other image processing algorithms to manifold-valued signal and images, like interpolation, diffusion, and restoration of missing data (extrapolation). This is the case for instance of diffusion tensor imaging for which manifold-valued image processing was pioneered in Pennec et al. (2006).

9.2 An Affine Symmetric Space Structure for Lie Groups

A classical way to perform statistics on shapes in computational anatomy is to estimate or assume a template shape and then to encode other shapes by diffeomorphic transformations of that template. This lifts the problem from statistics on manifolds to statistics on smooth transformation groups, i.e., Lie groups. The classical Riemannian methodology consists in endowing the Lie group with a left (or right) invariant metric which turns the transformation group into a Riemannian manifold. This means that the metric at a point x of the group is obtained by the left translation $L_x(y) = x \circ y$ of the metric at identity, or in a more computational way, that the scalar product of two tangent vectors at x is obtained by left-translating them back to identify using $DL_{x^{-1}}$ and taking the scalar product there. A right-invariant metric is obtained if we use the differential of the right translation $R_x(y) = y \circ x$ to identify the tangent space at x to the tangent space at identity. However, this Riemannian approach is consistent with the inversion operation of the group only if the metric is both left- and right-invariant. This is the case for compact or commutative groups, such as rotations or translations. But as soon as the Lie group is a non-direct product of simpler compact or commutative ones, such as rigid-body transformations in 2D or 3D, there does not exist a bi-invariant metric: left-invariant metrics are not right-invariant. Since the inversion exchanges left and right, such metrics are not inverse consistent either. This means that the Fréchet mean for a left (resp. right) invariant metric is not consistent with inversion and right (resp. left) composition. In particular, the mean of inverse transformations is not the inverse of the mean.

One can wonder if there exists a more general framework, obviously non-Riemannian, to realize consistent statistics on these Lie groups. Indeed, numerous methods in Lie groups are based on pure group properties, independent of the action of transformations on objects. These methods rely in particular on one-parameter subgroups, realized in finite-dimensional matrix Lie groups by the matrix exponential. There exist particularly efficient algorithms to compute the matrix exponential like the scaling and squaring procedure (Higham 2005) or for integrating differential equations on Lie groups in geometric numerical integration theory (Hairer et al. 2002; Iserles et al. 2000). In infinite dimension, one-parameter subgroups are deformations realized by the flow of stationary velocity fields (SVFs), as we will see in Sect. 9.3.1. Parametrizing diffeomorphisms with SVFs was proposed for medical image registration by Arsigny et al. (2006) and very quickly adopted by many other authors (Ashburner 2007; Vercauteren et al. 2008; Hernandez et al. 2009; Modat et al. 2010). The group structure was also used to obtain efficient low-dimensional parametric locally affine diffeomorphisms as we will see in Sect. 9.5.1.

In fact, these one-parameter subgroups (matrix exponential and flow of SVF) are the geodesics of the Cartan–Schouten connection, a more invariant and thus more consistent but non-metric structure on transformation groups. We detail in this section the extension of the computing and statistical framework to Lie groups endowed with the affine symmetric connection. In the medical imaging and geometric statistics communities, these notions were first developed in (Pennec and Arsigny 2012; Lorenzi and Pennec 2013). A more complete account on the theory appeared recently in Pennec and Lorenzi (2020). We refer the reader to this chapter for more explanations and mathematical details.

9.2.1 Affine Geodesics

Geodesics, exponential, and log maps are among the most fundamental tools to work on differential manifolds. In order to define a notion of geodesics in non-Riemannian spaces, we cannot rely on the shortest path as there is no Riemannian metric to measure length. The main idea is to define straight lines as curves with vanishing acceleration, or equivalently curves whose tangent vectors remain parallel to themselves (auto-parallel curves). In order to compare vectors living in different tangent spaces (even at points which are infinitesimally close), we need to provide a notion of parallel transport from one tangent space to the other. Likewise, computing accelerations require a notion of infinitesimal parallel transport that is called a connection.

In a local coordinate system, a connection is completely determined by its coordinates on the basis vector fields: $\nabla_{\partial_i} \partial_j = \Gamma_{ij}^k \partial_k$. The n^3 coordinates Γ_{ij}^k of the connection are called the Christoffel symbols. A curve $\gamma(t)$ is a geodesic if its tangent vector $\dot{\gamma}(t)$ remains parallel to itself, i.e., if the covariant derivative $\nabla_{\dot{\gamma}} \dot{\gamma} = 0$ of γ is zero. In a local coordinate system, the equation of the geodesics is thus $\ddot{\gamma}^k + \Gamma_{ij}^k \dot{\gamma}^i \dot{\gamma}^j = 0$, exactly as in a Riemannian case. The difference is that the affine connection case starts with the Christoffel symbols, while these are determined by the metric in the Riemannian case, giving a natural connection called the Levi-Civita connection. Unfortunately, the converse is not always possible; many affine connection spaces do not accept a compatible Riemannian metric. Riemannian manifolds are only a subset of affine connection spaces.

What is remarkable is that we conserve many properties of the Riemannian exponential map in affine connection spaces. For instance, the geodesic $\gamma_{(x,v)}(t)$ starting at any point x with any tangent vector v is defined for a sufficiently small time, which means that we can define the affine exponential map $\exp_x(v) = \gamma_{(x,v)}(1)$ for a sufficiently small neighborhood. Moreover, there exists at each point a *normal convex neighborhood* (NCN) in which any couple of points (x, y) is connected by a unique geodesic $\gamma(t)$ entirely contained in this neighborhood. We can thus define the log-map locally without ambiguity.

9.2.2 An Affine Symmetric Space Structure for Lie Groups

In the case of Lie groups, the Symmetric Cartan-Schouten (SCS) connection is a canonical torsion free connection introduced by Cartan and Schouten (1926) shortly after the invention of the notion of connection by Cartan. This is also the unique affine connection induced by the canonical symmetric space structure of the Lie groups with the symmetry $s_g(h) = gh^{(-1)}g$. The SCS connection exists on all Lie groups, and it is left- and right-invariant. When there exists a bi-invariant metric on the Lie group (i.e., when the group is the direct product of Abelien and compact groups), the SCS connection is the Levi-Civita connection of that metric. However, the SCS connection still exists when there is no bi-invariant metric.

Geodesics of the SCS connection are called group geodesics. The ones going through the identity are the flow of left-invariant vector fields. They are also called one-parameter subgroups since $\gamma(s + t) = \gamma(s) \circ \gamma(t)$ is an isomorphism of Lie groups, which is a mapping that preserves the Lie group structure. In matrix Lie groups, one-parameter subgroups are described by the exponential $\exp(M) = \sum_{k=0}^{\infty} M^k/k!$ of square matrices. Conversely, if there exists a square matrix M such that $\exp(M) = A$, then M is said to be a logarithm of the invertible square matrix A. In general, the logarithm of a real invertible matrix is not unique and may fail to exist. However, when this matrix has no (complex) eigenvalue on the (closed) half line of negative real numbers, then it has a unique real logarithm $\log(M)$, called the principal logarithm whose (complex) eigenvalues have an imaginary part in $(-\pi, \pi)$ (Kenney and Laub 1989; Gallier 2008). Moreover, matrix exp and log can be very efficiently numerically computed with the 'Scaling and Squaring Method' of Higham (2005) and 'Inverse Scaling and Squaring Method', see Hun Cheng et al. (2001).

Group geodesics starting from other point can be obtained very simply by left or right translation: $\gamma(t) = A \exp(tA^{(-1)}M) = \exp(tMA^{(-1)})A$ is the geodesic starting at A with tangent vector M. In finite dimension, the group exponential is a chart around the identity. In infinite-dimensional Fréchet manifolds, the absence of an inverse function theorem prevents the straightforward extension of this property to general groups of diffeomorphisms and one can show that there exists diffeomorphisms as close as we want to the identity that cannot be reached by one-parameter subgroups (Khesin and Wendt 2009). In practice, though, the diffeomorphisms that we cannot reach have not yet proved to be of practical use for any real-world application.

Thus, everything looks very similar to the Riemannian case, except that group geodesics are defined from group properties only and do not require any Riemannian metric. One should be careful that they are generally different from the Riemannian exponential map associated to a Riemannian metric on the Lie group.

9.2.3 Statistics in Affine Connection Spaces

In order to generalize the Riemannian statistical tools to affine connection spaces, the Fréchet/Karcher means have to be replaced by the weaker notion of exponential barycenters, which are the critical points of the variance in Riemannian manifolds. In an affine connection space, the exponential barycenters of a set of points $\{x_1 \ldots x_n\}$ are implicitly defined as the points x for which the tangent mean field vanishes:

$$\mathfrak{M}(x) = \frac{1}{n} \sum_{i=1}^{n} \log_x(x_i) = 0. \tag{9.1}$$

While this definition is close to the Riemannian center of mass (Gallier 2004), it uses the logarithm of the affine connection instead of the Riemannian logarithm.

For sufficiently concentrated distributions with compact support, typically in a normal convex neighborhood, there exists at least one exponential barycenter. Moreover, exponential barycenters are stable by affine diffeomorphisms (connection preserving maps). For distributions whose support is too large, exponential barycenters may not exist. This should be related to the classical non-existence of the mean for heavy tailed distributions in Euclidean spaces. The uniqueness of the exponential barycenter can be shown with additional assumptions, either on the derivatives of the curvature as in Buser and Karcher (1981) or on a stronger notion of convexity (Araudon and Li 2005).

Higher order moments can also be defined locally. For instance, the empirical covariance field is the two-contravariant tensor $\Sigma(x) = \frac{1}{n} \sum_{i=1}^{n} \log_x(x_i) \otimes \log_x(x_i)$ and its value $\Sigma = \Sigma(\bar{x})$ at the exponential barycenter $\bar{x}$ is called the empirical covariance. Notice that this definition depends on the chosen basis and that diagonalizing the matrix makes no sense since we do not know what are orthonormal unit vectors. Thus, tangent PCA is not easily generalized. Despite the absence of a canonical reference metric, the Mahalanobis distance of a point y to a distribution can be defined locally as in the Riemannian case with the inverse of the covariance matrix. This definition is independent of the basis chosen for the tangent space and is actually invariant under affine diffeomorphisms of the manifold. This simple extension of the Mahalanobis distance suggests that it might be possible to extend much more statistical definitions and tools on affine connection spaces in a consistent way.

9.2.4 The Case of Lie Groups with the Canonical Cartan–Schouten Connection

Thanks to the bi-invariance properties of the SCS connection, the exponential barycenters of Eq. (9.1) define bi-invariant group means. Let $\{A_i\}$ be a set of transformations from the group (we can think of matrices here). Then a transformation $\bar{A}$ verifying $\sum_i \log(\bar{A}^{(-1)} A_i) = \sum_i \log(A_i \bar{A}^{(-1)}) = 0$ is a group mean which exists and is unique for sufficiently concentrated data $\{A_i\}$. Moreover, the fixed point iteration $\bar{A}_{t+1} = \sum_i \log(\bar{A}_t^{(-1)} A_i)$ converges to the bi-invariant mean at least linearly (still under a sufficient concentration condition), which provides a very useful algorithm to compute it in practice.

The bi-invariant mean turns out to be globally unique in a number of Lie groups which do not support any bi-invariant metric, for instance, nilpotent or some specific solvable groups (Pennec and Arsigny 2012; Lorenzi and Pennec 2013; Pennec and Lorenzi 2020). For rigid-body transformations, the bi-invariant mean is unique when the mean rotation is unique, so that we do not lose anything with respect to the Riemannian setting. Thus, the group mean appears to be a very general and natural notion on Lie groups.

9.3   The SVF Framework for Shape and Deformation Modeling

In the context of medical image registration, diffeomorphic registration was introduced with the "Large Deformation Diffeomorphic Metric Mapping (LDDMM)" framework (Trouvé 1998; Beg et al. 2005), which parametrizes deformations with the flow of *time-varying velocity fields* $v(x, t)$ with a right-invariant Riemannian metric (see Younes (2010) for a complete mathematical description). In view of reducing the computational and memory costs, Arsigny et al. (2006) subsequently proposed to restrict this parametrization to the subset of diffeomorphisms parametrized by the flow of stationary velocity fields (SVFs), for which efficient image registration methods like log-Demons have been developed with a great success from the practical point of view. The previous theory of statistics on Lie groups with the canonical symmetric Cartan–Schouten connection provides strong theoretical bases for the use of these one-parameter subgroups.

9.3.1   Diffeomorphisms Parametrized by Stationary Velocity Fields

To construct our group of diffeomorphisms, one first restricts the Lie algebra to sufficiently regular velocity fields according to the regularization term of the SVF registration algorithms (Vercauteren et al. 2008; Hernandez et al. 2009) or to the spline parametrization of the SVF in Ashburner (2007); Modat et al. (2010). The flow of these stationary velocity fields and their finite composition generates a group of diffeomorphisms that we endow with the affine symmetric Cartan–Schouten connection. The geodesics starting from identity are then exactly the one-parameter subgroups generated by the flow of SVFs: the deformation $\phi = \exp(v)$ is parametrized by the Lie group exponential of a smooth SVF $v : \Omega \to \mathbb{R}^3$ through the ordinary differential equation (ODE) $\frac{\partial \phi(x,t)}{\partial t} = v(\phi(x, t))$ with initial condition $\phi(x, 0) = x$. It is known that not all diffeomorphisms can be reached by such a one-parameter subgroup (we might have to compose several ones to reach them all) but in practice this does not seem to be a limitation.

Many of the techniques developed for the matrix case can be adapted to SVFs. This is the case of the scaling and squaring algorithm, which integrates the previous ODE very effectively thanks to the iterative composition of successive exponentials: $\exp(v) = \exp(v/2) \exp(v/2) = (\exp(v/2^n))^n$. Inverting a deformation is usually quite difficult or at least computationally intensive as we have to find ψ such that $\psi(\phi(x)) = \phi(\psi(x)) = x$. This is generally performed using the least-square minimization of the error on the above equation integrated over the image domain. In the SVF setting, such a computation can be performed seamlessly since $\phi^{(-1)} = \exp(-v)$.

In order to measure volume changes induced by the deformation, one usually computes the Jacobian matrix $d\phi = \nabla\phi^{\mathsf{T}}$ using finite differences, and then takes its determinant. However, finite-differences schemes are highly sensitive to noise. In the SVF framework, the log-Jacobian can be reliably estimated by finite differences for the scaled velocity field $v/2^n$, and then recursively computed thanks to the chain rule in the scaling and squaring scheme and thanks to the additive property of the one-parameter subgroups. The Jacobian determinant that we obtain is, therefore, fully consistent with the exponential path taken to compute the diffeomorphism.

Last but not least, one often needs to compose two deformations, for instance, to update the current estimation in an image registration algorithm. The Baker Campbell Hausdorff (BCH) formula is a series expansion that approximates the SVF

$$BCH(v, u) = \log(\exp(v) \exp(u)) = v + u + \frac{1}{2}[v, u] + \frac{1}{12}[v, [v, u]] + \dots$$

as a power series in the two SVFs u and v. In this formula, the Lie bracket of vector fields is v: $[v, u] = dv\,u - du\,v = \partial_u v - \partial_v u$. In the context of diffeomorphic image registration, this trick to do all the computations in the Lie algebra was introduced by Bossa et al. (2007).

9.3.2 SVF-Based Diffeomorphic Registration with the Log-Demons

The encoding of diffeomorphisms via the flow of SVF of Arsigny et al. (2006) inspired several SVF-based image registration algorithms (Vercauteren et al. 2007, 2009; Bossa et al. 2007; Ashburner 2007; Hernandez et al. 2009; Modat et al. 2010, 2011; Lorenzi et al. 2013). Among them, the log-demons registration algorithm (Vercauteren et al. 2008; Lorenzi et al. 2013) found a considerable interest in the medical image registration community with many successful applications to clinical problems (Peyrat et al. 2008; Mansi et al. 2011; Lorenzi et al. 2011; Seiler et al. 2011).

Given a pair of images $I, J : \mathbb{R}^3 \mapsto \mathbb{R}$, the log-demons algorithm aims at estimating a SVF v parametrizing diffeomorphically the spatial correspondences that minimize a similarity functional $Sim[I, J \circ \exp(v)]$. A classically used similarity criterion is the sum of square differences (SSD) $Sim[I, J] = \int (I(x) - J(x))^2 dx$. In order to symmetrize the criterion and ensure inverse consistency, one can add the symmetric similarity criterion $Sim[I \circ \exp(-v), J]$ as in Vercauteren et al. (2008) or more simply measure the discrepancy at the mid-deformation point using $Sim[I \circ \exp(-v/2), J \circ \exp(v/2)]$. This last formulation allows to easily symmetrize a similarity functional that is more complex than the SSD, such as the local correlation coefficient (LCC) (Lorenzi et al. 2013).

In order to prevent overfitting, a regularization term that promotes spatially more regular solutions is added to the similarity criterion. In the log-demons framework, this regularization is naturally performed on the SVF v rather than on the deformation $\phi = \exp(v)$. A feature of the demons' type algorithms is also to introduce an auxiliary variable encoding for the correspondences, here a SVF v_c, in addition to the SVF v encoding for the transformation (Cachier et al. 2003). The two variables are linked using a coupling criterion that prevents the two from being too far away from each other. The criterion optimized by the log-demons is then

$$E(v, v_c, I, J) = \tfrac{1}{\sigma_i^2} Sim(I, J, v_c) + \tfrac{1}{\sigma_x^2} \|v_c - v\|_{L_2}^2 + \tfrac{1}{\sigma_T^2} Reg(v). \qquad (9.2)$$

The interest of the auxiliary variable is to decouple a non-linear and non-convex optimization problem into two simpler optimization problems that are, respectively, local and quadratic. The classical criterion is obtained at the limit when the typical scale of the error σ_x^2 between the transformation and the correspondences tends to zero.

The minimization of (9.2) is alternatively performed with respect to the correspondence SVF v_c and the transformation SVF v. The first step is a non-convex but purely local problem which is usually optimized via gradient descent using Gauss–Newton or Levenberg–Marquardt algorithms. To simplify the second step, one can choose $Reg(\cdot)$ to be an isotropic differential quadratic form (IDQF, see Cachier and Ayache (2004)), which leads to a closed form solution by convolution. In most cases, one chooses this convolution to be Gaussian: $v = G_\sigma * v_c$, which can be computed very efficiently using separable recursive filters.

9.4 Modeling Longitudinal Deformation Trajectories in Alzheimer's Disease

With the log-demons algorithm, we can register two longitudinal images of the same subject. When more images are available at multiple time-points, we can regress the geodesic that best describes the different registrations to obtain a longitudinal deformation trajectory encoded by a single SVF (Lorenzi et al. 2011; Hadj-Hamou et al. 2016). We should notice that while such a geodesic is a linear model in the space of SVFs, it is a highly non-linear model on the displacement field and on the space of images.

However, follow-up imaging studies usually require to transport this subject-specific longitudinal trajectories in a common reference for group-wise statistical analysis. A typical example is the analysis of structural brain changes with aging in Alzheimer's disease versus normal controls. It is quite common in neuroimaging to transport from the subject to the template space a scalar summary of the changes over time like the Jacobian or the log-Jacobian encoding for local volume changes. This is easy and numerically stable as we just have to resample the scalar map. However, this does not allow to compute the "average" group-wise deformation and its variability, nor to transport it back at the subject level to predict what will be the future deformation. To realize such a generative model of the longitudinal deformations, we should normalize the deformations as a geometric object and not just its components independently. This involves defining a method of transport of the longitudinal deformation parameters along the inter-subject change of coordinate system.

9.4.1 Parallel Transport in Riemannian and Affine Spaces

Depending on the considered parametrization of the transformation (displacement fields, stationary velocity fields, initial momentum field…), different approaches have been proposed in the literature to transport longitudinal deformations. In the Riemannian and affine connection space setting, where longitudinal deformations are encoded by geodesics parametrized by their initial tangent vector, it is natural to consider the parallel transport of this initial tangent vector (describing the longitudinal deformation) along the inter-subject deformation curve. Parallel transport is an isometry of tangent spaces in the Riemannian case, so that the norm is conserved. In the affine connection case, this is an affine transformation of tangent spaces. Instead of defining properly the parallel transport in the continuous setting and approximating it in an inconsistent discrete setting, it was proposed in Lorenzi et al. (2011) to rely on a carefully designed discrete construction that intrinsically respects all the symmetries on the problem: the Schild's Ladder. This algorithm was initially introduced in the 1970s by the physicist Alfred Schild Ehlers et al. (1972) in the field of the general relativity. The method was refined with the pole ladder in Lorenzi and Pennec (2013) to minimize the number of steps when the transport is made along geodesics. Schild's and pole ladders only require the computation of exponentials and logarithms, and thus can easily and consistently be implemented for any manifold provided that we have these basic algorithmic bricks.

In this process, the numerical accuracy of parallel transport algorithm is the key to preserve the statistical information. The analysis of pole ladder in Pennec (2018) actually showed that the scheme is of order three in general affine connection spaces with a symmetric connection, an order higher than expected. Moreover, the fourth-order error term vanishes in affine symmetric spaces since the curvature is covariantly constant. In fact, the error terms vanish completely in a symmetric affine connection space: one step of pole ladder realizes a transvection, which is an exact parallel transport (provided that geodesics and mid-points are computed exactly of course), see Pennec (2018). These properties make pole ladder a very attractive alternative for parallel transport in the framework of diffeomorphisms parametrized by SVFs. In particular, parallel transport has a closed form expression $\Pi^v(u) = \log(\exp(v/2)\exp(u)\exp(-v/2))$ as shown in Lorenzi and Pennec (2013). In practice, the symmetric reformulation of the pole ladder scheme using the composition of two central symmetries (a transvection) gives numerically more stable results and was recently shown to be better than the traditional Euclidean point distribution model on cardiac ventricular surfaces (Jia et al. 2018).

9.4.2 Longitudinal Modeling of Alzheimer's Progression

Parallel transport allows us to compute a mean deformation trajectory at the group level and to differentiate populations on the basis of their full deformation features and not only according to local volume change as in traditional tensor-based morphometry (TBM). We illustrate in this section an application of this framework to the statistical modeling of the longitudinal changes in a group of patients affected by Alzheimer's disease (AD). In this disease, it was shown that the brain atrophy that one can measure using the registration of time sequences of magnetic resonance images (MRI) is strongly correlated to cognitive performance and neuropsychological scores. Thus, deformation-based morphometry provides an interesting surrogate image biomarker for the progression of the disease from pre-clinical to pathological stages.

The study that we summarize here was published in Lorenzi and Pennec (2013). We took 135 Alzheimer's subjects of the ADNI database with images at baseline and one year later. The SVFs v_i parametrizing the longitudinal deformation trajectory $\phi_i = \exp(v_i)$ between the two time-points was estimated with the LCC log-demons. These SVFs were then transported with the pole ladder from their subject-specific space to the template reference T along the subject-to-template geodesic, also computed using the LCC log-demons. The mean $\bar{v}$ of the transported SVFs in the template space parametrizes our model of the group-wise longitudinal progression $\exp(t\bar{v})$. The spatial localization of significant longitudinal changes (expansion or contraction) was established using one-sample t-test on the log-Jacobian scalar maps after parallel transport. In order to compare with the traditional method used in tensor-based morphometry, another one-sample t-test was computed on the subject-specific log-Jacobian scalar maps resampled in the template space.

Results are presented in Fig. 9.2. Row A illustrates the mean SVF of the transported one-year longitudinal trajectories. It shows a pronounced enlargement of the ventricles, an expansion of their temporal horns and a consistent contracting flow in the temporal areas. It is impressive that the extrapolation of the deformation along the geodesic from 1 year to 15 years produces a sequence of very realistic images going from a young brain at $t = -7$ years to a quite old AD brain with very large

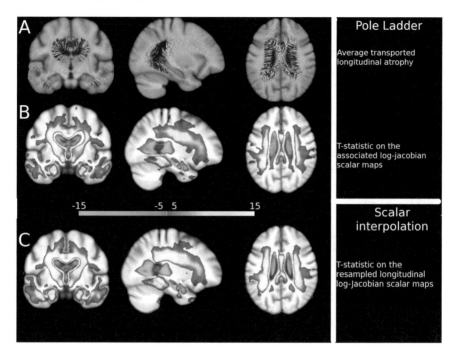

Fig. 9.2 One year structural changes for 135 Alzheimer's patients. **A** Mean of the longitudinal SVFs transported in the template space with the pole ladder. We notice the lateral expansion of the ventricles and the contraction in the temporal areas. **B** T-statistic for the corresponding log-Jacobian values significantly different from 0 ($p < 0.001$ FDR corrected). **C** T-statistic for longitudinal log-Jacobian scalar maps resampled from the subject to the template space. Blue color: significant expansion, Red color: significant contraction (Figure reproduced from Lorenzi and Pennec (2013) with permission)

ventricles and almost no hippocampus at $t = 8$ years. This shows that a linear model in a carefully designed non-linear manifold of diffeomorphisms can handle realistically very large shape deformations. Such a result is definitely out of sight with a statistical model on the displacement vector field or even with a classical point distribution model (PDM), as is often done in classical medical shape analysis.

Evaluating the volumetric changes (here computed with the log-Jacobian) leads to areas of significant expansion around the ventricles with a spread in the Cerebrospinal Fluid (CSF, row B). Areas of significant contraction are located as expected in the temporal lobes, hippocampi, parahippocampal gyrus, and in the posterior cingulate. These results are in agreement with the classical resampling of the subject-specific log-Jacobian maps done in TBM presented in row C. It is striking that there is no substantial loss of localization power for volume changes by transporting SVFs instead of resampling the scalar log-Jacobian maps. In contrast to TBM, we also preserve the full multidimensional information about the transformation, which allows to make more powerful multivariate voxel-by-voxel comparisons than the ones obtained with the classical univariate tests. For example, we could show for the first time in Lorenzi et al. (2011) a statistically significant different brain shape evolutions depending on the level of $A\beta_{1-42}$ protein in the CSF. As the level of $A\beta_{1-42}$ is sometimes considered as pre-symptomatic of Alzheimer's disease, we could be observing the very first morphological impact of the disease. More generally, a normal longitudinal deformation model allows to disentangle normal aging component from the pathological atrophy even with one time-point only per patient (cross-sectional design) (Lorenzi et al. 2015).

The SVF describing the trajectory can also be decomposed using Helmholtz' decomposition into a divergent part (the gradient of a scalar potential) that encodes the local volume changes and a divergence free reorientation pattern, see Lorenzi et al. (2015). This allows to consistently define anatomical regions of longitudinal brain atrophy in multiple patients, leading to improved measurements of the quantification of the longitudinal hippocampal and ventricular atrophy in AD. This method provided very reliable results during the MIRIAD atrophy challenge for the regional atrophy quantification in the brain, with really state-of-the-art performances (first and second rank on deep structures, Cash et al. 2015).

9.5 The SVF Framework for Cardiac Motion Analysis

Cardiac motion plays an important role in the function of the heart, and abnormalities in the cardiac motion can be the cause of multiple diseases and complications. Modeling cardiac motion can, therefore, provide precious information. Unfortunately, the outputs from cardiac motion models are complex. Therefore, they are hard to analyze, compare, and personalize. The approach described below relies on a polyaffine projection applied to the whole cardiac motion and results in a few parameters that are physiologically relevant.

9.5.1 Parametric Diffeomorphisms with Locally Affine Transformations

The polyaffine framework assumes that the image domain is divided into regions defined by smooth normalized weights ω_i (i.e., summing up to one over all regions). The transformation of each region is modeled by a locally affine transformation expressed by a 4×4 matrix A_i in homogeneous coordinates. Using the principal logarithm $M_i = \log(A_i)$ of these matrices, we compute the SVF at any voxel x (expressed in homogeneous coordinates) as the weighted sum of these locally affine transformation (Arsigny et al. 2005, 2009):

$$v_{poly}(x) = \sum_i \omega_i(x) \, M_i \, x.$$

The polyaffine transformation is then obtained by taking the flow of SVF using the previous scaling and squaring algorithm for the exponential. This leads to a very flexible locally affine diffeomorphism parametrized by very few parameters. In this formulation, taking the log in homogeneous coordinates ensures that the inverse of the polyaffine transformations is also a polyaffine transformation. This property is necessary to create generative motion models.

As shown in Seiler et al. (2012), the log affine matrix parameters M_i can be estimated explicitly by a linear least squares projection of an observed velocity field $v(x)$ into the space of Log-Euclidean Polyaffine Transformations (LEPT's). Denoting $\Sigma_{ij} = \int_\Omega \omega_i(x)\omega_j(x)xx^\mathsf{T}dx$ and $B_i = \int_\Omega \omega_i(x)v(x)x^\mathsf{T}dx$, the optimal matrix of log-affine transformation parameters $M = [M_1, M_2, ... M_n]$, minimizing the criterion $C(M) = \int_\Omega \| \sum_i \omega_i(x)M_i x - v(x)\|^2 dx$ is given by $M = B\Sigma^{(-1)}$. The solution is unique when the Gram matrix Σ of the basis vectors of our polyaffine SVF is invertible. This gives rise to the polyaffine log-demons algorithm where the estimated SVF at each step of the log-demons algorithm is projected into this low-dimensional parameter space instead of being regularized.

A cardiac-specific version of this model was proposed in Mcleod et al. (2015b) by choosing regions corresponding to the standard American Heart Association (AHA) regions for the left ventricle (Fig. 9.3). The weights ω_i are normalized Gaussian functions around the barycenter of each regions. Furthermore, an additional regularization between neighboring regions was added to account for the connectedness of cardiac tissue among neighboring regions, as well as an incompressibility penalization to account for the low volume change in cardiac tissue over the cardiac cycle.

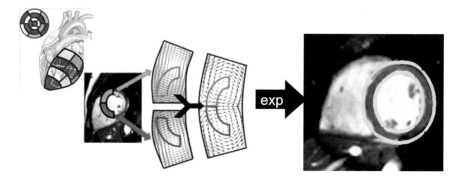

Fig. 9.3 A low-dimensional parametrization of diffeomorphisms for tracking cardiac motion in cine-MRI: the flow of an affine transformation with 12 parameters (middle) is generating a local velocity field around each of the 17 AHA regions (on the left). The weighted average of these 17 affine velocity fields produces a global velocity field whose flow (the group exponential) is parametrizing the heart deformation (on the right). In this context, motion tracking consists in optimizing the 12*17 = 204 regional parameters, which is easily done in the log-demons framework

9.5.2 Toward Intelligible Population-Based Cardiac Motion Features

The interpretability of the affine parameters of each region can be considerably increased by expressing the local affine transformations in a local coordinate system having radial, longitudinal vector and circumferential axes for each region. The resulting parameters can be related to physiological deformation: the translations parameters correspond to the motion along the radial, longitudinal, and circumferential axes while the linear part of the transformation encodes the circumferential twisting, radial thickening, and longitudinal shrinking. In a first study, the parameters were further reduced by assuming the linear part of each matrix M_i to be diagonal, thus reducing the number of parameters to 6 per region. These intelligible parameters were then used by supervised learning algorithms to classify a database of 200 cases with equal number of infarcted and non-infarcted subjects (the STACOM statistical shape modeling). A tenfold cross-validation showed that the method was achieving more than 95% of correct classification on yet-unseen data (Rohé et al. 2015).

In (Mcleod et al. 2015a, b, 2018), relevant factors discriminating between the motion patterns of healthy and unhealthy subjects were identified, thanks to a Tucker decomposition on Polyaffine motion parameters with a constraint on the sparsity of the core tensor (which essentially defines the loadings of each mode combination). The key idea is to consider that the parameters resulting from the tracking of the motion over cardiac image sequences of a population can be stacked in a 4-way tensor along motion parameters × region × time × subject. Performing the decomposition on the full tensor directly using 4-way Tucker Decomposition has the advantage of describing how all the components interact (as opposed to matricising the tensor and performing 2-way decomposition using classical singular value decomposition). The Tucker tensor decomposition method is a higher order extension of PCA which computes orthonormal subspaces associated with each axis of the data tensor. Thus, we get modes that independently describe a reduced basis of transformations (common to all regions, all time-points of the sequence and all subjects); a spatial basis (region weights) that localize deformations on the heart; a set of modes along time that triggers the deformation; and discriminative factors across clinical conditions. In order to minimize the number of interactions between all these modes along all the tensor axes, sparsity constraints were added on the core tensor. The sparsity of the discriminating factors and their individual intelligibility appears to be a key for a clear and intuitive interpretation of differences between populations in order to gain insight into pathology-specific functional behavior.

The method was applied to a dataset of 15 healthy subjects and 10 Tetralogy of Fallot patients with short-axis cine MRI sequences of 12 to 16 slices (slice thickness of 8mm) and 15 to 30 image frames. The decomposition was performed with 5 modes per axis and the core tensor loadings for each subject were averaged for the different groups. This showed that the two groups share some common dominant loadings. As expected, the Tetralogy of Fallot group also has some additional dominant loadings representing the abnormal motion patterns in these patients.

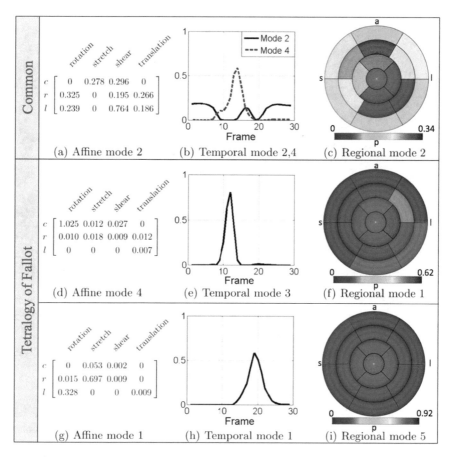

Fig. 9.4 Dominant mode combinations common to healthy and ToF cohorts: affine mode 2 (**a**), temporal modes 2 and 4 (**b**), and regional mode 2 (**c**). Key - a: anterior, p: posterior, s: septal, l: lateral. Figure reproduced from Mcleod et al. (2015a) with permission

The common dominant mode combinations are plotted in Fig. 9.4 (top row). The affine mode for the dominant mode combinations (Fig. 9.4a) shows predominant stretching in the circumferential direction related to the twisting motion in the left ventricle. The temporal modes (Fig. 9.4b) show a dominant pattern around the end- and mid-diastolic phases for mode 2, which may be due to the end of relaxation and end of filling. The dominant regions for these mode combinations are anterior (Fig. 9.4c). The dominant mode combinations for the Tetralogy of Fallot group are plotted in Fig. 9.4. The affine mode for the first dominant combination (Fig. 9.4d) indicates little longitudinal motion. The corresponding temporal mode (Fig. 9.4e) represents a peak at the end systolic frame (around one-third of the length of the cardiac cycle). The corresponding regional mode (Fig. 9.4, f) indicates that there is a dominance in the motion in the lateral wall. This is an area with known motion

Healthy controls   Tetralogy of Fallot

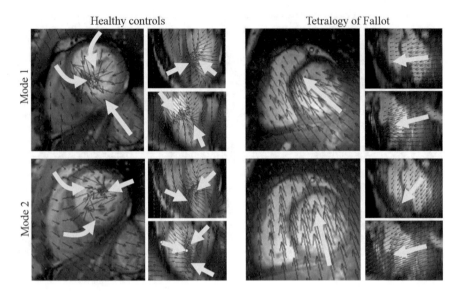

Fig. 9.5 Three views of the first (top row) and second (bottom row) spatial modes for the healthy controls (left) and for the Tetralogy of Fallot patients (right). The modes for the healthy controls represent the radial contraction and circumferential motion, whereas the modes for the Tetralogy of Fallot patients represent the translation toward the right ventricle. Yellow arrows indicate the general direction of motion. Figure ©IEEE 2015, reproduced from Mcleod et al. (2015b) with permission

abnormalities in these patients given that the motion in the free wall of the left ventricle is dragged toward the septum. The temporal mode for the second dominant mode (Fig. 9.4h) has instead a peak around mid-systole, with corresponding regional mode (Fig. 9.4i), indicating dominance around the apex, which may be due to poor resolution at the apex. The SVF corresponding to the first two spatial modes are shown in Fig. 9.5. The first mode for the healthy controls appears to capture both the radial contraction and the circumferential motion (shown in block yellow arrows). The Tetralogy of Fallot modes, on the other hand, appear to capture a translation of the free wall and septal wall toward the right ventricle (RV). This abnormal motion is evident in the image sequences of these patients.

9.6 Conclusion

We have presented in this chapter an overview of the theory of statistics on non-linear spaces and of its application to the modeling of shapes and deformations in medical image analysis. When the variability of shapes becomes important, linear methods like point distribution models for shapes or linear statistics on displacement vector fields for images and deformations become ill-posed as they authorize self-intersections. Considering non-linear spaces that are locally Euclidean (i.e., Riemannian manifolds) solves this issue. The cost to pay is that we have to work locally with tangent vectors and geodesics. However, once the exponential and log maps are implemented at any point of our shape space, many algorithms and statistical methods can be generalized quite seamlessly to these non-linear spaces.

For statistics on deformations, we have to consider smooth manifolds that have an additional structure: transformations form a Lie group under composition and inversion. One difficulty to use the Riemannian framework is that there often does not exist a metric which is completely invariant with respect to all the group operations (composition on the left and on the right, inversion). As a consequence, the statistics that we compute with left- or right-invariant metrics are not fully consistent with the group structure. We present in this chapter an extension of the Riemannian framework to affine connection spaces that solves this problem. In this new setting, all the computations continue to be related to geodesics using the exp and log maps. Here, geodesics are defined with the more general notion of straight lines (zero acceleration or auto-parallel curves) instead of being shortest paths. Every Riemannian manifold is an affine connection space with the Levi-Civita connection, but the reverse is not true. This is why we can find a canonical connection on every Lie group (the symmetric Cartan–Schouten connection) that is consistent with left and right composition as well as inversion while there is generally no bi-invariant Riemannian metric.

We have drafted the generalization of the statistical theory to this affine setting, and we have shown that it can lead to a very powerful framework for diffeomorphisms where geodesics starting from the identity are simply the flow of stationary velocity fields (SVFs). Very well-known non-linear registration algorithms based on this parametrization of diffeomorphisms are the log-demons (Vercauteren et al. 2008), Ashburner's DARTEL toolbox in SPM8 (Ashburner 2007), and the NiftyReg registration package (Modat et al. 2010, 2011). The combination of these very efficient algorithm with the well-posed geometrical and statistical framework allows to develop new methods for the analysis of longitudinal data. Furthermore, the affine symmetric structure of our group of deformation provides parallel transport algorithms that are numerically more stable and efficient than in the Riemannian case. We showed on two brain and cardiac applications that this allows to construct not only statistically more powerful analysis tools, but also generative models of shape motion and evolution.

With polyaffine deformations in the cardiac example, we have also shown that the deformation parameters can be localized and aligned with biophysical reference frames to produce a diffeomorphism parametrized by low-dimensional and intelligible parameters. Such a sensible vectorization of deformations is necessary for sparse decomposition methods: each parameter has to make sense individually as an atom of deformation if we want to describe the observed shape changes with an extremely low number of meaningful variables. This opens the way to very promising factor analysis methods dissociating the influence of the type of local deformation, the localization, the time trigger, and the influence of the disease as we have shown with the sparse Tucker tensor decomposition. There is no doubt that these methods will find many other applications in medical image analysis.

For efficiency, the medical applications shown in this chapter were implemented using C++ software dedicated to 3D image registration parameterized by SVFs. An open-source implementation of the symmetric log-demons integrated into the Insight Toolkit (ITK) is available at http://hdl.handle.net/10380/3060. A significant improvement of this software including the more robust LCC similarity measure and symmetric confidence masks is available at https://team.inria.fr/epione/fr/software/lcclogdemons/ (Lorenzi et al. 2013), along with additional standalone tools to work on SVFs including the pole ladder algorithm (Lorenzi and Pennec 2013). The code for the polyaffine log-demons is also available as an open-source ITK package at https://github.com/ChristofSeiler/PolyaffineTransformationTrees (Seiler et al. 2012). For Riemannian geometric data which are less computationally demanding than the very large 3D images, it is more comfortable to work in python. The recent *Geomstat* Python toolbox https://github.com/geomstats/geomstats provides an efficient and user-friendly interface for computing the exponential and logarithmic maps, geodesics, parallel transport on non-linear manifolds such as hyperbolic spaces, spaces of symmetric positive definite matrices, Lie groups of transformations, and many more. The package provides methods for statistical estimation and learning algorithms, clustering and dimension reduction on manifolds with support for different execution backends, namely NumPy, PyTorch, and TensorFlow, enabling GPU acceleration (Miolane et al. 2020).

Acknowledgements This work has received funding from the European Research Council (ERC) under the European Union's Horizon 2020 research and innovation programme (grant G-Statistics No 786854) and from the French Government through the 3IA Côte d'Azur Investments in the Future project (National Research Agency ANR-19-P3IA-0002).

References

Araudon M, Li X-M (2005) Barycenters of measures transported by stochastic flows. Ann Probab 33(4):1509–1543

Arsigny V, Commowick O, Ayache N, Pennec X (2009) A fast and log-euclidean polyaffine framework for locally linear registration. J Math Imaging Vis 33(2):222–238

Arsigny V, Commowick O, Pennec X, Ayache N (2006) A log-Euclidean framework for statistics on diffeomorphisms. In: Proceedings of the 9th international conference on medical image computing and computer assisted intervention (MICCAI'06), Part I, number 4190 in LNCS, pp 924–931, 2-4 October 2006

Arsigny V, Pennec X, Ayache N (2005) Polyrigid and polyaffine transformations: a novel geometrical tool to deal with non-rigid deformations - application to the registration of histological slices. Med Image Anal 9(6):507–523

Ashburner J (2007) A fast diffeomorphic image registration algorithm. NeuroImage 38(1):95–113

Bauer M, Bruveris M, Michor PW (2014) Overview of the geometries of shape spaces and diffeomorphism groups. J Math Imaging Vis 50(1–2):60–97

Beg MF, Miller MIMI, Trouvé A, Younes L (2005) Computing large deformation metric mappings via geodesic flows of diffeomorphisms. Int J Comput Vis 61(2):139–157

Bossa M, Hernandez M, Olmos S (2007) Contributions to 3D diffeomorphic atlas estimation: application to brain images. In: Ayache N, Ourselin S, Maeder A (eds), Proceedings of medical image computing and computer-assisted intervention (MICCAI 2007), volume 4792 of *LNCS*, pp 667–674. Springer

Buser P, Karcher H (1981) Gromov's almost flat manifolds. Number 81 in Astérisque. Société mathématique de France

Cachier P, Ayache N (2004) Isotropic energies, filters and splines for vector field regularization. J Math Imaging Vis 20(3):251–265

Cachier P, Bardinet E, Dormont D, Pennec X, Ayache N (2003) Iconic feature based nonrigid registration: the pasha algorithm. Comput Vis Image Underst 89(2-3):272–298. Special Issue on Nonrigid Registration

do Carmo M (1992) Riemannian geometry. Mathem Theory Appl, Birkhäuser, Boston

Cartan E, Schouten JA (1926) On the geometry of the group-manifold of simple and semi-simple groups. Proc Akad Wekensch, Amsterdam 29:803–815

Cash DM, Frost C, Iheme LO, Ünay D, Kandemir M, Fripp J, Salvado O, Bourgeat P, Reuter M, Fischl B, Lorenzi M, Frisoni GB, Pennec X, Pierson RK, Gunter JL, Senjem ML, Jack CR, Guizard N, Fonov VS, Collins DL, Modat M, Cardoso MJ, Leung KK, Wang H, Das SR, Yushkevich PA, Malone IB, Fox NC, Schott JM, Ourselin S (2015) Assessing atrophy measurement techniques in dementia: results from the MIRIAD atrophy challenge. NeuroImage 123:149–164

Dryden IL, Mardia KV (2016) Statistical shape analysis with applications in R. Wiley series in probability and statistics. Wiley, Chichester, UK; Hoboken, NJ, second edition edition

Ehlers J, Pirani F, Schild A (1972) The geometry of free fall and light propagation, in O'Raifeartaigh. Papers in Honor of J. L. Synge. Oxford University Press, General Relativity

Fletcher P, Lu C, Pizer S, Joshi S (2004) Principal geodesic analysis for the study of nonlinear statistics of shape. IEEE Trans Med Imaging 23(8):995–1005

Gallier J (2008) Logarithms and square roots of real matrices. arXiv:0805.0245 [math]

Groisser D (2004) Newton's method, zeroes of vector fields, and the Riemannian center of mass. Adv Appl Math 33:95–135

Hadj-Hamou M, Lorenzi M, Ayache N, Pennec X (2016) Longitudinal analysis of image time series with diffeomorphic deformations: a computational framework based on stationary velocity fields. Front Neurosc 10(236):18

Hairer E, Lubich C, Wanner G (2002) Geometric numerical integration: structure preserving algorithm for ordinary differential equations, volume 31 of Springer series in computational mathematics. Springer

Hernandez M, Bossa MN, Olmos S (2009) Registration of anatomical images using paths of diffeomorphisms parameterized with stationary vector field flows. Int J Comput Vis 85(3):291–306

Higham NJ (2005) The scaling and squaring method for the matrix exponential revisited. SIAM J Matrix Anal Appl 26(4):1179–1193

Hun Cheng S, Higham NJ, Kenney CS, Laub AJ (2001) Approximating the logarithm of a matrix to specified accuracy. SIAM J Matrix Anal Appl 22(4):1112–1125

Iserles A, Munthe-Kaas HZ, Norsett SP, Zanna A (2000) Lie-group methods. Acta Numer 9:215–365

Jia S, Duchateau N, Moceri P, Sermesant M, Pennec X (2018) Transport Parallel, of Surface Deformations from Pole Ladder to Symmetrical Extension. In ShapeMI MICCAI, Workshop on Shape in Medical Imaging. Sept, Granada, Spain, p 2018

Kenney CS, Laub AJ (1989) Condition estimates for matrix functions. SIAM J Matrix Anal Appl 10:191–209

Khesin BA, Wendt R (2009) The geometry of infinite dimensional lie groups, volume 51 of Ergebnisse der Mathematik und ihrer Grenzgebiete. 3. Folge / A Series of Modern Surveys in Mathematics. Springer

Lorenzi M, Ayache N, Frisoni GB, Pennec X (2013) LCC-Demons: a robust and accurate symmetric diffeomorphic registration algorithm. NeuroImage 81(1):470–483

Lorenzi M, Ayache N, Pennec X (2011) Schild's ladder for the parallel transport of deformations in time series of images. In Szekely G, Hahn H (eds) IPMI—22nd international conference on information processing in medical images-2011, volume 6801, pp 463–474, Kloster Irsee, Germany, July 2011. Springer

Lorenzi M, Ayache N, Pennec X (2015) Regional flux analysis for discovering and quantifying anatomical changes: an application to the brain morphometry in Alzheimer's disease. NeuroImage 115:224–234

Lorenzi M, Frisoni GB, Ayache N, Pennec X (2011) Mapping the effects of $A\beta_{1-42}$ levels on the longitudinal changes in healthy aging: hierarchical modeling based on stationary velocity fields. In: Fichtinger G, Martel A, Peters T (eds) Medical image computing and computer-assisted intervention - MICCAI 2011, vol 6892. LNCS. Springer, Heidelberg, pp 663–670

Lorenzi M, Pennec X (2013) Efficient parallel transport of deformations in time series of images: from schild's to pole ladder. J Math Imaging Vis 50(1–2):5–17

Lorenzi M, Pennec X (2013) Geodesics, parallel transport & one-parameter subgroups for diffeomorphic image registration. Int J Comput Vis 105(2):111–127

Lorenzi M, Pennec X, Frisoni GB, Ayache N (2015) Disentangling normal aging from Alzheimer's disease in structural MR images. Neurobiol Aging 36:S42–S52

Mansi T, Pennec X, Sermesant M, Delingette H, Ayache N (2011) iLogDemons: A demons-based registration algorithm for tracking incompressible elastic biological tissues. Int J Comput Vis 92(1):92–111

Mcleod K, Sermesant M, Beerbaum P, Pennec X (2015a) Descriptive and intuitive population-based cardiac motion analysis via sparsity constrained tensor decomposition. Medical Image Computing and Computer Assisted Intervention (MICCAI, (2015) volume 9351 of Lecture notes in computer science (LNCS). Munich, Germany, pp 419–426

Mcleod K, Sermesant M, Beerbaum P, Pennec X (2015b) Spatio-temporal tensor decomposition of a polyaffine motion model for a better analysis of pathological left ventricular dynamics. IEEE Trans Med Imaging 34(7):1562–1675

Mcleod K, Tøndel K, Calvet L, Sermesant M, Pennec X (2018) Cardiac motion evolution model for analysis of functional changes using tensor decomposition and cross-sectional data. IEEE Trans Biomed Eng 65(12):2769–2780

Miolane N, Le Brigant A, Mathe J, Hou B, Guigui N, Thanwerdas Y, Heyder S, Peltre O, Koep N, Zaatiti H, Hajri H, Cabanes Y, Gerald T, Chauchat P, Shewmake C, Kainz B, Donnat C, Holmes S, Pennec X (2020) Geomstats: a Python Package for Riemannian Geometry in Machine Learning. J Mach Learn Res 21(223):1–9

Modat M, Ridgway GR, Daga P, Cardoso MJ, Hawkes DJ, Ashburner J, Ourselin S (2011) Log-euclidean free-form deformation. In: Proceedings of SPIE medical imaging 2011. SPIE

Modat M, Ridgway GR, Taylor ZA, Lehmann M, Barnes J, Hawkes DJ, Fox NC, Ourselin S (2010) Fast free-form deformation using graphics processing units. Comput Methods Programs Biomed 98(3):278–284

Pennec X (1996) L'incertitude dans les problèmes de reconnaissance et de recalage—applications en imagerie médicale et biologie moléculaire. Phd thesis, Ecole Polytechnique

Pennec X (1999) Probabilities and statistics on riemannian manifolds: basic tools for geometric measurements. In: Cetin AE, Akarun L, Ertuzun A, Gurcan MN, Yardimci Y (eds) Proceedings of nonlinear signal and image processing (NSIP'99), volume 1, pages 194–198, June 20-23, Antalya, Turkey, Turkey, 1999. IEEE-EURASIP

Pennec X (2006) Intrinsic statistics on Riemannian manifolds: Basic tools for geometric measurements. J Math Imaging Vis 25(1):127–154

Pennec X (2018) Barycentric subspace analysis on manifolds. Ann Statist 46(6A):2711–2746

Pennec X (2018) Parallel transport with pole ladder: a third order scheme in affine connection spaces which is exact in affine symmetric spaces. arXiv:1805.11436

Pennec X, Arsigny V (2012) Exponential barycenters of the canonical Cartan connection and invariant means on Lie groups. In: Barbaresco F, Mishra A, Nielsen F (eds) Matrix information geometry, pp 123–168. Springer

Pennec X, Ayache N (1998) Uniform distribution, distance and expectation problems for geometric features processing. J Math Imaging Vis 9(1):49–67

Pennec X, Fillard P, Ayache N (2006) A Riemannian framework for tensor computing. Int J Comput Vis 66(1):41–66

Pennec X, Lorenzi M (2020) Beyond Riemannian: the affine connection setting for transformation groups. In Pennec X, Sommer S, Fletcher T (eds) Riemannian Geometric Statistics in medical image analysis, number Chap. 5, pages 169–229. Academic Press

Pennec X, Sommer S, Fletcher PT (2020) Riemannian geometric statistics in medical image analysis. Elsevier

Peyrat J-M, Delingette H, Sermesant M, Pennec X (2008) Registration of 4D time-series of cardiac images with multichannel diffeomorphic demons. In: Metaxas D, Axel L, Fichtinger G, Székely G (eds) Medical image computing and computer-assisted intervention - MICCAI 2008, vol 5242. LNCS. Springer, Heidelberg, pp 972–979

Rohé M-M, Duchateau N, Sermesant M, Pennec X (2015) Combination of polyaffine transformations and supervised learning for the automatic diagnosis of LV infarct. In: Statistical atlases and computational modeling of the heart (STACOM 2015), Munich, Germany

Seiler C, Pennec X, Reyes M (2011) Geometry-aware multiscale image registration via OBBTree-based polyaffine log-demons. In: Fichtinger G, Martel A, Peters T (eds) Medical image computing and computer-assisted intervention–MICCAI 2011, vol 6893. LNCS. Springer, Heidelberg, pp 631–638

Seiler C, Pennec X, Reyes M (2012) Capturing the multiscale anatomical shape variability with polyaffine transformation trees. Med Image Anal 16(7):1371–1384

Trouvé A (1998) Diffeomorphisms groups and pattern matching in image analysis. Int J Comput Vis 28(3):213–221

Vercauteren T, Pennec X, Perchant A, Ayache N (2007) Non-parametric diffeomorphic image registration with the Demons algorithm. In: Ayache N, Ourselin S, Maeder A (eds) Proceedings of medical image computing and computer-assisted intervention (MICCAI 2007), pp 319–326, Berlin, Heidelberg, 2007. Springer, Berlin, Heidelberg

Vercauteren T, Pennec X, Perchant A, Ayache N (2008) Symmetric Log-domain diffeomorphic registration: a Demons-based approach. In: Metaxas D, Axel L, Fichtinger G, Szekely G (eds) Proceedings medical image computing and computer-assisted intervention - MICCAI, (2008), vol 5241. Lecture Notes in Computer Science - LNCS. Springer, Heidelberg, pp 754–761

Vercauteren T, Pennec X, Perchant A, Ayache N (2009) Diffeomorphic demons: efficient non-parametric image registration. NeuroImage 45(1):S61–S72

Younes L (2010) Shapes and diffeomorphisms. Number v. 171 in Applied mathematical sciences. Springer, Heidelberg; New York

Chapter 10
High Fidelity 3D Anatomical Visualization of the Fibre Bundles of the Muscles of Facial Expression as In situ

Zhi Li, John Tran, Jacobo Bibliowicz, Azam Khan, Jeremy P. M. Mogk, and Anne Agur(ID)

Abstract The ability to express emotion through facial gestures impacts social and mental health. The production of these gestures is the result of the individual function and complex synergistic activities of the muscles of facial expression. Visualization and modelling techniques provide insight into how the muscles individually and collectively contribute to the shaping and stiffening of facial soft tissues. However, due to lack of detailed anatomical data, modellers are left to heuristically define the inner structure of each muscle, often resulting in a relatively homogeneous distribution of muscle fibres, which may not be accurate. Recent technological advances have enabled the reconstruction of entire muscles in 3D space as In situ using dissection, digitization and 3D modelling at the fibre bundle/aponeurosis level. In this chapter, we describe the use of this technology to visualize the muscles of facial expression and mastication at the fibre bundle level. The comprehensive 3D model provides novel insights into the asymmetry and complex interrelationships of the individual muscles of facial expression. These data possess great value to improve the anatomical fidelity of biomechanical models, and subsequently simulations, of facial gestures. Furthermore, these data could advance imaging and image processing techniques that are used to derive models.

Z. Li · J. Tran · J. P. M. Mogk · A. Agur (✉)
Musculoskeletal and Peripheral Nerve Anatomy Laboratory, Division of Anatomy, Department of Surgery, University of Toronto, Toronto, ON, Canada
e-mail: anne.agur@utoronto.ca

J. Bibliowicz
Autodesk Research, Autodesk Inc., Toronto, ON, Canada

A. Khan
Department of Computer Science, University of Toronto, Toronto, ON, Canada

© The Author(s), under exclusive license to Springer Nature Switzerland AG 2021
J.-F. Uhl et al. (eds.), *Digital Anatomy*, Human–Computer Interaction Series,
https://doi.org/10.1007/978-3-030-61905-3_10

185

10.1 Introduction

Facial expression is important in both verbal and non-verbal communication. The ability to express emotion through facial gestures impacts social and mental health (Swearingen et al. 1999). It has been previously described that more than 10 muscles of facial expression play a biomechanical role in the deformation of tissues surrounding the lips, which is essential for smiling and frowning (Chabanas et al. 2003). The production of facial gestures is the result of the individual function and complex synergistic activities of the muscles of facial expression.

Visualization and modelling techniques provide insight into how the muscles individually and collectively contribute to the shaping and stiffening of facial soft tissues during functional activities. Computer models used in early facial animation efforts contained parameterized representations of the muscles, their dimensions and geometric lines of action defined according to images found in anatomy textbooks (Platt and Badler 1981; Waters 1987; Terzopoulos and Waters 1990). An increasing number of studies use medical imaging to reconstruct the geometry of the face, muscles, and bones (Gladilin et al. 2004; Sifakis et al. 2005; Kim and Gomi 2007; Barbarino et al. 2009; Beldie et al. 2010; Nazari et al. 2010; Wu et al. 2014; Flynn et al. 2017). However, the thinness and complexity of the facial muscles can make delineating muscle boundaries in serial images a challenge (Som et al. 2012; Hutto and Vattoth 2015). Often, images are segmented from only one side of the face and either used to create a half-facial model reflected across the sagittal plane to create a full-facial model with right-left symmetry. Moreover, without more involved imaging techniques (e.g., diffusion tensor imaging, micro-computed tomography with iodine staining), modellers are left to heuristically define the inner structure of each muscle. This typically takes the form of some relatively homogeneous distribution of muscle fibres, which may not accurately reflect the architecture of a particular muscle.

An important determinant of muscle function is its architecture, the arrangement of contractile and connective tissue elements within the muscle volume (Zajac 1989; Gans and Gaunt 1991). The contractile elements consist of muscle fibre bundles, whereas the connective tissue elements that passively transmit forces include aponeuroses and tendons. Few studies have implemented techniques to reconstruct the architecture of select facial muscles (Liu et al. 2016; Wu and Yin 2016; Falcinelli et al. 2018; Sun et al. 2018). Thus, the 3D internal architecture of the muscles of facial expression remains largely unknown, impeding the development of high fidelity 3D visualization models and simulation of facial muscle contraction. Importantly, modelling muscle actuated facial deformation is challenging and requires an accurate representation of the internal anatomic elements and their interactions (Mazza and Barbarino 2011).

10.2 Methods

Recent technological advances have enabled the reconstruction of entire muscles in 3D space, as In situ, using dissection, digitization and 3D modelling at the fibre bundle/aponeurosis level (Ravichandiran et al. 2009; Li et al. 2015). More recently, collaborators incorporated the volumetric musculoaponeurotic 3D digitized data of the human masseter into a prototype finite element model that would better represent larger attachment areas and internal stresses of the muscle during contraction, i.e., chewing (Sánchez et al. 2017). This was the first reported application of digitized volumetric 3D data in the development of simulation models. Results demonstrated that the incorporation of 3D digitized data into their prototype finite element model "increase simulated maximum bite forces to more realistic levels" (Sánchez et al. 2017). Although presently available, this 3D digitization methodology has not been used to quantify morphology and architectural parameters of the muscles of facial expression, nor have the data been used to construct more detailed finite element models at the fibre bundle level.

The dissection, digitization and 3D modelling protocol has been developed in our laboratory over the last 20 years and has been used to model the musculoaponeurotic architecture of upper limb, lower limb and masticatory muscles (Agur et al. 2003; Li et al. 2015; Castanov et al. 2019). Approval was received from the University of Toronto Health Sciences Research Ethics Board (Protocol Reference #27210 and #28530). The protocol was adapted to study the muscles of facial expression. To enable digitization of the musculoaponeurotic elements of the muscles of facial expression, the head was first stabilized in a casing of polyurethane foam (Great Stuff™, Dow Chemical Co, Midland, Michigan, USA), leaving the face exposed. Next, three screws were inserted into the frontal, right and left temporal bones of the skull to provide a reference frame for reconstructing digitized data into 3D models. The skin of the face and neck was meticulously removed to expose the superficial muscles of facial expression. Fibre bundles of each muscle were exposed and digitized throughout the muscle volume, from superficial to deep, using a MicroScribe G2X Digitizer (0.05 mm accuracy; Immersion Corporation, San Jose, CA).

A custom software program, named "Fibonacci," was created to digitize the fibre bundles of each muscle (Fig. 10.1). This program models the captured fibre bundle data as a hierarchy of muscles, muscle heads, layers, fibre bundles, and points. The first three levels of the hierarchy were created at the user's discretion, whereas the latter two were defined using the MicroScribe digitizer. The MicroScribe pedal attachments send commands to the software. Specifically, pressing one pedal captures the current position of the digitizer and adds a point to the current fibre bundle. Pressing the other pedal completes the current fibre bundle and begins a new one. The digitizer was calibrated using the three screws embedded in the bones of the skull.

To capture a 3D representation of a fibre bundle, the user first placed the tip of the digitizer on the fibre bundle and pressed the pedal to capture the current location as a point. This process was repeated while incrementally advancing along the entire

Fig. 10.1 Fibonacci user interface

length of the fibre bundle. Once digitized, the fibre bundle was carefully removed from the cadaver, exposing the deeper fibre bundles for digitization.

Fibonacci provides a simple user interface comprised of three panels. The main panel shows the captured 3D fibre bundle data. One smaller panel controls the connection to the MicroScribe scanner, and another panel displays a tree-view representation of the hierarchy of muscle data. In this latter panel, the user can modify attributes of each fibre bundle, such as the display colour, or add comments. Users can also use this panel to hide parts of the hierarchy or reorganize various components. Once the muscle data have been captured, it can be exported into the Autodesk® Maya® format for further processing and visualization.

The following muscles of facial expression were digitized bilaterally: risorius, platysma, zygomaticus major/minor, nasalis, levator labii superioris alaeque nasi, levator labii superioris, orbicularis oculi, levator anguli oris, depressor anguli oris, depressor labii inferioris, mentalis, orbicularis oris, frontalis, procerus, buccinator and corrugator. Additionally, digitized were the muscles of mastication including masseter, temporalis, and medial/lateral pterygoids. Digitized data of individual fibre bundles of each muscle, collected using Fibonacci, were imported into Autodesk® Maya® (Autodesk Inc., San Rafael, California) and reconstructed into a 3D model. The three-step modelling process is summarized below (Fig. 10.2):

1. Digitized fibre bundles were represented as points connected by line segments (Fig. 10.2b, c). Using the built-in function available in Maya®, the trajectory of each fibre bundle was approximated using a cubic uniform B-Spline curve for

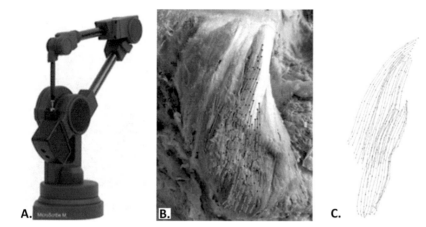

Fig. 10.2 Muscle fibre bundle digitization. **a** Microscribe™ digitizer. **b** Cadaveric dissection of masseter muscle with digitized fibre bundles (data points are blue, and points are connected by polylines making up the fibre bundle). **c** Digitized fibre bundle of masseter as seen on Fibonacci software used to capture muscle fibre bundle trajectory. Reprinted with permission from Musculoskeletal and Peripheral Nerve Anatomy Laboratory, University of Toronto

smoothness and stored as a non-uniform rational basis spline (NURBS) curve (Ravichandiran et al. 2009). Each NURBS curve (representation of a fibre bundle) was reconstructed into a cylindrical tube for visualization purposes.

2. High resolution computed tomography (CT) scans of the skull and cervical spine were used to volumetrically reconstruct the skeletal elements to act as the base for the 3D models.

3. All of the muscle models were combined and registered to the CT-based skeletal model using the digitized reference frames.

Using this process, a complete, 3D model of the muscles of facial expression and mastication was generated.

10.3   Results

In total, 22 muscles were digitized and modelled. The number of fibre bundles digitized ranged from 22 to 1265, depending on the size of the muscle (Table 10.1). The smallest number of fibre bundles digitized was in risorius and the largest in masseter.

The high-fidelity 3D model enables visualization of the relationships of the musculotendinous elements of the muscles and their attachment sites to the underlying skeleton, precisely as found in the specimen. The muscles of facial expression are superficial to the more deeply located muscles of mastication (Figs. 10.3 and 10.4). The muscles of facial expression, as seen in the 3D model, can be divided into five main groups: (1) muscles of the orbit; (2) muscles of the nose; (3) muscles of the scalp

Table 10.1 Number of digitized fibre bundles in each muscle modelled

Muscles of facial expression	nFB	Muscles of facial expression	nFB
Frontalis	63	Depressor anguli oris	122
Corrugator supercilii	43	Platysma	144
Procerus	35	Auricularis	56
Orbicularis oculi	119	Buccinator	262
Nasalis	45	Mentalis	144
LLSAN	72	Orbicularis Oris	263
Levator labii superioris	124	–	–
Levator anguli oris	71	Muscles of mastication	–
Zygomaticus major	45	Masseter	1265
Zygomaticus minor	47	Temporalis	890
Risorius	22	Medial pterygoid	789
Depressor labii inferioris	225	Lateral pterygoid	349

LLSAN, levator labii superioris alaeque nasi; nFB, number of fibre bundles

and ear (auricle); (4) muscles of the mouth; and (5) platysma. Generally, the main function of each muscle can be determined by the direction and spatial arrangement of the fibre bundles. For example, the main muscle of the orbit, the orbicularis oculi, has three main functions depending on the location of the part of the muscle that contracts. The palpebral, orbital and lacrimal parts of the orbicularis oculi function to gently close the eye, tightly close the eye and compress the lacrimal sac, respectively (Fig. 10.5). Muscles of the nose will compress (nasalis) or dilate (levator labii superioris alaeque nasi) the nostrils. The anterior muscles of the scalp, the frontalis, elevate the eyebrows and wrinkles the skin of the forehead. Muscles that attach to the upper lip (levator labii superioris) will raise it, while muscles that attach to the lower lip (depressor labii inferioris) will depress the lip. Muscles that attach to the angle of the mouth superiorly (levator anguli oris) will contribute to smiling, whereas muscles (depressor anguli oris) that attach inferiorly will contribute to frowning. The orbicularis oris is a sphincter-like muscle that closes the mouth, whereas the transversely orientated fibres of the buccinator, inserting into the angle of the mouth, are antagonists that retract the angle of the mouth. The platysma is the most extensive of the muscles of facial expression, spanning between the neck and the mandible, and functions to tense the skin of the neck (e.g., when shaving).

More specifically, the comprehensive 3D model provides novel insights into the asymmetry and complex interrelationships of the individual muscles of facial expression. When comparing the right and left pairs of the muscles of facial expression, notable asymmetry existed in their size and shape. For example, the fibre bundles of

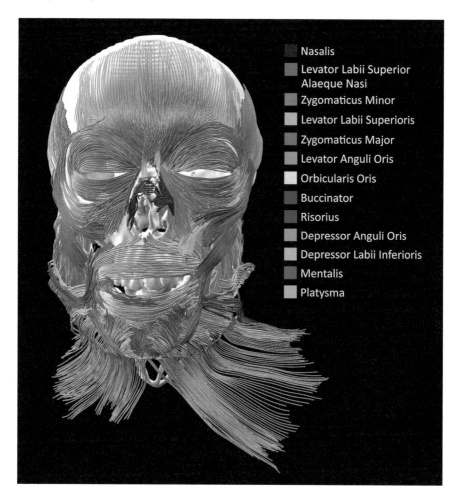

Fig. 10.3 Anterior view of 3D model of the individual muscles of facial expression and mastication at the fibre bundle level. Reprinted with permission from Musculoskeletal and Peripheral Nerve Anatomy Laboratory, University of Toronto

the platysma were visibly longer on the left than the right (Figs. 10.3 and 10.4). Also, the fibre bundles of depressor anguli oris extended from the superior margin of the orbicularis oris and coursed inferiorly as far as the inferior border of the mandible on the right side. In comparison, the fibre bundles on the left side were shorter, extending from the angle of the mouth to the inferior border of the mandible (Fig. 10.4).

Furthermore, most of the muscles of facial expression exhibited complex inter-relationships of surrounding muscle bellies and interdigitation of fibre bundles. For example, the fibre bundles of levator labii superioris lay deep to the oribularis oculi and zygomaticus minor as they course inferiorly to their attachment to the connective tissue superficial to the superior part of the orbicularis oris on the left side.

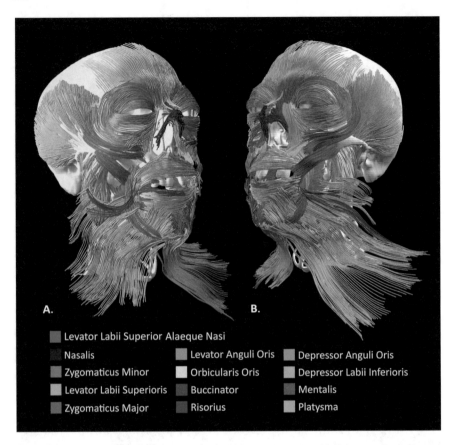

Fig. 10.4 Anterolateral views of 3D model of the individual muscles of facial expression and mastication. **a** Right side. **b** Left side. Reprinted with permission from Musculoskeletal and Peripheral Nerve Anatomy Laboratory, University of Toronto

On the right side, the fibre bundles were markedly shorter with a longer connective tissue attachment (Fig. 10.4). Interdigitation of fibre bundles of the muscles of facial expression was common. Some examples of fibre bundle interdigitation are listed below:

- Peri-oral musculature at the angle of the mouth: the buccinator, levator anguli oris, depressor anguli oris, risorius and zygomaticus major all interdigitated with the orbicular oris (Fig. 10.6).
- Buccinator was found to have extensive interdigitations with the inferior part of orbicularis oris (Fig. 10.6b). Whereas the superior part of orbicularis oris lay superficial to buccinator.
- Inferior part of levator anguli oris interdigitated with depressor anguli oris to form a continuous band of muscle at the angle of the mouth (Fig. 10.6d).

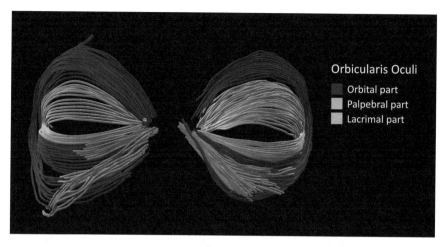

Fig. 10.5 Anterior view of 3D model of the 3 parts of orbicularis oculi muscle. Reprinted with permission from Musculoskeletal and Peripheral Nerve Anatomy Laboratory, University of Toronto

10.4   Discussion

Currently, no complete datasets exist that contain digital representations of the 3D internal musculoaponeurotic architecture of, and intermuscular relationships among, the muscles of facial expression. Studies to date have been more general and limited to numerical data about the average length and width of selected muscles (Balogh et al. 1988; Happak et al. 1997). In contrast, the current 3D model generated from digitized data could function as a digital atlas that provides 3D coordinate data of the musculotendinous anatomy of the muscles of facial expression and mastication. By digitizing fibre bundles from the same specimen, any architectural differences that existed In situ were real differences and not due to interspecimen variation. Furthermore, this model enabled quantification and visualization of the intricacies of the 3D spatial relationships among muscles, tendons, and aponeuroses, including the arrangement of fibre bundles within each muscle. These data possess great value to improve the anatomical fidelity of biomechanical models, and subsequently simulations, of facial gestures. Additionally, such data could prove useful for advancing imaging and image processing techniques that are used to derive models.

The current data revealed important geometrical information regarding asymmetry that exists between homologous muscles of the right and left sides of the face. Using the 3D model, morphological comparison of the left and right muscles of facial expression demonstrated marked asymmetry. The asymmetric features included differences in fibre bundle length, attachment sites, and interdigitation patterns between the right and left muscles. In the modelling literature, facial models constructed from medical images sometimes assume muscular symmetry (Beldie et al. 2010; Wu et al. 2014). This reduces the time required to construct models, as segmented images from one side of the face are either used to create a half-facial

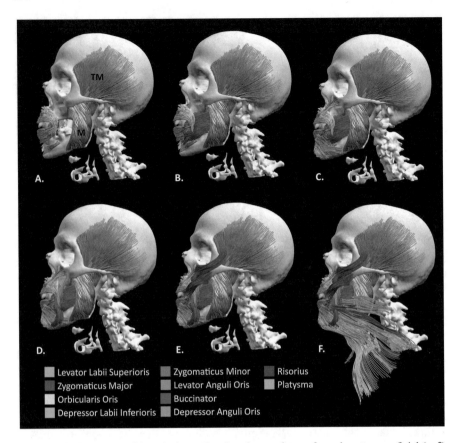

Fig. 10.6 Serial images of lateral views of peri-oral musculature, from deep to superficial (**a–f**). All images include temporalis (TM) and masseter (M) muscles. Reprinted with permission from Musculoskeletal and Peripheral Nerve Anatomy Laboratory, University of Toronto

model or are duplicated and reflected to represent the other side in a full-facial model. Accentuation of muscle asymmetries could occur due to use patterns, pathology, injury or deformation of the underlying skeleton. Thus, creating 3D models that assume symmetry of the muscles of facial expression would likely produce inaccuracies in the model's predictive capabilities. The architectural asymmetry demonstrated in the current 3D model provides a unique opportunity to further investigate the functional impact of facial muscle asymmetry, and to potentially move toward more realistic simulation.

As expected, interdigitating fibres existed in many of the muscles modelled. Due to the 3D data collection, the current model contains the precise location, morphology, interrelationships, and interdigitations of the muscles of facial expression, as In situ. While previous studies have reported such crossing of fibres, the specific geometry of the interdigitations remained undefined (Shim et al. 2008; Al-Hoqail and Abdel Meguid 2009; Yu et al. 2013; Kim et al. 2015). As determined in studies of the brain and tongue, standard diffusion tensor imaging fails to reconstruct crossing fibres (Wiegell et al. 2000; Ye et al. 2015; Voskuilen et al. 2019; Yang et al. 2019). Thus, the current model indicates the need to implement multi-tensor diffusion models when imaging the muscles of the face. As noted with the tongue, the fibre orientations and interdigitation greatly influence real and simulated tissue motion, particularly in the absence of internal joints (Gomez et al. 2018). Like the tongue, much of facial expression involves stiffening and movement of soft tissues. Including realistic architecture can also significantly alter muscle force predictions and resulting geometries of contraction (Sánchez et al. 2014; Sánchez et al. 2017). Consequently, accounting for interdigitation and heterogeneous fibre orientations are likely essential for improving the realism of facial appearance in animations and simulations.

10.5   Conclusions

By generating 3D representations of the musculoaponeurotic geometry of the face, we can move beyond the often idealized 2D depictions presented in anatomy texts; representations that are commonly used for learning and/or interpreting clinical and experimental findings. The current framework enables 3D structure and relationships to be explored and quantified, rather than simply interpreted from collections of images. High fidelity computerized representations of the human anatomy, as In situ, present the opportunity to develop advanced and interactive visualization techniques (e.g., augmented/virtual reality). Researchers and clinicians require resources to which image-based models can be compared, particularly when validating new techniques and/or understanding and handling the appearance of potential image artefacts. Advanced understanding of how musculoaponeurotic anatomy attaches to, and interacts with, neighbouring hard and soft tissues is fundamental for modellers looking to improve the realism of animations and simulations. The current 3D modelling process provides a common framework upon which these and other advances can occur, in a unified way.

References

Agur AM, Ng Thow Hing V, Ball KA, Fiume E, McKee NH (2003) Documentation and three dimensional modelling of human soleus muscle architecture. Clinical Anat 16: 285–293

Al-Hoqail RA, Abdel Meguid EM (2009) An anatomical and analytical study of the modiolus: enlightening its relevance to plastic surgery. Aesth Plast Surg 33(2):147–152

Balogh B, Wessig M, Millesi W, Millesi H, Firbas W, Fruhwald F (1988) Photogrammetry of the muscles of facial expression. Acta Anatom 133:183–187

Barbarino GG, Jabareen M, Trzewik J, Nkengne A, Stamatas G, Mazza E (2009) Development and validation of a three-dimensional finite element model of the face. J Biomech Eng 131(4):041006

Beldie L, Walker B, Lu Y, Richmond S, Middleton J (2010) Finite element modelling of maxillofacial surgery and facial expressions – a preliminary study. Int J Med Rob Comput Assis Sur 6(4):422–430

Castanov V, Hassan SA, Shakeri S, Vienneau M, Zabjek K, Richardson D, McKee NH, Agur AM (2019) Muscle architecture of vastus medialis obliquus and longus and its functional implications: a three dimensional investigation. Clinical Anat 32:515–523

Chabanas M, Luboz V, Payan Y (2003) Patient specific finite element model of the face soft tissues for computer-assisted maxillofacial surgery. Med Image Anal 7(2):131–151

Falcinelli C, Li Z, Lam WW, Stanisz GJ, Agur AM, Whyne CM (2018) Diffusion-tensor imaging versus digitization in reconstructing the masseter architecture. J Biomech Eng 140(11):111010

Flynn C, Nazari MA, Perrier P, Fels S, Nielsen PMF, Payan Y (2017) Computational modeling of the passive and active components of the face. In: Payan Y, Ohayon J (eds) Biomechanics of living cells: hyperelastic constitutive laws for finite element modeling. Academic Press/Elsevier, San Diego, CA, pp 377–394

Gans C, Gaunt AS (1991) Muscle architecture in relation to function. J Biomech 24:53–65

Gladilin E, Zachow S, Deuflhard P, Hege H-C (2004) Anatomy- and physics-based facial animation for craniofacial surgery simulations. Med Biol Eng Comput 42(2):167–170

Gomez AD, Elsaid N, Stone ML, Zhuo J, Prince JL (2018) Laplace-based modeling of fiber orientation in the tongue. Biomech Model Mechanobiol 17(4):1119–1130

Happak W, Liu J, Burggasser G, Flowers A, Gruber H, Freilinger G (1997) Human facial muscles: dimensions, motor endplate distribution, and presence of muscle fibres with multiple motor endplates. Anatom Rec 249:276–284

Hutto JR, Vattoth S (2015) A practical review of the muscles of facial mimicry with special emphasis on the superficial musculoaponeurotic system. AJR Am J Roentgenol 204(1):W19–W26

Kim K, Gomi H (2007) Model-based investigation of control and dynamics in human articulatory motion. J Syst Des Dyn 1(3):558–569

Kim HS, Pae C, Bae JH, Hu KS, Chang BM, Tansatit T, Kim HJ (2015) An anatomical study of the risorius in Asians and its insertion at the modiolus. Surg Radiol Anat 37(2):147–151

Li Z, Mogk JPM, Lee D, Bibliowicz J, Agur AM (2015) Development of an architecturally comprehensive database of forearm flexors and extensors from a single cadaveric specimen. Comput Methods Biomech Biomed Eng Imag Visual 3(1):3–12

Liu S, Wang M, Ai T, Wang Q, Wang R, Chen W, Pan C, Zhu W (2016) In vivo morphological and functional evaluation of the lateral pterygoid muscle: a diffusion tensor imaging study. British J Radiol 89(1064):20160041

Mazza E, Barbarino GG (2011) 3D mechanical modeling of facial soft tissue for surgery simulation. Facial Plast Surg Clin North Am 19(4):623–637

Nazari MA, Perrier P, Chabanas M, Payan Y (2010) Simulation of dynamic orofacial movements using a constitutive law varying with muscle activation. Comput Methods Biomech Biomed Eng 13(4):469–482

Platt SM, Badler NI (1981) Animating facial expressions. ACM Siggraph Computer Graph 15(3):245–252

Ravichandiran K, Ravichandiran M, Oliver ML, Singh KS, McKee NH, Agur AM (2009) Determining physiological cross-sectional area of extensor carpi radialis longus and brevis as a whole and by regions using 3D computer muscle models created from digitized fiber bundle data. Comput Methods Progr Biomed 95(3):203–212

Shim KS, Hu KS, Kwak HH, Youn KH, Koh KS, Kim HJ (2008) An anatomical study of the insertion of the zygomaticus major muscle in humans focused on the muscle arrangement at the corner of the mouth. Plastic Reconstr Surg 121(2):466–473

Sifakis E, Neverov I, Fedkiw R (2005) Automatic determination of facial muscle activations from sparse motion capture marker data. ACM Trans Graph 24(3):417–425

Sánchez CA, Lloyd JE, Fels S, Abolmaesumi P (2014) Embedding digitized fibre fields in finite element models of muscles. Comput Methods Biomech Biomed Eng Imag Visual 2(4):223–236

Sánchez CA, Li Z, Hannam AG, Abolmaesumi P, Agur A, Fels S (2017) Constructing detailed subject-specific models of the human masseter. In: Cardoso MJ et al. (eds) Imaging for patient-customized simulations and systems for point-of-care ultrasound. BIVPCS 2017, POCUS 2017. Lecture Notes in Computer Science, vol 10549. Springer

Som PM, Ng SA, Stuchen C, Tang CY, Lawson W, Laitman JT (2012) The MR imaging identification of the facial muscles and the subcutaneous musculoaponeurotic sustem. Neurographics 2(1):35–43

Sun M, Chen GC, Wang YQ, Song T, Li HD, Wu D, Yin NB (2018) Anatomical characterization and three-dimensional modeling of the muscles at the corner of the mouth: an iodine staining technique based on micro-computed tomography. Plastic Reconstr Surg 142(3):782–785

Terzopoulos D, Waters K (1990) Physically-based facial modeling, analysis, and animation. J Visual Comput Animat 1(2):73–80

van Swearingen JM, Cohn JF, Bajaj-Luthra A (1999) Specific impairment of smiling increases the severity of depressive symptoms in patients with facial neuromuscular disorders. Aesth Plast Surg 23(6):416–423

Voskuilen L, Mazzoli V, Oudeman J, Balm AJM, van der Heijden F, Froeling M, de Win MML, Strijkers GJ, Smeele LE, Nederveen AJ (2019) Crossing muscle fibers of the human tongue resolved in vivo using constrained spherical deconvolution. J Mag Reson Imag 50(1):96–105

Waters K (1987) A muscle model for animating three-dimensional facial expression. ACM Siggraph Comput Graph 21(4):17–24

Wiegell MR, Larsson HB, Wedeen VJ (2000) Fiber crossing in human brain depicted with diffusion tensor MR imaging. Radiology 217(3):897–903

Wu J, Yin N (2016) Detailed anatomy of the nasolabial muscle in human fetuses as determined by micro-CT combined with iodine staining. Annals Plastic Surg 76(1):111–116

Wu TF, Hung A, Mithraratne K (2014) Generating facial expression using an anatomically accurate biomechanical model. IEEE Trans Visual Comput Graph 20(11):1519–1529

Yang S, Ghosh K, Sakaie K, Sahoo SS, Carr SJA, Tatsuoka C (2019) A simplified crossing fiber model in diffusion weighted imaging. Front Neurosci 13:492

Ye C, Murano E, Stone M, Prince JL (2015) A Bayesian approach to distinguishing interdigitated tongue muscles from limited diffusion magnetic resonance imaging. Comput Med Imag Graph 45:63–74

Yu SK, Lee MH, Kim HS, Park JT, Kim HJ, Kim HJ (2013) Histomorphologic approach for the modiolus with reference to reconstructive and aesthetic surgery. J Craniof Surg 24(4):1414–1417

Zajac FE (1989) Muscle and tendon: Properties, models, scaling, and application to biomechanics and motor control. Critical Rev Biomed Eng 17(4):359–411

Chapter 11
Collaborative VR Simulation for Radiation Therapy Education

Haydn Bannister, Ben Selwyn-Smith, Craig Anslow⊙, Brian Robinson⊙, Paul Kane⊙, and Aidan Leong⊙

Abstract Cancer is the cause of over 16% of deaths globally. A common form of cancer treatment is radiation therapy; however, students learning radiation therapy have limited access to practical training opportunities due to the high demand upon equipment. Simulation of radiation therapy can provide an effective training solution without requiring expensive and in-demand equipment. We have developed LINACVR, a Virtual Reality radiation (VR) therapy simulation prototype that provides an immersive training solution. We evaluated LINACVR with 15 radiation therapy students and educators. The results indicated that LINACVR would be effective in radiation therapy training and was more effective than existing simulators. The implication of our design is that VR simulation could help to improve the education process of learning about domain-specific health areas such as radiation therapy.

H. Bannister · B. Selwyn-Smith · C. Anslow (✉)
School of Engineering and Computer Science, Victoria University of Wellington,
PO Box 600, Wellington 6140, New Zealand
e-mail: craig@ecs.vuw.ac.nz

H. Bannister
e-mail: haydn.j.bannister@gmail.com

B. Selwyn-Smith
e-mail: benselwynsmith@googlemail.com

B. Robinson
School of Nursing, Midwifery, and Health Practice, Victoria University of Wellington,
PO Box 600, Wellington 6140, New Zealand
e-mail: brian.robinson@vuw.ac.nz

P. Kane · A. Leong
Department of Radiation Therapy, University of Otago, Wellington,
PO Box 7343, Wellington South 6242, New Zealand
e-mail: paul.kane@otago.ac.nz

A. Leong
e-mail: aidan.leong@otago.ac.nz

11.1   Introduction

Cancer was responsible for an estimated 9.6 million deaths in 2018, accounting for about 16% of all deaths globally (World Health Organization 2018), and it is estimated that 40% of people will have cancer at some stage of their life (The Economist 2017). Radiation therapy is a common form of treatment and is in high demand. The Royal College of Radiologists found that the average wait time from diagnosis of cancer to the beginning of radiation treatment in the UK was 51 days, with some waiting as long as 379 days (Faculty of Clinical Oncology 1998). Radiation therapy requires highly trained operators; however, these operators have limited access to practical training due to the cost of, and demand for, specialized equipment.

Medical Linear Particle Accelerator (LINAC) machines (Fig. 11.1) are used by radiation therapists to deliver targeted radiation to tumors for the treatment of cancers. For this procedure, a patient is positioned on a motorized platform called a treatment couch, and once the patient is in place radiation is delivered from a part of the machine called the gantry (Tello 2018). These two pieces of radiation equipment are important for therapists to learn to position correctly.

Patients undergoing radiation therapy treatment often experience severe psychosocial stress (Sehlen et al. 2003) and psychological distress (Sostaric and Lilijana 2004). Rainey (1985) found that radiation therapy patients who had undergone a patient education program providing them with more information about the upcoming procedure experienced significantly less emotional distress from the procedure. They found that 85% of patients reported that they would like to learn more about radiation therapy.

Carlson (2014) discusses six reported errors during radiation therapy. Five errors involved delivering radiation to the wrong area, and one used considerably higher levels of radiation than the treatment plan listed. Events like these can put patients at risk of serious harm. They found that an important way to minimize the risk of incidents such as these is to have clear procedures that the therapists are thoroughly trained in. Kaczur (2013) reported that the medical radiation accidents with most severe consequences, such as severe burns or internal damage, were related to miscalibration of radiation therapy equipment. Similarly to Carlson, Kaczur found that the cause of these accidents was usually poor radiation education and training.

Radiation therapy students must train extensively in the use of Medical LINAC equipment. A LINAC machine can cost several million dollars (USD) to purchase and up to a half million dollars in annual operational costs, with a lifespan of approximately 10 years (Van Der Giessen et al. 2004). This makes it financially infeasible for an educational facility to have a dedicated LINAC machine for students to train with. It is vital, however, that students are able to train sufficiently before they interact with real equipment or assist with the treatment of real patients. Students typically gain their first experiences with LINAC machines during observational placements within the hospital, and later through practical placements. Opportunities for inexperienced students to actually practice with real equipment are limited due to the high demand for radiation therapy treatment. A training simulation in which inexperienced students can familiarize themselves with LINAC machine operation would enable them to go into the work force with a more comprehensive understanding of the equipment and environment. This would allow them to gain practical experience earlier. Additionally, a virtual simulation can allow students to experience, and to train in dealing with, negative scenarios such as errors causing miscalibration and misalignment (Beavis and Ward 2014). An effective simulation would also allow experienced students to further develop their skills in a risk-free environment.

In order to increase the effectiveness and reduce the cost of radiation therapy training, we have developed *LINACVR* which is able to simulate a LINAC machine and environment in VR. The application involves two simulation scenarios. The first simulation allows radiation therapists to learn and practice the operation of a LINAC machine. The second simulation shows the experience of the radiation treatment procedure from the perspective of a patient.

11.2 Background

Medical LINACs treat cancer by delivering targeted high-precision ionizing radiation to the tumor. This is done across multiple regular treatment sessions which vary depending on the cancer being treated, but is often between 10 and 40 sessions. Figure 11.1a shows a labeled Vairan TrueBeam LINAC machine (Varian Medical Systems 2021), the same model of machine that is common in radiation therapy departments at hospitals. The machine being operated in Fig. 11.1b is also a Varian TrueBeam.

The radiation is generated within either the stand or the gantry, and is directed out of the emitting collimator head of the gantry and through the patient (Podgorsak 2005). The exact path and shape of the radiation can be finely tuned by the radiation therapist based on a treatment plan specific to each patient. In Fig. 11.1a, the yellow diverging triangle coming out of the gantry head represents the radiation. The isocenter is the intersection of the center of the radiation and the horizontal axis of the gantry. This is where the center of the tumor must be in order for the radiation to properly irradiate the cancerous cells. Indicators of the location of the tumor are

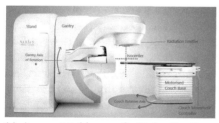

(a) An annotated image of a Varian TrueBeam LINAC machine [34].

(b) Radiation therapy students positioning a patient in a LINAC environment.

Fig. 11.1 Medical linear particle accelerator (LINAC) for radiation therapy

(a) Educator exploring digital anatomy.   (b) Students exploring a virtual LINAC.

Fig. 11.2 VERT simulation of radiation therapy (Beavis et al. 2009; Phillips et al. 2008; Vertual Limited 2021; Ward et al. 2011)

marked with tattoos externally on the body of the patient. These tattoos are placed based on a digitized plan created using a 3D scan of the patient, and are used by the radiation therapists to triangulate the internal location of the tumor. The tattoos are lined up with a laser grid projected onto the patient, allowing correct repeatable positioning of a tumor in the isocenter. While radiation is being emitted, the gantry rotates around the horizontal axis, passing through the space under the end of the couch. Ensuring that the gantry does not collide with the couch is vital.

To position the patient so that the tumor is at the radiation isocenter, the treatment couch can be moved. This is done with a couch movement controller, known as the "pendant," where the student on the right of Fig. 11.1b is using one. The couch on a modern treatment couch is motorized and can move and rotate in almost all directions, although some older designs support less of these. The couch is able to both move and tilt up, down, left, and right. It can also rotate around a circular path on the ground, labeled as "Couch Rotation Axis" in Fig. 11.1a.

In order to allow students to learn and practice LINAC operations without access to an actual machine they must initially be trained using simulation. VERT (Fig. 11.2) is the only available training simulation for radiation therapy (Beavis et al. 2009; Phillips et al. 2008; ?; Vertual Limited 2021; Ward et al. 2011). VERT involves projecting imagery onto a large screen in front of a student to represent the 3D LINAC environment, and a controller resembling those used for controlling a treatment couch. To provide depth perception to the imagery, 3D glasses are worn. This means that users cannot interact with the simulation as though they are present in the treatment room. Due to this, students cannot practice manually positioning the patient by hand, an important skill to learn. Instead they are limited to performing alignment by moving the treatment couch. VERT only supports one user at a time, but typically there are at least two radiation therapists working together. A simulation that supports collaborative operation by multiple simultaneous users would allow students to practice in a way that more closely resembles the actual environment, and develop more relevant operational and communication skills. VERT can cost up to $1,000,000 (NZD) which makes it very expensive for teaching institutions. VERT is not fully immersive so students may not be able to fully experience a realistic environment, and may therefore have difficulty applying what they have learned to a real-life treatment. A fully immersive low-cost VR simulation would give students a more affordable and easier way to familiarize themselves with LINAC operation in a way that is more directly applicable to the real world, which is what we propose with LINACVR.

Kane (2017) conducted a review of the current state, effectiveness, and usage of VERT. He found that VERT is the only widely used training solution, and is generally considered effective compared to traditional non-interactive media. A primary limitation is the inability to manually position a patient on the couch and is an important skill to learn. Kane (2015) further explores the impact that VERT has had upon the radiation therapy teaching program at Otago University and found that the integration of VERT as a training tool had difficulties but the simulation in the training of radiation therapy had significant potential. Leong et al. (2018) studied the effects of using VERT in the training of the planning of treatment, and found that it increased the perceived conceptual understanding of the procedure for students.

Collaboration is important in radiation therapy and various studies have examined this aspect in virtual environments. Cai et al. (2000) developed a multi-user application to simulate the planning of radiation therapy. In their application, physicians in remote locations were able to collaboratively perform a radiation treatment planning procedure. This allowed them to successfully collaboratively develop patient treatment plans without requiring the actual physical equipment. While this application intended to develop treatment plans for patients rather than to train therapists for machine operation, this still demonstrates the advantage of collaborative simulation when access to radiation equipment is limited. Churchill and Snowdon (1998) examined a range of designs for collaboration in a virtual environment. The primary focus is the affordance of communication. Real-world collaboration involves large amounts of both implicit and explicit communication. In order to achieve this, a collaborative virtual environment should be designed in a way that allows many channels of communication. Fraser et al. (2000) examined the differences between collaboration in VR and the real world. One example is the limited field of view in VR. Humans have a horizontal angle of vision of around 210° including peripheral vision, whereas VR headsets are around 110°. This can lead to confusion in communication in VR, as users will often think that another user can see something that is in fact outside of their vision. They suggest visually indicating the field of view of other users, in order to explicitly show users the limitations of the system.

Simulation training in healthcare has been widely adopted including the use of VR. Cook et al. (2011) conducted a systematic review of 609 studies evaluating the effectiveness of simulation for the education of health professionals. They found that simulation training consistently provided participants with large positive effects in terms of knowledge, skills, and behaviors, and moderate positive effects for patients. Across this review, only four percent of the studies did not show any benefit. Mantovani et al. (2003) reviewed and discussed the current state and usefulness of VR in the training of healthcare professionals. They found that VR provided significant benefits over traditional training and education methods such as print and film media. Many knee injuries can be treated through arthroscopic surgery; however, most training tools have issues due to cost, maintenance, or availability. Arthroscopy surgery involves using tools and sensors inserted through small incisions, and so the tools cannot be seen by the surgeon while they are using them. Hollands and Trowbridge (1996) provided surgical training simulation for knee surgeries where they used 3D representations of the geometry of a knee to allow surgeons to practice the operation in VR. VR allowed these surgical tools to be made visible, so that the surgeon can learn the correlation between manipulation of these tools and how the knee moves internally. Davies et al. (2018) evaluated the effectiveness of using VR simulation for clinical X-ray imaging with 17 healthcare students. The study found that most students were both more confident with being present for the X-ray procedure, and had a better understanding of where to stand during the procedure.

A recent state-of-the-art report is given by Schlachter et al. (2019) on visual computing in radiation therapy planning. They concluded that it is not easy to introduce new approaches in the clinic but with the exception of applications for training and education. They also stated that radiation therapy is a patient-oriented process and that visual computing and radiation therapy people need to collaborate more, especially in the interface between hospitals, universities, and companies. Our work builds upon these areas by developing a novel VR system for radiation therapy education using the latest VR headset technology.

11.3   LINACVR for Radiation Therapy Education

Simulation and VR simulation have shown to provide effective training benefits and transferable skills in healthcare education. We present *LINACVR* which is the first VR simulation radiation therapy treatment tool for both therapist training and patient education (Fig. 11.3). LINACVR includes a multi-user simulation for both patient education and for therapist training, and a portable headset version for the patient-perspective simulation.

11.3.1   User Interface

Figure 11.3a shows a user with the VR headset and controllers interacting with the patient and treatment couch. A 3D representation of a patient is constructed from Digital Imaging and Communications in Medicine (DICOM) data files. The patient model can be manually moved by interacting with it directly using the VR controllers. The treatment couch can be moved using a series of slider bars within a virtual menu panel (Fig. 11.3b). The patient model and individual organs can be made transparent using similar slider bars in order to allow therapists to see the internal location of the

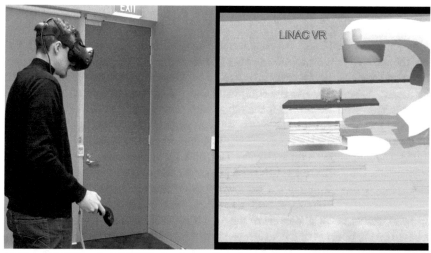

(a) User with VR headset (HTC Vive) and controllers viewing the gantry, couch, and patient model.

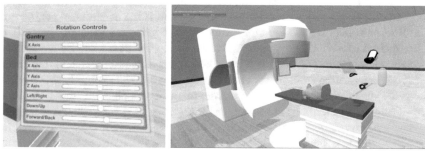

(b) The slider bar system for adjusting treatment couch position.

(c) A remote user (VR controllers and headset) standing next to the LINAC equipment and using a menu, as seen by a local user.

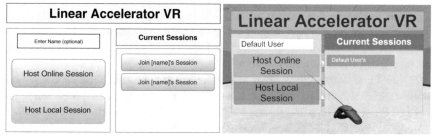

(d) Networking user interface. Host Local Session button is being selected (green).

(e) Networking user interface. Host Online Session button is being selected (green).

Fig. 11.3 LINACVR for simulation of radiation therapy

isocenter. A projected laser grid indicating the location of the isocenter can also be activated.

When a user first loads the collaborative simulation and equips the headset, they find themselves in the LINAC treatment room facing a user interface, giving them the option to either host a session or join an existing one. Once an option is chosen, the user is placed closer to the equipment and patient model and can now interact with them. From here they can see other users who already joined or can wait for others to join the session. The users can then perform the LINAC procedure, with the actions of each also occurring in the simulations of the others.

Figure 11.3c shows a remote user ("Default User" represented by controllers and headset) standing next to the LINAC equipment and using a menu as they appear to a local user. Users joining this session will follow the same process as in the collaborative simulation. Each user is visible to others through a set of models representing the head, hands, body, and name of the user. The position of the head and the hands is based on the position of the headset and controllers for that user, while the position of the body and name is calculated based on the position and angle of the head. The reason that the body position is extrapolated rather than tracked is that the VR sensors can only detect specific tracking devices present in the headset and controllers. The VR controllers are used to represent where the hands are located. The body is a transparent capsule shape which represents the spatial area filled by a user than an accurate location. The head is represented as a VR headset which is influenced by a recommendation from Fraser et al. (2000), who suggest explicitly showing the limits of field of view of other users in order to avoid miscommunication. The head representation also reminds users that the other person is wearing the same headset as they are, serving as a further reminder of the angle of vision. For example, by looking at this headset, we can tell that the other user is looking at their menu, and that the view of the other user is slightly outside of their field of view.

Figure 11.3d shows the network selection menu. From this menu, users can choose to host an online session, host a local network session, or join an existing session. A session name can also be entered, which is important for differentiating between sessions if there are multiple running. The session name is also displayed over the head of a user, identifying them to others in a session. The buttons are gray by default and turn green when currently selected.

Figure 11.3e shows the network user interface for the collaborative simulation. In this example, there is one session currently being hosted, and this is shown in the panel on the right. The patient-perspective user interface shares the same layout and design, but uses slightly different text. The menu takes the form of a wall-sized set of panels extending from slightly above the floor to slightly below the ceiling. It is interacted with by the user via a laser pointer that extends from the end of one of the controllers. When a button is pointed at, as Host Online Session is in Fig. 11.3e, it is highlighted green. By pulling the trigger on the controller, the highlighted option is selected. The reason for the large size of the menu comparative to the user is that it aids the ease with which they can correctly point the laser at a button and pull the trigger on the controller. After selecting the text box in the left panel containing the placeholder text "Default User," a user can then type on their keyboard the name

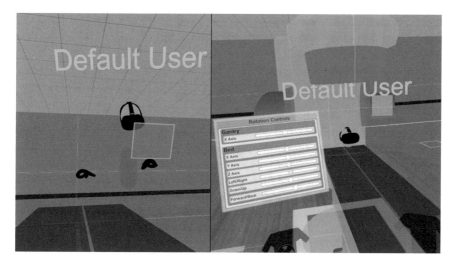

Fig. 11.4 Left: Patient-perspective view. Right: Patient-perspective therapist view

they want for their label and session. This requires the user to temporarily remove the headset, but could be implemented using a virtual keyboard in the future.

The patient-perspective simulation functions in the same way as the multi-user except that the user hosting the session will find themselves placed on the treatment couch in the perspective of a patient while the other user is the therapist. Figure 11.4 shows the views of the patient (left) and therapist (right) who is adjusting the treatment couch using the movement controls and has turned on the laser grid for patient alignment. This means that the user can get used to the room and environment before being joined in the simulation by a therapist. This order is important, as the therapist will generally need to be observing the patient in the real world as they acclimatize to the simulation before they can join. This also means that the patient-perspective simulation can be used by just one user. To ensure that the patient user sees the simulation from the perspective of someone who is lying on the couch, the translational movement of the headset is locked. This means that if the patient moves within the physical space they will not move within the virtual space. Rotational movement is allowed, and so the user can look around within the simulation as they would be able to during the actual procedure. The controller models for the patient user are hidden in the view but they can use them to control the movement of the gantry. As the treatment procedure shown to the patient is performed by an actual radiation therapist, they can tailor the experience to the exact treatment plan that the patient will undergo. This gives the patient a much more accurate preparatory experience, as the patient's experience for different treatments can vary significantly. The therapist user has been deployed to an HTC Vive while the patient user has been deployed to an Occulus Go (wireless headset). This wireless VR headset allows us to demonstrate the patient perspective to patients in isolation or in a distributed environment where they can communicate with a radiation therapist at a different site. To further enhance the scenario, patients can lay on a physical table while the therapist can physically walk around the environment.

11.3.2 Implementation and Architecture

The simulations were developed in Unity3D (Unity Technologies 2021c). A Unity application is constructed from objects placed in a 3D space, with scripts containing code attached to them. These scripts were written in C#. The multi-user applications are exported as executable build files. The application uses SteamVR library (Valve Corporation 2021) which acts as a bridge between the application and the HTC Vive. The portable patient-perspective simulation for the Oculus Go was also developed with Unity3D, but uses a combination of Android Studio (2021) code libraries and the Oculus Core Utilities Unity library.

As the users of LINACVR may not necessarily be proficient with technology, it is important that the multi-user simulation runs without any manual network configuration. The network design goal has been to make launching a multi-user simulation no more difficult than launching a single-user version. For this reason, the client–server architecture has been designed to not require a dedicated server. In many multi-user programs, there is a dedicated server program, which is then connected to by all of the users (the clients), using a separate client program. This server is then responsible for all communication between the clients. The server, running as a separate application, would receive input from the clients, process it, and then distribute the current program state back to each client. The issue with this design is that setting up a server creates another layer of tasks that a user must complete in order to start the simulation. Addressing this, and reducing the amount of expertise required to run a multi-user simulation, the server functionality is contained within the LINACVR program rather than being a separate program. This network design is known variously as both "peer-to-peer hosted" and "listen server" architecture. It is worth noting that in some strict peer-to-peer designs there is no server at all, and all clients share the responsibilities of a server. This, however, can cause large issues and delayed feedback when users are performing simultaneous actions, as each user must wait for an update on the actions of all other users after every network refresh (Fiedler 2010). To avoid this, the chosen design uses some of the distributed processing of the peer-to-peer design, with the authoritative server of most listen server designs. As seen in Fig. 11.5, in this peer-hosted system design, the program of the host user simultaneously and automatically acts as a server and a client. This means that there is only one program version needed for any user, and this version is able to act as both a host server and a local client, or just as a remote client. This design is easier to run than a dedicated server, and unlike a full peer-to-peer system the number of network connections per user is not exponential.

The network architecture has been developed using the Unity Multiplayer High-Level API (HLAPI) (Unity Technologies 2021a). The core of this implementation is a script called LinacNetworkManager, a class that extends the UnityEngine Net-

Fig. 11.5 LINACVR
peer-hosted client/server
architecture, showing a host
computer running a local
client and a server, which is
connected to by two remote
client computers

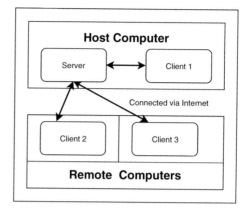

workManager class. Through this, LinacNetworkManager is able to interface with, extend, and control the HLAPI functionalities. LinacNetworkManager manages the stopping and starting of server hosting, the creation and connections of clients, and the creation and joining of sessions. The setting up and maintenance of connections between remote clients and the server have been implemented using the Unity Internet Services platform (Unity Technologies 2021b). An important advantage of this service, and a primary reason that it was chosen, is that users do not need to know or enter the IP address of the other users that they are connecting to. This gives greater convenience and accessibility to less technically experienced users, and drastically reduces the time taken to set up the simulation. Unity Internet Services allows the connection of up to 20 concurrent users across all sessions at any one time without any hosting costs. Sessions hosted on a local network using the "Host Local Session" option are not effected by this limit, and have the added advantage of considerably lower latency due to not using an external match hosting service. The limit refers not to the number of actual users but to the number of potential users at any time. This means that if one online session containing only two users is running that has a user capacity of 20 users, then no other sessions can be hosted. For this reason, the network is currently configured to allow online sessions a maximum capacity of four users. This limit was decided due to normal collaborative usage of LINAC equipment involving low number of operators. If the use of this needs to change to require larger sessions, for example, if an institution acquired many HTC Vive units and wishes to use this simulation in a lecture-type format with many simultaneous users, this limit can be easily raised using a "Max Connections" field in the Unity Editor which interfaces with the LinacNetworkManager script. This LinacNetworkManager script is the main interface between the system and the matchmaking functionality given by the UnityEngine NetworkMatch class.

11.4   Evaluation

To evaluate the effectiveness of LINACVR for simulation of radiation therapy education we conducted a user study. The simulation is designed to be used by radiation therapy students and radiation therapy educators, and so these people were the target participants for this study. The aim of the study was to evaluate the multi-user and patient-perspective features and the simulation, in general, by addressing the following questions:

- How easily do users learn how to operate the simulation, controls, and interface?
- How effective is manually positioning the patient?
- How effective are the couch controls for positioning the patient?
- How effective is the multi-user feature?
- How effective is the patient-perspective simulation?
- How effective would this simulation be in training to use LINAC machines?
- How does this simulation compare to existing LINAC simulation programs?
- In what ways could this simulation be improved in the future?

Participants were recruited from the Department of Radiation Therapy at the University of Otago, Wellington. Any student was considered eligible for the study if they had used, or experienced any computer simulations of LINAC machines. Similarly, any educator was considered eligible if they have taught the use of LINAC machines in any sort of educational capacity. The participants were recruited through email mailing lists, and the researchers advertising the study in person during student lectures at the university. Upon completion of the study, participants were awarded an honorarium in the form of a $10 supermarket voucher. We recruited 15 participants, 11 of these being students and 4 being educators.

This study was conducted in a clinical examination room at the Victoria University of Wellington hospital campus which is right next to the Department of Radiation Therapy. The edges of the room were set up within the headset as a border for the SteamVR Chaperone system. This created a grid in the virtual space to prevent participants from walking outside of the cleared space and colliding with real-world objects. This was essential for when the experimenter was also in the virtual space, as they cannot be watching the participant at all times.

The user study was conducted over 3 days. Each user study was a one-on-one session between the participant and the experimenter, and lasted approximately 60 minutes. The study was a within-subjects study, where all participants were exposed to all study conditions (Nielsen 1993). Participants were given an information sheet, consent form to sign, and a pre-study questionnaire. Each participant was screened for nausea via verbal questioning. Participants were then given some training time with LINACVR where the features and control options were demonstrated. Participants then completed the study tasks for each of the scenarios: individual and then collaboratively with the experimenter acting as another educator. Participants then experienced the collaborative and the portable version of the patient perspective. The study was then concluded with a post-study questionnaire and follow-up interview.

During the study participants were regularly asked whether they are experiencing any motion sickness.

The study tasks are as follows and repeated for both times the participant performed the two scenarios:

1. Navigate to the equipment, either by teleporting or by walking.
2. Manually adjust the patient on the bed.
3. Turn on the laser grid to aid positioning.
4. Move the bed using the menu controls.
5. Line up, through preferred combination of manual adjustment and bed positioning, the isocenter with the indicators.
6. Use transparency controls.
7. Initiate radiation delivery.

Data collection took the form of two questionnaires and observation notes taken by the experimenter during the session. The first questionnaire was a pre-study questionnaire in order to determine background factors such as experience with the various technologies involved. The second was a post-study questionnaire about the participant's experiences with the simulations. The first 11 questions in this survey were Likert scale questions, recording the participant's perceived effectiveness through ratings from 1 (Very Ineffective) to 5 (Very Effective). The remaining 4 questions sought qualitative responses via free text answers.

11.5 Results

We now present the results from the user study from a quantitative and qualitative perspective based on responses from the post-study questionnaire.

11.5.1 Quantitative Data

Figure 11.6 shows the ratings for each Likert scale question in the post-session questionnaire for all participants. The color representation is green (Very Effective) to red (Very Ineffective). From this data, we can see some interesting features. From this, we can see that the median rating for every question was Effective, and that no question received less than two-thirds of its ratings being positive. Some participants gave generally higher or lower ratings than others. Comparing participants number two and three we can see this trend, with two giving consistently higher ratings. Reasons for these differences can often be explained in some form by the information given in the pre-study questionnaire. In this specific case, participant two was a first-year student with 6 months practical experience and little exposure to virtual reality or to VERT. Participant three, however, was an experienced radiation therapy educator and specialist practitioner. This could indicate that those with higher experience with

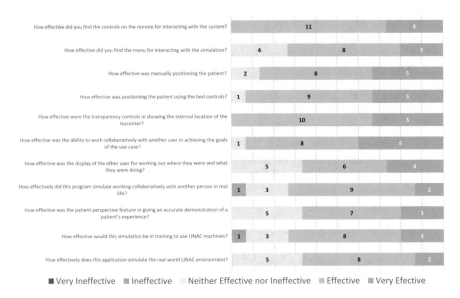

How effective did you find the controls on the remote for interacting with the system?   11   4

How effective did you find the menu for interacting with the simulation?   4   8   3

How effective was manually positioning the patient?   2   8   5

How effective was positioning the patient using the bed controls?   1   9   5

How effective were the transparency controls in showing the internal location of the Isocenter?   10   5

How effective was the ability to work collaboratively with another user in achieving the goals of the use case?   1   8   6

How effective was the display of the other user for working out where they were and what they were doing?   5   6   4

How effectively did this program simulate working collaboratively with another person in real life?   1   3   9   2

How effective was the patient perspective feature in giving an accurate demonstration of a patient's experience?   5   7   3

How effective would this simulation be in training to use LINAC machines?   1   3   8   3

How effectively does this application simulate the real world LINAC environment?   5   8   2

■ Very Ineffective   ■ Ineffective   ▨ Neither Effective nor Ineffective   ■ Effective   ■ Very Efective

Fig. 11.6 Likert scale results from post-study questionnaire, 11 questions

LINAC machines have higher expectations of functionality or realism due to their increased experience with the real environment. This is corroborated by the fact that participants six, seven, and ten were the other educators involved in this study, with both six and ten giving relatively lower scores than most participants. However, participant seven gave relatively high scores, however, so this relationship likely bares further investigation. We now discuss the results from the four open-ended questions.

11.5.2 Question 1: What differences did you notice between this simulation and the real-world LINAC environment?

The most frequent reported difference given was sound. Participants pointed out that in real life the LINAC machines make a considerable amount of noise, mainly when they are activated and emitting radiation. Many verbally noted that this was particularly important for the patient-perspective simulations, as the noise made by a LINAC machine was likely to be one of the most frightening aspects of treatment. As both headsets support binaural audio, this could make a valuable and effective addition to the project in the future.

> No radiation sound.—PID 1
> Lack of sound.—PID 3
> Lack of sound made a difference as often LINACs can be relatively loud.—PID 4

Sound the machine makes as it delivers RT.—PID 6
First thing to notice is sound—LINACs are noisy!—PID 10

An aspect commonly mentioned differences was the LINAC control pendant. While the menu system and Vive controller have the same functionality as a pendant controller, it seems that it is still very different to use.

Hand controls different to LINAC pendants.—PID 1
Controls of the LINAC are important, touch controls take some getting used to.—PID 3
LINAC controls are more easier to get mm movements for accuracy.—PID 8
Controllers feels very different.—PID 11

The positioning of the patient using bed controls received overall quite positive ratings, indicating that this is not an issue based on difficulty of use in the simulation, but rather of the quality and transferability of training with the actual controls.

Manual couch movements.—PID 7
We move the couch manually more often than using the pendants of the couch controls to make the required adjustments.—PID 8
Scale of couch is similar, the machine feels smaller.—PID 10
Different way of using couch controls.—PID 13

A common comment was that in the real world it is much more difficult to manually position a patient. Seemingly due to how hands on it is to physically move a patient in real life, operators are less likely to choose to do so instead of moving the treatment couch compared to how likely they are in this simulation.

Much more hands on in the real world in terms of handling the patient.—PID 5
Equipment to position immobilise a patient.—PID 7
Unable to roll patient's skin when positioning. Aside from the using the remote to adjust couch position, RTs commonly use controls underneath the couch and on the floor next to the couch. This component is missing from the VR simulation.—PID 9
It was easier to setup the patient with the VR simulation.—PID 12

A further difference that was mentioned by multiple people was the terminology used to refer to planes of direction, where, in this simulation, the directions are referred to as forward/backward and up/down, the equivalent terms used by therapists would be superior/inferior and anterior/posterior. The directions of left and right use the same words, but in the medical context the directions should be from the perspective of the patient, not the operator as in this simulation. The directions are also used in reference to the laser positioning lines. For example, a therapist might say to their colleague: "Please move the patient superior from the horizontal laser," rather than "Please move the couch forwards."

Different terminology forwards/backwards vs superior/inferior.—PID 13

Many participants reported that the motorized movement of the couch and gantry is a lot slower in the real world, making it considerably easier to accurately position a patient with higher precision. Apparently, some treatment couches have a manual, non-motorized mode for large movements and then slow motorized controls for precise positioning. One participant suggested having more gradual movements, but a feature to speed movement up if required.

Slower gantry.—PID 1
In the real world the bed moves slower, more momentum, e.g. with heavy patient.—PID 15

There were also many other small differences related to visual feedback such as tattoo marks being crosses so that they can line up with the laser grid, tattoo marks being on both sides of the patient, lasers being thinner in real life, viewing patient organs, real room being darker, and a light field being emitted from the gantry head for some treatments.

The lasers are only 1-2mm thick.—PID 2
We can view certain organs on the machine rather than imaging it.—PID 5
Darker room, X tattoo marks to align to.—PID 6
Markings on skin.—PID 7
Skin marks are important to visualize during patient positioning.—PID 13
Light field can be helpful for breast setups.—PID 14

As a tool for training purposes, some participants thought LINACVR could be quite useful, helpful, and complementary to existing tools and techniques.

The detail of the machine were minimal, however it was beneficial for giving an overall impression of a LINAC bunker and could be quite useful for training.—PID 4
LINACVR could be complementary to several current training programs.—PID 10
Some aspects of LINACVR could be helpful for training.—PID 14

11.5.3   Question 2: How did this application compare to any other LINAC simulation programs you have used (e.g., VERT)?

It is worth noting that in the pre-study survey no participants reported having used a LINAC simulation other than VERT in the past. Similarly, all answers to this question compared LINACVR to VERT as the participants had no other simulation experiences.

Comparisons to VERT were almost exclusively positive toward LINACVR.

Really great but I have not used VERT much before.—PID 2
LINACVR was a lot more user friendly than my experience with VERT.—PID 4
Fills many of the gaps in experience offered by VERT.—PiD 7
Would be way more useful than VERT in preparing students for the clinical environment.—PID 13
So much better than VERT.—PID 14
Much more immersive + similar to real world application.—PID 15

A common comparison was that LINACVR gave a simulation that was more interactive, tactile, and realistic.

Really liked the tactile component. I think more practice with it you could become more proficient and able to make fine adjustments.—PID 6
Ability to interact with patients in LINACVR is fabulous.—PID 7
I like that LINACVR is more interactive and that you can do a lot more with it. I like

that it provided an experience in the role of a radiation therapist whereas VERT is more observational.—PID 8
LINACVR was more interactive and hands on which is an advantage.—PID 12

The interaction capabilities in LINACVR made for a more effective teaching and collaborative experience. One participant also stated that the controllers were similar in feel to LINAC pendants, contrary to the feedback of several other participants.

LINACVR is much more hands on application, allowing a more realistic and more useful teaching experience.—PID 9
Nice to be able to have collaborative work that makes it a bit closer to an authentic experience. The controllers were similar in feel to the LINAC pendants.—PID 10
LINACVR way more interactive and cool that you can do two users to work together at the same time.—PID 13

Some aspects of LINACVR which are not present in VERT include the patient perspective which were an advantage for educational aspects.

Wireless headset is a convenient option for introducing patients to the clinical environment.—PID 3
LINACVR would be beneficial for patient education to give an idea of what a LINAC machine actually looks like.—PID 4

The freedom to move around and to interact with patients were both also reported as positive comparisons.

The ability to move through 3D space is a valuable feature compared to VERT.—PID 3
Similar to VERT but LINACVR gives us the freedom to move around and we feel like we are in the clinic when we are not.—PID 5

Some participants mentioned there were disadvantages with LINACVR due to the lack of a LINAC pendant remote for moving the bed.

VERT includes real LINAC pendants which is an important component for training.—PID 3
VERT has a pendant with a lot more buttons on the trigger (controller) than in the virtual world.—PID 5
VERT uses real life LINAC equipment such as the pendant which makes the patient movement more like the LINAC machine.—PID 11

Another disadvantage of LINACVR compared to VERT is that it does not support different radiation treatment modes such as electron therapy, a type of radiation therapy (but not very common) that targets cells near the skin rather than inside the patient.

Would be good to add electron option like VERT has as electrons can be hard for students to learn and not as common so heaps of practice would be great.—PID 14

11.5.4   Question 3: Are there any improvements you think could be made to this VR simulation?

There were some additional improvements participants suggested. Sound of the real LINAC machine was a key important aspect that is needed to be included.

> Put in sounds the LINACs make.—PID 1
> For the patient perspective you could add sound as the sound of the LINAC and the room alarms aren't that pleasant.—PID 2
> Include sound.—PID 3
> The addition of sound a LINAC makes could be beneficial for patient education.—PID 4
> Addition of sound if possible.—PID 6

One of these that was suggested by several participants is to have props in the background of the treatment room and more detail on the LINAC model and features in order to make the environment more realistic, particularly for the patient-perspective simulations.

> Add in more details specific to LINACS e.g. light fields, collimators, laws. A simulation of a CBCT scan.—PID 3
> More in the surrounding environment to the treatment room so it more lifelike.—PID 6
> Permanent representation of the isocentre.—PID 7

The pendant and controls could be improved so they resembled closer to what the pendant is like in a real environment similar to what is available in the VERT simulator that the participants are more familiar with.

> The pendant could mimic the real controller in the department when doing couch movements for a more representative idea of what it is like in the clinic.—PID 8
> The controls on the remote were a bit too sensitive at times especially the up, down, left, and right arrow buttons. I opened the wrong menu multiple times by accidentally tapping the down arrow and the centre button. A more gradual/slower sliding tool when adjusting the couch could be useful. Maybe double tap to speed things up. Add another controller at the couch.—PID 9
> The controls for scrolling through the menus seemed tricky.—PID 15

Some participants would have liked to have seen more data about the simulated patient such as the complete model and avatars of other users.

> Whole body for realism.—PID 3
> Datasets that showed the whole patient anatomy rather than a torso to be more lifelike.—PID 6
> Maybe a body of the person using the VR simulation.—PID 12

A suggested further improvement to the realism of the patient perspective is to include the panels that fold out from the gantry on either side of the patient's head. This would help the patients prepare themselves for the experience, as the panels can come quite close to the patients' head for some treatments and can cause feelings of claustrophobia.

> Panels coming out for the patient mode as can come quite close to the patient.—PID 14

The multi-user and collaboration with patient features were particularly useful but there was some feedback on how to improve these aspects and not all participants were comfortable with that simulation scenario.

> Maybe when the partner is working we could see what they are doing instead of their hands/trigger/controller dissolving into the couch.—PID 5
>
> It's good to have the multi-user function. The next step in really effective VR for radiation therapy is mannequins to provide feedback to the users. Currently it helps in teaching the steps through which a team sets up a patient but the most variable part of the setup is the patient.—PID 10
>
> The use of another user was hard - being unfamiliar with the controls of this technology. Felt with the other user a bit lost as I worked a lot slower not understanding the controls.—PID 11

Some participants would like to use LINACVR with treatment plans they have designed in other tools and import them into LINACVR.

> Making use of treatment plan data from DICOM files.—PID 7

Some participants would have liked more time with the tool, but if we were to deploy LINACVR in an educational setting this would be possible. This could help to alleviate some of the interaction usability issues with participants having more time to learn the controller options.

> More time to familiarise with the environment and controls.—PID 3

11.5.5   Question 4: Do you have any other feedback about the simulation that you would like to give?

Most participants' feedback, in general, was positive and encouraging about supporting a VR radiation therapy simulator experience.

> Was cool to use.—PID 1
>
> Really advanced and hope to see it in clinical when I start working.—PID 5
>
> Really impressive.—PID 6
>
> Very impressed.—PID 7
>
> Really enjoyed the experience.—PID 8
>
> It was very effective, a good representation of a real life LINAC.—PID 12
>
> Really cool!—PID 13
>
> This is awesome. Well done!—PID 14

Several participants specifically mentioned that the patient perspective seemed like it would be very beneficial for patient education and preparation. Another mentioned that while actual clinical experience is still more important for training that LINACVR would be a good way to educate beginners and to introduce staff to new concepts.

> It would be very beneficial to nervous patients/children receiving treatment.—PID 1
>
> I think the use of VR could be very helpful for patient education and easing anxiety of

patients. Clinical experience will still be more valuable I think for education, however I can see the value in VR use for educating beginners, or introducing new techniques for staff."—PID 4

I like the idea of having it from the patient's perspective and it being portable.—PID 6

I think it could be real beneficial for both staff and patients.—PID 8

It was very cool, would be a great learning and teaching tool!—PID 15

11.6   Discussion

For those who have never used VR before, it can be an exciting and novel experience. In a user study involving VR, it is important to isolate the effects of this novelty upon the results. For example, a user who had never used VR before may give positive feedback because the VR paradigm is very effective compared to virtual systems they have used before, not necessarily because the simulation being studied is effective. Comparing mean rated effectiveness across all scale-based questions between those who had experienced virtual reality before and those who had not gives us a mean rating of 3.964 for no prior experience and of 4.005 for prior experience. This shows that for this study the bias most likely did not have a significant effect on rating.

While this tool would in theory be used in an educational setting and so students would initially be taught in the use of it by an expert, having the researcher present and assisting may still have influenced the results. Due to the unconventional nature of the VR control scheme and interface, users were not left to figure controls out by themselves but were rather initially guided through them by the experimenter. This may have had an impact on the participants' impression of the usability and learnability of the user interface.

The comparison by users of the single- and multi-user use cases may have been impacted by several factors. The first is that counterbalancing was not used. Counterbalancing should normally be used in a comparative VR user study to control the effects of user learning and fatigue (Edward Swan et al. 2004). This was not included in the design for this study as it was necessary for the experimenter to observe and assist the participant learning how to use the system, which would not have been possible with the experimenter also using the simulation for the multi-user scenario. Likewise, the fact that the other user that participants worked collaboratively with in the multi-user use case was the experimenter who could also have impacted results. Ideally, this would have been another radiation therapy student or educator.

Additionally, the small participant sample size may not have been representative of the radiation therapy education industry as a whole. In particular, while most students were in their first or second year of training, the third-year students were all on practical placements during this study and so only one of them participated, meaning that results may not be representative of all experience levels within an educational institute. All participants were from the same institution which may also mean that results from this study may not be applicable to other institutions.

Overall, the results of the study indicate that LINACVR provides an effective training solution. 11 out of the 15 participants responded that this solution would be either effective or very effective for the training of radiation therapy. For the remaining four, only one considered it ineffective. A similar majority also considered the collaboration and patient-perspective features effective or very effective. It was found that the simulations developed have distinct advantages over the existing alternative VERT system which includes interactivity, immersion, and collaboration features. The study also produced a set of useful potential future improvements to realism and effectiveness that could be implemented in the future.

11.7   Conclusions

Cancer is one of the leading global causes of death and requires treatment by highly trained medical professionals, some of whom currently have limited access to effective training tools. To provide better access to this training, our goal was to provide the experience of radiation therapy using a Linear Accelerator (LINAC) Machine from the perspective of both the radiation therapist and the patient through the use of VR simulation. The therapist perspective would provide effective low-cost training to radiation therapy students. The patient perspective would give patients thorough education and preparation, reducing the psychological stress of treatment.

LINACVR is the first collaborative VR tool which represents a radiation therapy environment without needing to use actual LINAC equipment. LINACVR provides an effective and immersive simulation of radiation therapy treatment for both therapist training and patient education. The inclusion of multi-user functionality increases realism, accuracy, and therefore training effectiveness for students. The multi-user functionality allows customized treatment experiences for patient education, increasing patient preparation effectiveness. LINACVR features a completely portable version of the patient perspective.

We conducted a user study of LINACVR to evaluate the usability and effectiveness of both the patient-perspective and training simulations. We found that the patient-perspective simulation gave an effective representation of the patient experience which would be beneficial for patient education. We found that the training simulation was easy to learn, very effective compared to the existing alternative (e.g., VERT), and effective in the training of radiation therapy. The main disadvantage of LINACVR was the lack of a real-life physical treatment couch remote control. The development and integration of this pendant hardware would be a valuable addition, and would further increase the applicability of the simulation training to the real-world procedure. There was a significant difference between the virtual simulation and the real world in manually positioning the body of a patient. Future work would be to explore a mixed reality solution using a combination of a physical mannequin and table but a virtual LINAC machine as this would allow for an even greater level of training accuracy and realism.

Acknowledgements We thank the participants who performed in the user study and collaborators at the Department of Radiation Therapy at the University of Otago, Wellington.

References

Android Studio (2021) Build for android. https://developer.android.com/. Accessed 01 Mar 2021

Beavis AW, Page L, Phillips R, Ward J (2009) Vert: Virtual environment for radiotherapy training. In: Dössel O, Wolfgang CS (eds) World congress on medical physics and biomedical engineering, September 7–12. Munich, Germany. Springer, pp 236–238

Beavis AW, Ward JW (2014) The use of a virtual reality simulator to explore and understand the impact of linac mis-calibrations. J Phys Conf Series 489

Cai W, Walter S, Karangelis G, Sakas G (2000) Collaborative virtual simulation environment of radiotherapy treatment planning. Comput Graph Forum 19:379–390

Carlson AL (2014) Medical errors in radiation therapy. Division of Emergency Preparedness and Community Support Bureau of Radiation Control Florida Department of Health

Churchill EF, Snowdon D (1998) Collaborative virtual environments: an introductory review of issues and systems. Virtual Reality

Cook D, Hatala R, Brydges R, Zendejas B, Szostek J, Wang A, Erwin P, Hamstra S (2011) Technology-enhanced simulation for health professions education a systematic review and meta-analysis. JAMA J Am Med Assoc 306(9):978–988

Davies AG, Crohn NJ, Treadgold LA (2018) Can virtual reality really be used within the lecture theatre? BMJ Simul Technol Enhan Learn 4

Faculty of Clinical Oncology (1998) A national audit of waiting times for radiotherapy. Technical report. The Royal College of Radiologists May

Fiedler G (2010) What every programmer needs to know about game networking: A short history of game networking techniques. https://gafferongames.com/post/what_every_programmer_needs_to_know_about_game_networking/. Accessed 01 Mar 2021

Fraser M, Glover T, Vaghi I, Benford S, Greenhalgh C (2000) Revealing the realities of collaborative virtual reality. In: Proceedings of the third international conference on collaborative virtual environments. San Francisco, CA USA

Van Der Giessen PH, Alert J, Badri C, Bistrovic M, Deshpande D, Kardamakis D, Van Der Merwe D, Motta N, Pinillos L, Sajjad R, Tian Y, Levin V (2004) Multinational assessment of some operational costs of teletherapy. Radiotherapy Oncol J Eur Soc Therap Radiol Oncol 71(07):347–355

Hollands RJ, Trowbridge EA (1996) A virtual reality training tool for the arthroscopic treatment of knee disabilities. In: 1st European conferences disability, virtual reality and association technology

Edward Swan II J, Gabbard J, Hix D, Ellis S, Adelstein B (2004) Conducting human-subject experiments with virtual and augmented reality. In: IEEE virtual reality conference, VR, IL. USA, March, Chicago, p 266

Kaczur R (2013) Safety is not an accident: lessons to be learned from catastrophic events. Am Assoc Phys Med. http://chapter.aapm.org/pennohio/2013FallSympPresentations/FI2_Ray_Kaczur.pdf. Accessed 01 Mar 2021

Kane P (2015) The impact of a virtual reality system on teaching and learning. In: AIR- NZIMRT scientific meeting. Wellington, New Zealand, p 07

Kane P (2017) Simulation-based education: a narrative review of the use of vert in radiation therapy education. J Med Radiat Sci 65:09

Leong A, Herst P, Kane P (2018) Vert, a virtual clinical environment, enhances understanding of radiation therapy planning concepts. J Med Radiat Sci 65:03

Mantovani F, Castelnuovo G, Gaggioli A, Riva G (2003) Virtual reality training for health-care professionals. Cyberpsychol Behav Impact Inter Multim virtual Reality behav Society 6(09):389–395

Nielsen J (1993) Usability engineering. Morgan Kaufmann Publishers Inc., San Francisco, CA, USA

Phillips R, Ward JW, Page L, Grau C, Bojen A, Hall J, Nielsen K, Nordentoft V, Beavis AW (2008) Virtual reality training for radiotherapy becomes a reality. Stud Health Technol Inform 132:366–371

Podgorsak EB (2005) Treatment machines for external beam radiotherapy. Department of medical physics. McGill University Health Centre

Rainey LC (1985) Effects of preparatory patient education for radiation oncology patients. Cancer 56:1056–1061

Schlachter M, Raidou RG, Muren LP, Preim B, Putora PM, Bühler K (2019) State-of-the-art report: visual computing in radiation therapy planning. Comput Grap Forum 38(3):753–779

Sehlen S, Hollenhorst H, Schymura B, Herschbach P, Aydemir U, Firsching M, Dühmke E (2003) Psychosocial stress in cancer patients during and after radiotherapy. Radiol Oncol 179(3):175–180

Sostaric M, Lilijana S (2004) Psychological distress and intervention in cancer patients treated with radiotherapy. Radiother Oncol J Eur Soc Therap Radiol Oncol 38(3):193–203

Tellow VM (2018) Medical linear accelerators and how they work. Healthphysics society. https://web.archive.org/web/20180403121129/http:/hpschapters.org/florida/13PPT.pdf https://web.archive.org/web/20180403121129/http://hpschapters.org/florida/13PPT.pdf. Accessed 01 May 2021

The Economist. Cancer drugs are getting better and dearer (2017). https://www.economist.com/business/2017/05/04/cancer-drugs-are-getting-better-and-dearer. Accessed 01 Mar 2021

Unity Technologies (2021a) The multiplayer high level api. https://docs.unity3d.com/Manual/UNetUsingHLAPI.html. Accessed 01 Mar 2021

Unity Technologies (2021b) UNet internet services overview. https://docs.unity3d.com/540/Documentation/Manual/UNetInternetServicesOverview.html. Accessed 01 Mar 2021

Unity Technologies (2021c) Unity. https://unity.com/. Accessed 01 Mar 2021

Valve Corporation (2021) Steam vr. https://steamcommunity.com/steamvr. Accessed 01 Mar 2021

Varian Medical Systems (2021) Truebeam radiotherapy system. https://www.varian.com. Accessed 01 Mar 2021

Vertual Limited (2021) Vertual limited. https://www.vertual.co.uk/. Accessed 01 Mar 2021

Ward JW, Phillips R, Boejen A, Grau C, Jois D, Beavis AW (2011) A virtual environment for radiotherapy training and education—VERT. In: Buehler K, Vilanova A (eds) Eurographics 2011—Dirk Bartz prize, the eurographics association

World Health Organization (2018) Cancer fact sheet. http://www.who.int/news-room/fact-sheets/detail/cancer

Chapter 12
Multi-Touch Surfaces and Patient-Specific Data

Anders Ynnerman⓪, **Patric Ljung**⓪, **Anders Persson**⓪, **Claes Lundström**⓪, and **Daniel Jönsson**⓪

Abstract While the usefulness of 3D visualizations has been shown for a range of clinical applications such as treatment planning it still had difficulties in being adopted in widespread clinical practice. This chapter describes how multi-touch surfaces with patient-specific data have contributed to breaking this barrier, paving the way for adoption into clinical practice and, at the same time, also found widespread use in educational settings and in communication of science to the general public. The key element identified for this adoption is the string of steps found in the full imaging chain, which will be described as an introduction to the topic in this chapter. Emphasis in the chapter is, however, visualization aspects, e.g., intuitive interaction with patient-specific data captured with the latest high speed and high-quality imaging modalities. A necessary starting point for this discussion is the foundations of and state-of-the-art in volumetric rendering, which form the basis for the underlying theory part of the chapter. The chapter presents two use cases. One case is focusing on the use of multi-touch in medical education and the other is focusing on the use of touch surfaces at public venues, such as science centers and museums.

12.1 Motivation and Background

The rapid development of modalities for medical imaging opens up possibilities for introduction of new workflows in medical research and practice. Making use of digital representations of digital anatomy, and use of patient-specific data creates new opportunities, but is also demanding as the uniqueness of the data requires high level of detail and freedom in interaction to enable exploration of features of interest. Thus, the availability of data is only the first step towards uptake and widespread use of patient-specific digital representations. This chapter presents how use of multi-touch surfaces is one of these enabling components and how it is placed in the underpinning

A. Ynnerman (✉) · P. Ljung · C. Lundström · D. Jönsson
Department of Science and Technology, Linköping University, 60174 Norrköping, Sweden
e-mail: anders.ynnerman@liu.se

A. Persson
CMIV, Linköping University, 58183 Linköping, Sweden

Fig. 12.1 The use of multi-touch surfaces has opened up for widespread use of patient-specific and interactive medical visualization in the medical workflow, and in particular uptake in educational settings, such as teaching of medical students, has been rapid. Image copyright Sectra AB, reprinted with permission

chain of advancement of technology and methodology. The starting point is found in the advances and technology trends that have made this possible:

- Rapid development of medical imaging modalities such as computer tomography (CT), see Sect. 12.2, leading to availability of high-quality patient-specific data at unprecedented levels and amounts.
- General availability of graphics processing units (GPUs) with capability, in terms of processing power and memory sizes, of supporting direct volume rendering (DVR).
- Development of methods for high-quality volumetric rendering and illumination at speeds supporting high-fidelity interaction, see Sect. 12.3.
- Availability of multi-touch technology on a variety of screen formats and interaction design approaches enabling intuitive interaction with complex 3D environments.

The chapter will describe each of these trends in some detail. We will start by describing the use of the technology in medical applications and also in science communication and education. We then provide a description of the imaging chain from data capture to rendering and touch interaction with a focus on state-of-the-art practise. Our ambition is to provide as much information as possible in the context of applications and we thus give detailed information on the use of medical visualization tables in the context of the chosen scenarios.

12.1.1 Medical Application Scenarios

3D visualization of medical data has been pursued almost as long as the capacity for volumetric data acquisition has been available (Herman and Liu 1979). The main producer of volumetric data in medicine is of course the radiology department. The dominating visualization of 3D data that radiologists use is, however, not volume rendering but 2D slice views (Lundström 2007; Preim and Bartz 2007). Since 2D slices are considered necessary to convey the full diagnostic details, radiologists commonly find that 3D visualizations do not add substantial value. While there are of course exceptions where the overview provided by volume rendering is essential, such as vessel visualizations, radiologists mostly rely on the mental 3D models they construct from browsing slices. In contrast, other physicians have a mindset centered on the physical meeting with the patient, such as during surgery, which makes 3D visualizations the natural information carrier for radiology data.

The context of this chapter is touch interaction. We can begin by concluding that this is, in view of the analysis above, less relevant for the image specialists in Radiology. The rich and advanced toolbox radiologists require fits better in a traditional workstation with primarily mouse-based input. For other physicians and medical students, however, the critical factor for adopting advanced visualization is ease of use. The advent of touch interfaces thus removed an important barrier for adoption, and also induced a user experience redesign for 3D visualization suitable for wider audiences.

Outside of the daily diagnostic workflow in Radiology, there are several scenarios where exploration of patient data through multi-touch surfaces can be effective and valuable. In clinical care, the main need resides with the physicians having referred the patient for a radiological examination, in particular surgeons (Volonte et al. 2016). One such well-documented use case is found in planning of orthopedic surgery (Lundström et al. 2011), for which an evaluation clearly showed the positive implications of touch table-based visualization. In surgery workflows, an early activity is often a multi-disciplinary team meeting (MDT). At the MDT an initial assessment of injuries and treatment options is made during an image-centered session with a radiologist and one or more surgeons. The collaborative opportunities of a large touch interface can be very beneficial during such a session (Isenberg et al. 2012). A second stage in the surgeons' workflows, during which visualization can play a central role, is the pre-operative planning, where strategies for quick and effective surgery are formed. Depending on the sub-specialty, the planning can be more or less image-guided. Touch interaction is particularly relevant when several colleagues discuss, and this is also true for post-operative follow-up. Intra-operative use of touch interaction to dynamically produce informative visualizations is attractive due to ease of use, but typically the demands on sterile equipment pose practical challenges.

Outside clinical care, a very important application area for multi-touch technology is medical education (Darras et al. 2018; Baratz et al. 2019; Preim and Saalfeld 2018). In anatomy teaching, virtual dissections provide additional and complementary opportunities compared to the traditional dissection of human cadavers. Multi-touch technology has low needs of application training, and interaction similarity across devices from large table-tops to tablets improves ease of use. Training on data from several clinical patients instead of a single cadaver or generic anatomical model inherently provides the students with the insight that the notion of normality spans many indiviual variations. Furthermore, cadavers are an expensive resource that is not always available, and they cannot be reused.

In the case of postmortem imaging (PMI), cadavers are the source of the imaging data. The autopsy is a long-standing cornerstone for quality improvement in healthcare, and PMI has been shown to be a valuable complement to the physical procedure (Burton and Underwood 2007; Lundström 2012). PMI is also a key component in forensic investigations, and a multi-touch 3D visualization application has been specifically tailored for this application (Ljung et al. 2006), see Fig. 12.1. In contrast to the regular radiology workflow touch technology is an excellent fit for use by the forensic pathologist before or during the autopsy, and in collaborative settings with the radiologist, and is a useful complement as it currently cannot fully replace the autopsy. In Fig. 12.2 a use case showing potential impact of touch technology in the forensic workflow is shown.

12.1.2   Science Communication Scenarios

The last decades of rapidly increasing digitization of artefacts and subjects (Wayne 2013) and touch device technology have paved the way for interactive exploration-based communication of science at museums and science centers (Ynnerman et al.

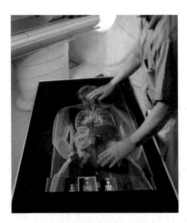

Fig. 12.2 Postmortem forensic imaging. During an autopsy it is normally difficult to map gas distribution in the body. After examination with computed tomography, gas distribution can be visualized in all directions with the help of the visualization table. Important information can be obtained about the direction of a possible gas-filled injury channel prior to forensic autopsy. Image courtesy of CMIV, Linköping University

2016). The primary aims of this type of science communication can be divided into three parts. First, communicating the scientific knowledge and advances related to the displayed artefact. Second, communicating the scientific advances in technology and methods enabling interactive exploration of the underlying data. Third, education for learning natural science. These three aims are all included in the following three science communication scenarios.

A visit to a museum or science center is perhaps one of the most common science communication scenarios for interactive touch surfaces. Here, typically only a few people interact with an installation at any given time and interactive multi-touch surfaces displaying scanned artefacts effectively allow the visitors to become explorers themselves. In addition, the multi-touch surface can promote communication among a group of visitors since gathering around the installation to discuss and interact both verbally and physically at the same time (Ynnerman et al. 2015) is a natural behavioral pattern. Visitors to museums and science centers typically only spend a short amount of time at each installation. Therefore, it is essential that the installation has an intriguing design, to first attract the visitor and once attracted, is easy to understand and interact with. The overarching goal for museums and science centers is the learning outcomes of a visit. An indirect measure of this is the visitor dwell time, i.e. the time spent in a gallery. Naturally, if an installation is engaging and builds on user-driven exploration, it can also increase the dwell time and the learning experience. In Sect. 12.4.2, we present such engaging and exploratory aspects that have led to the success of multi-touch surfaces in museums and science centers. This scenario, with short sessions and few people engaging at the same time, is mainly addressing the first two aims, i.e., communication of the application content and the technology behind it. In Fig. 12.3 we show a volumetric rendering on a touch table of a mummy at the British Museum. The work behind the image is described in detail in Ynnerman et al. (2016).

Facilitated presentations is another type of science communication scenario, involving an audience, a presenter/facilitator and possibly a controller/pilot who steers the experience while the presenter communicates (Sundén et al. 2014). For smaller groups, it is commonly included in a guided tour where more in-depth knowledge is communicated by a facilitator. For larger groups, a large display, which mirrors the image on the touch display, is commonly used to augment the multi-touch surface interaction. In this case, it is important that the audience can see both displays since the physical interaction with the touch surface provides cues that help the audience to appreciate the connection between interaction and visual representation. These types of presentations can include more advanced features that are not available to the visitor otherwise, due to the prior usage experience of the presenter and/or controller. Depending on the presenter and topic, focus can be put on the technology, the content, or educational aspects. Therefore, this scenario spans all of the three science communication aims.

While educational aspects are inherent to a visit at a museum or science center and guided tours, they can be enhanced through a more produced and facilitated experience. The experience can for example involve the combination of a lecture on the human body with interactive exploration of its anatomy. By integrating interactive

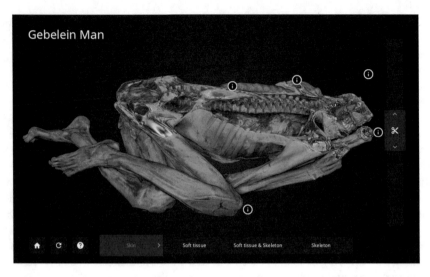

Fig. 12.3 An interactive multi-touch table installation at the British Museum showing the 5500 year old Gebelein mummy. Visitors can perform a virtual autopsy of the mummy and find the cause of death. Data is courtesy of the British Museum

teaching elements, such as placing human organs at their corresponding location, more of the educational aspects can be taught using multi-touch surfaces. It is our experience that using multi-touch surfaces for natural science communication has great potential in reaching and engaging also a younger audience. While this scenario has yet not received much research attention it is an area emerging in studies of visual learning and communication as well as in interaction research. The above-described scenarios all require the content to be specific to the theme of the exhibition or the topic being taught during education providing specific examples. Thus, this patient-specific aspect is essential also to the science communication scenarios.

12.2   Data Acquisition

Computer tomography (CT) is one of the critical imaging modalities available today and is a driving force in patient diagnosis and management. CT has been shown to improve the ability to detect disease in numerous studies, determine its etiology, and guide management. CT can also be used for the collection of data from a range of non-human objects. A CT scan makes use of computer-processed combinations of many X-ray measurements taken from different angles to produce cross-sectional images of specific areas of a scanned object, allowing the user to see inside the object without cutting.

The latest development enables extremely fast volume scans that can generate two-dimensional images in all possible orientations, as well as three-dimensional volume reconstructions. New CT technology also enables functional imaging of the body. Today CT technology allows significant reduction of radiation dose and increased tissue differentiation through the use of two simultaneous X-ray energies, the so-called "dual-energy technology". As before, each examination must be clinically motivated and the radiation dose as low as possible while maintaining image quality to be able to answer the clinical question.

In CT-scanning image quality depends mainly on two factors: CT technique (scan protocol, image reconstruction methods from raw data) and data visualization technique (image post processing). It is of crucial importance that data collection is adapted to the type of object being scanned. Modern CT modalities allow for customization of a large number of scan parameters such as temporal resolution (1–0.25 sec), spatial resolution (0.24–1.0 mm), radiation energy—(70–150 kV), the amount of radiation (mAs) and number of energies used (Single or Dual-energy), etc.

To obtain satisfactory 3D images from computed tomography (CT) several other key parameters need to be optimized depending on the examined object and which issue or problem that is in focus. It is important to understand the principles of CT to appreciate the new capabilities it affords and also to be aware of its limitations. One of the keys to high-quality 3D imaging is the use of data sets that have been reconstructed at close inter-slice spacing to obtain isotropic volume elements (voxels). It is also important to attain maximum difference in gray-scale value attenuation between the structure of interest and the surrounding tissue.

Computed tomography is rapidly evolving and will be further enhanced with photon-counting detectors that deliver high-resolution images with multispectral information at very low radiation doses. Typical CT scanners use energy integrating detectors (EID); the X-rays are converted into visible light in a scintillator layer and emitted light converted into electric signal pulses. This technique is susceptible to noise and other factors which can affect the image quality. Photon-counting detectors (PCDs) are able to convert X-rays directly into electric signal pulses. PCDs are still affected by noise but it does not change the measured counts of photons. PCDs have several potential advantages, including improving signal (and contrast) to noise ratios, reducing doses, improving spatial resolution spatial resolution (0.1–0.2 mm), and through use of several energies, distinguishing multiple contrast agents. PCDs have only recently become feasible in CT scanners due to improvements in detector technologies which can handle the volume and rate of data required (Yu et al. 2016). This promising new technology poses challenges to the user as the amount of generated high-resolution and multivariate image data will increase dramatically.

12.3   Rendering Volumetric Data

The process of computing visual representations of volumetric data is referred to as volume rendering, or more generally, rendering with participating media. Realistic rendering of volumetric phenomena, using physically based principles, is extremely computationally demanding, every interaction between light and media recursively generates additional light-media interactions, making it of exponential complexity. See Fig. 12.4 for an illustration of the three major interaction events: absorption, emission, and scattering. To make volume rendering feasible and support interactivity in data exploration, on reasonably powerful GPUs, several simplifications are necessary. In essence, it is necessary to reduce the exponential complexity into a linear scheme with respect to the volume size and rendering viewport size. More demanding operations also need to be amortized over multiple frames. Traditional raycasting approaches to volume rendering ignore the multiple scattering events and often employs a surface-based shading method at each sample point. In raycasting, rays are cast from the viewer into the volumetric data and throughout the ray traversal typically only absorption and a single scattering event per volume sample is afforded. In the following subsections we will briefly describe the raycasting method and recent techniques to enhance lighting and shadowing along with the principles of translating volumetric data into appearance parameters, basically colors and transparencies, referred to as Transfer Functions (TFs).

12.3.1 Volume Raycasting

Light can be simulated as particles traveling along a ray through a medium. These particles can be absorbed (transformed into heat), reflected into another direction, emitted (transformed from heat), or travel unobstructed through the medium, as illustrated in Fig. 12.4. The simplified physical process of emission and absorption, simulating the amount of light starting at position $\mathbf{x}_0$ and going in the direction towards the camera, $\vec{\omega} = \|\mathbf{x}_c - \mathbf{x}_0\|$, is given by Max (1995):

$$L(\mathbf{x}_c, \vec{\omega}) = T(\mathbf{x}_0, \mathbf{x}_c) \underbrace{L_0(\mathbf{x}_0, \vec{\omega})}_{\text{background}} + \int_0^D \underbrace{T(\mathbf{x}_c, \mathbf{x})}_{\text{attenuation}} \underbrace{\sigma_a(\mathbf{x}) L_e(\mathbf{x}, \vec{\omega})}_{\text{scattering+emission}} ds, \qquad (12.1)$$

where the first term describes the attenuation of the initial light and the second term describes the amount of light $L_e(\mathbf{x}, \vec{\omega})$ emitted and absorbed along the ray. $D = \|\mathbf{x}_c - \mathbf{x}_0\|$ is the length of the ray and $\mathbf{x} = \mathbf{x}_0 + s\vec{\omega}$, $s \in [0, D]$, is the position along the ray. The amount of light absorbed at a position is described by the

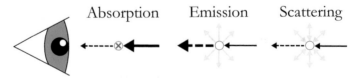

Fig. 12.4 Illustration of three different types of light-media interactions occurring during light transport in volumetric data. Image courtesy of Jönsson (2016)

absorption coefficient $\sigma_a(\mathbf{x})$. The absorption and scattering coefficients at a given position provides the extinction coefficient σ_t, which in turn can be used to compute the attenuation between two points $\mathbf{x}_1$ and $\mathbf{x}_2$:

$$T(\mathbf{x}_1, \mathbf{x}_2) = e^{-\int_0^{\|\mathbf{x}_2-\mathbf{x}_1\|} \sigma_t(\mathbf{x}_1+s\vec{\omega})ds}. \tag{12.2}$$

Solving the two equations above is computationally expensive and there have been several works trying to do so efficiently without introducing artifacts or limiting the number, type or location of light sources (Hernell et al. 2010; Ropinski et al. 2008; Hernell et al. 2008; Díaz et al. 2010; Kronander et al. 2012). While more computationally expensive, more accurately simulated advanced effects such as light scattering can be achieved (Jönsson et al. 2012; Jönsson and Ynnerman 2017), see Fig. 12.5. For a full survey on illumination techniques for interactive volume rendering we refer to Jönsson et al. (2014).

While techniques exist to capture all significant light interaction in a scene, such as Monte Carlo Path Tracing Lafortune (1995), that can be adopted for volume rendering, like the Exposure Renderer by Kroes et al. (2012); these techniques typically require hundreds of samples per pixel, to reduce the noise from stochastic sampling of ray paths and yield high-quality images, and are thus currently not supporting interactivity. Near interactive performance can be achieved using multiple GPUs, or generating lower quality previews, and accumulate the results over multiple frames for a final high-quality image, as demonstrated by Siemens' Cinematic VRT feature (Paladini et al. 2015; Comaniciu et al. 2016), see Fig. 12.6.

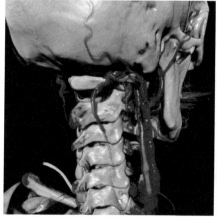
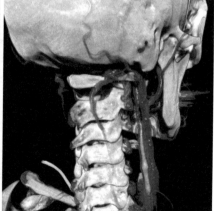

(a) Without multiple scattering.   (b) With multiple scattering.

Fig. 12.5 Example of high-quality illumination using methods for interactive photon mapping (Jönsson et al. 2012), to account for multiple scattering of light in volume raycasting of a CT-scan of a human head, as seen from its back. Data is courtesy of CMIV, Linköping University

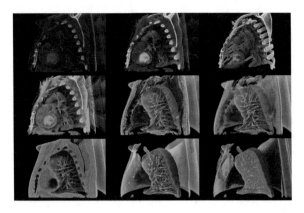

Fig. 12.6 Image from Siemens' Cinematic VRT feature in Syngo via that uses a Monte Carlo Path Tracing approach. Rendering and dataset courtesy of Anders Persson, CMIV

12.3.2 Transfer Function

In the context of volume rendering, a transfer function (TF) serves two purposes: to classify samples of the volume into materials or tissues, or in more generic terms, features; and second to map those features into visual appearances, at least a color and optical density (opacity). An illustration of these two aspects of the TF is shown in Fig. 12.7. In most cases the classification of features in the data is associated with a probability, or likelihood, of being a specific feature of interest. Often there are many possible interpretations of which feature is represented by the sampled data in the volume. This is clearly exhibited in CT data where many soft tissue organs have overlapping regions in the Hounsfield scalar domain. In one of the earliest publications on volume rendering by Drebin et al. (1988) the TF is defined as a dual operation of material presence and material appearance. Nevertheless, in much of the literature and practical implementations of volume rendering both of these

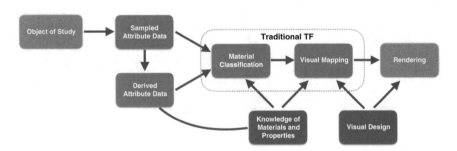

Fig. 12.7 The context of the traditional TF in volume rendering. For visualization tasks having the support for interactive updates to image generation and updates to the TF are essential

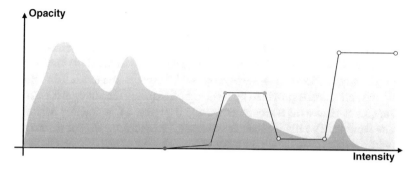

Fig. 12.8 A standard user interface for 1D TF settings. Color and opacity are assigned to data ranges using piecewise linear TF widgets. The background represents the histogram of binned scalar attribute data

operators are combined into a simple lookup table where probabilities and opacities are inseparable, as can be seen illustrated in Fig. 12.8.

We find in the literature that explicitly and separately consider material probabilities and the visual appearance mapping to be superior over the simplified inseparable approach, see, for instance, (Kniss et al. 2005; Lundström et al. 2007). In many practical applications adjusting the TF settings for each object of study is time-consuming and a complex task that most professional users do not have the time and experience to deal with. Development of automated methods that would apply templates of TF settings is still an active research topic and severely challenged by the fact that much data, outside of CT data, have inconsistent value domains. For example, MRI volumes and confocal microscopy images expose gradients and variations of the data throughout the volume domain that makes such automation highly problematic. In recent years attempts have been made to utilize Machine Learning techniques to assist with the classification of features in the data (Soundararajan and Schultz 2015). A comprehensive overview of the state-of-the-art of TFs in direct volume rendering is provided in Ljung et al. (2016).

12.4 Touch Interaction with Patient-Specific Volumetric Data

The applied interaction techniques for volumetric data need to be adapted depending on the target usage scenario. In this section we present two specific scenarios, medical education, and science communication. While these two scenarios are of quite different nature, they both need similar approaches when it comes to interaction, and there are indeed experiences, in terms of usability and user experiences, that can cross-fertilize between the two. Most notably, they both benefit from the collaborative social context arising from the ability to gather around a large multi-touch

surface, resembling a tabletop with objects of interest displayed. Since the participants all have different positions and orientation around the table, the effective use
of stereoscopic projection and viewing is restricted. It could be beneficial for a single
user but, the best results would require head tracking due to the short view distance.
An informative overview of the use of large-scale displays and touch technology for
scientific visualization is provided by Isenberg (2016), and the state-of-the-art report
by Besançon et al. provides a comprehensive overview of interaction schemes for
data visualization. The multi-touch functionality provides means for directly interacting with the data, introducing the sense of touching the object. However, the ability
to interact with many degrees of freedom can result in a steep learning curve and
there is, therefore, a need for simplified, and restricted, interaction schemes, which
is further discussed and analyzed below in the context of use cases. We first present
the case of using multi-touch surfaces for medical education followed by science
communication.

12.4.1   Case: Visualization Table for Medical Education

The medical education domain has seen a particularly wide adoption of touch-guided
3D visualizations. In this section we describe solution design and usage experiences
in this application area. Two widespread commercial products centered around volume visualization for medical education are the tables from Sectra (Ynnerman et al.
2015; Martín et al. 2018), both seen in Fig. 12.9. To aid in the creation of a sense
of realism, akin to having an actual patient or cadaver on a table, a large screen is
instrumental. This is embraced by both products, offering 86-inch screens in their
largest respective version. There are also versions where a screen is mounted on a
stand tiltable between horizontal and vertical positions.

As an exemplar case of solution design, we will next focus on the conclusions from
the design effort preceding the Sectra table, which we here refer to as the medical
visualization table (MVT).

The design process for the MVT consisted of recurring feedback sessions with
targeted end users in the form of interviews and prototype interaction at public exhi-
bitions. In their final form, the following list of design requirements was established
(Lundström et al. 2011).

R1 The learning threshold for basic interaction should be close to none.
R2 Advanced interaction should be available without disturbing the ease of use of
the basic interaction.
R3 The system should be alluring such that novice users are inspired to start inter-
acting.
R4 The users should experience similarity with real-world situations in patient care.
R5 The interaction should be highly responsive and robust.
R6 The interaction should be equally intuitive for all positions around the table.
R7 Swift but distinct changes of which person who is currently interacting should
be supported.

(a) The Sectra table for patient-specific medical visualization used in medical training. Image copyright Sectra AB, reprinted with permission.

(b) The Anatomage Virtual Dissection Table for education in medical imaging. Image copyright Custer *et al.* [40], Creative Commons Attribution License.

Fig. 12.9 Examples of two widespread commercial medical visualization products utilizing touch interaction for medical education

R8 The system should work in versatile environments without recurring maintenance efforts.

One of the key principles is that when in data exploration mode, the GUI is highly dominated by the visualization itself. This conveys simplicity to the novice user, and an alluring focus on the content at hand. For both effects, the touch control is a key component, and also for the ability to use the table from any position in a group setting. The basic volume navigation is operated through simple gestures in a six degree-of-freedom design. Apart from standard two-touch pinching for rotate-scale-translate movements, rotation around x- and y-axes is achieved by a single-touch movement.

A selected set of additional actions is reached through alternator objects. Since they are movable, obscuration is not a problem, and usable from any position around the screen. One alternator activates a transfer function (TF) preset gallery, and another enables TF adjustment. Clipping is controlled through the third alternator, in combination with a one-touch scribble to define the extent, a (rough) line for a clip plane, or a (rough) circle for a sphere.

The MVT functionality set has evolved to enrich the support for pedagogical use (Ynnerman et al. 2018). Central information carriers are the labels, see Fig. 12.10, through which educators can enrich the image data. A spatial point of the anatomy is connected by an arrow to a text label, also having a 3D position. The more the arrow is perpendicular to the screen plane when rotating, the more of the text is faded to mark its reduced relevance in that projection. There is also functionality where labels can be fully or partially hidden for quiz purposes. Case creation, including TF presets, labeling, and case descriptions, is a crucial component in supporting MVT use. Another central usefulness aspect of the MVT-centered ecosystem is that the user community actively readily can share their educational cases with each other.

Many positive usage experiences of large-screen visualization systems for virtual dissections have been reported. From the educator's perspective, there are several

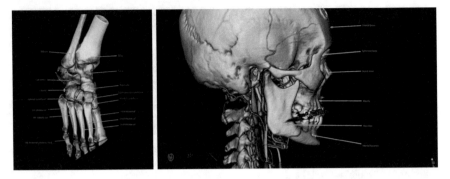

Fig. 12.10 Volumetric labeling is a key feature in medical education. Labels can be used in examination to document student performance and pose questions. Image copyright Sectra AB, reprinted with permission

types of benefits. Content-wise, it is seen as an advantage over cadavers to explore the anatomy of a living human (Darras et al. 2018). The organization of the curriculum becomes more flexible, as the educator is not restricted to the order necessary in a physical dissection, or to dissecting one system at a time (Darras et al. 2018; Munk et al. 2017).

From the student perspective, studies have shown distinctly improved learning results in some cases (Baratz et al. 2019; Mansoori et al. 2013), and no significant difference in others (Baratz et al. 2019; Mahindra and Singel 2014). An effect that is highly consistent, however, is that students report a positive subjective experience of increased learning attributed to the 3D visualization systems (Baratz et al. 2019; Mahindra and Singel 2014; Darras et al. 2019; Kazoka and Pilmane 2019).

12.4.2 Case: Visualization Table for Science Communication

Capitalizing on many of the experiences drawn from medical visualization on touch surfaces for medical professionals and students it is possible to create engaging installations for public venues targeting laymen, and still make use of the same data and underpinning methods. Simplifying the interaction and user interface is, however, key in a setting where a visitor, without prior experience of visualization tables or volume rendering, should be able to walk up to an installation and immediately start interacting. However, it should not be simplified to the point at which the exploratory experience is lost. Striking this balance is one of the main challenges that need to be met in science communication using interactive visualization. Jönsson et al. (2016) summarized a set of requirements for interactive installations at science centers and museums (Fig. 12.11):

R1 A new user should be able to interact with little additional meta information and/or instruction.

Fig. 12.11 The Inside Explorer Table by Interspectral AB used at science centers and museums to reveal the inside of human cadavers as well as animals. Image copyright Interspectral AB, reprinted with permission

R2 An intriguing design should attract the user to engage with the application.
R3 Interactions should give feedback such that the user quickly can understand the effect.
R4 The application should be responsive for smooth interaction.
R5 The installation should be engaging and lead to deep involvement.
R6 Users should be able to explore data beyond presets.

Here we present how the most important components in volume rendering can be exposed and how they can be restricted to allow visitors to immediately start interacting and exploring subjects and artifacts.

View point interaction design that matches the visual representations provides the starting point. In particular, the selection of view point/camera position must be intuitive and provide optimal views of selected data. One finger controls rotation around the focus point while two fingers control zooming. Translation is generally also controlled using one finger, which is possible since it is usually combined with rotational restrictions. Arbitrary view points can cause the displayed object to disappear, so they must be restricted to always show the subject or artifact. The user should not be able to excessively zoom either, as it results in a loss of orientation. In many cases it is beneficial to limit the axis along which rotation can occur, which is determined by the shape of the subject or artifact. For example, the human body is highly elongated so the rotation should be restricted to the transverse plane of the body, while translation is only allowed along the elongated direction. While these restrictions prevent the visitor from seeing the bottom of the feet and top of the head it makes it easier to get started and prevents the user from getting lost in views that are hard to recover from using standard interaction. On the contrary, the roughly spherically shaped brain does not need rotation constraints, but here the focus point can be fixed to its center and thus prevent translation. These concepts are illustrated in Fig. 12.12. Shape-dependent constraints for view point interaction are therefore key to providing instant interaction in interactive science communication.

The transfer function provides the means for highlighting different aspects of the artifact and is therefore one of the most central tools in the exploration of the volumetric artifact. Over the last decades there have been many transfer function methods and accompanying interaction techniques presented, see Sect. 12.3.2. However, most of these methods have been designed for expert users and are too complicated for laymen interaction at public venues. Instead, a set of predefined transfer functions can be designed by an expert (Ynnerman et al. 2015). The predefined transfer functions are then exposed through buttons or sliders, which allow the visitor to essentially peel off layers of skin or bring out contrast enhanced blood vessels through touch interaction. Multiple visitors can then collaborate in the exploration process by utilizing multi-touch interaction, where one can control the view point and another the transfer function. Experiments removing the restriction of predefined transfer functions have been conducted and it has been shown that the complexity of standard interfaces can potentially be reduced through the use of design galleries, where thumbnail images represent TF settings and guide the user in the exploration process (Jönsson et al. 2016; Marks et al. 1997).

Clipping is the concept of cutting away parts of the volume to reveal its interior. Medical applications often allow users to apply arbitrary cut planes and, for ultrasound visualization, it can even be possible for the user to use free hand to draw custom areas to remove. In the work described by Ynnerman et al. (2015) these types of interactions have been simplified into two ways of clipping. First, through

(a) Elongated shapes, such as the depicted body, are restricted to translation along the elongated axis and rotation around the transverse plane. Data is courtesy of CMIV, Linköping University.

(b) Full rotation can be allowed for spherically shaped objects while translation is disabled. Data is courtesy of Aubert-Broche et al. [48].

Fig. 12.12 Illustration of shape-dependent constrained camera interaction for decreasing the learning curve in science communication. Renderings have been made using techniques by Jönsson et al. (2012), Jönsson and Ynnerman (2017)

axis-aligned cut planes, which are exposed as movable scissors in the 3D scene. The axis-aligned cut planes provide means for clipping from each of the six sides of the artifact. Second is cutting aligned with the view direction, which is exposed through a slider in the user interface. View-aligned cutting, combined with moving the view point along the cutting direction, provides a sensation of moving into the artifact.

Unconstrained volume rendering provides powerful analysis capabilities, allowing for inspection from any view point and highlighting of any feature inside the volume. Our use case shows how these powerful features can be constrained in multi-touch visualization tables to allow instant interaction for science communication. The combination of these constrained concepts provides a good trade off between exploratory possibilities and ease of use.

12.5 Summary and Conclusions

This chapter provides an overview of the use of multi-touch surfaces to visualize data from CT-scanning. We have described how use of patient-specific visualization on touch tables has the potential to find many uses in the medical workflow in a range of medical sub-disciplines. These are, however, primarily found outside of the radiology department and in situations when intuitive 3D interaction can play an important role. Examples of this are surgical planning situations, and an early application was found in postmortem visualization in forensic investigations. However, the primary medical application is training and education and we have described how touch table interaction can be a valuable addition in educational contexts. We have also described how the captured data, visualization, and interaction techniques can be used in science communication, and how the explanatory aspect can be addressed in installations at science centers and museums. There is also a cross-fertilization between the medical use and the science communication scenarios based on the confluence of explanatory and exploratory visualization. We have also described the underlying concepts upon which medical touch table visualization is based with a starting point in state of the art in data capture and rendering. The work presented has resulted in two separate commercialization's. The medical market is addressed by Sectra AB targeting medical education setting. Full medical workstation capability is then provided on the table and a significant effort to curate data for education is a key factor in successful deployment. The other success case is based on the need for advanced visualization in science communication. Our spin-off company Interspectral AB is providing software and touch tables to science centers and museums worldwide. The key to success is here based on visual rendering quality, ease of interaction, and non-invasive storytelling of interesting subjects and objects, such as mummies and rare animals. A challenge for the future is to support hand-held devices and translate the positive experiences from large-scale multi-touch interfaces also to more generally available platforms.

References

Aubert-Broche N, Evans AC, Collins L (2006) A new improved version of the realistic digital brain phantom. NeuroImage 32(1):138–145

Burton JL, Underwood J (2007) Clinical, educational, and epidemiological value of autopsy. LANCET 369(9571):1471–1480

Comaniciu D, Georgescu B, Mansi T, Engel K (2016) Shaping the future through innovations: from medical imaging to precision medicine. Med Image Anal 33:19–26

Custer TM, Michael K (2015) The utilization of the Anatomage virtual dissection table in the education of imaging science students. J Tomogr Simul 1

Darras KE, Nicolaou S, Forster BB, Munk PL (2017) A golden opportunity for radiologists: bringing clinical relevance to undergraduate anatomy through virtual dissection. Can Assoc Radiol J 68(3):232–233

Darras KE, de Bruin ABH, Nicolaou S, Dahlström N, Persson A, van Merriënboer J, Forster BB (2018) Is there a superior simulator for human anatomy education? How virtual dissection can overcome the anatomic and pedagogic limitations of cadaveric dissection. Medical Teacher 40(7):752–753. PMID: 29569960

Darras KE, Forster BB, Spouge R, de Bruin ABH, Arnold A, Nicolaou S, Hu J, Hatala R, van Merriënboer J (2019) Virtual dissection with clinical radiology cases provides educational value to first year medical students. Acad Radiol

Díaz J, Vázquez P, Navazo I, Duguet F (2010) Real-time ambient occlusion and halos with summed area tables. Comput Graph 34:337–350

Drebin RA, Carpenter L, Hanrahan P (1988) Volume rendering. In: Computer Graphics (Proceedings of SIGGRAPH 1988), pp 65–74

Guy Baratz, Wilson-Delfosse Amy L, Singelyn Bryan M, Allan Kevin C, Rieth Gabrielle E, Rubina Ratnaparkhi, Jenks Brenden P, Caitlin Carlton, Freeman Barbara K, Susanne Wish-Baratz (2019) Evaluating the anatomage table compared to cadaveric dissection as a learning modality for gross anatomy. Med Sci Educat 29:499–506

Herman GT, Liu HK (1979) Three-dimensional display of human organs from computed tomograms. Comput Grap Image Process 9(1):1–21

Hernell F, Ljung P, Ynnerman A (2010) Local ambient occlusion in direct volume rendering. IEEE Trans Visualizat Comput Graph 16(4):548–559 July

Hernell F, Ljung P, Ynnerman A (2008) Interactive global light propagation in direct volume rendering using local piecewise integration. In: IEEE/EG Symposium on Volume and Point-Based Graphics, pp 105–112

Isenberg T (2016) Interactive exploration of three-dimensional scientific visualizations on large display surfaces. In: Anslow C, Campos P, Jorge J (eds) Collaboration Meets Interactive Spaces, chapter 6. Springer, Berlin/Heidelberg, pp 97–123

Isenberg P, Fisher D, Paul SA, Morris MR, Inkpen K, Czerwinski M (2012) Co-located collaborative visual analytics around a tabletop display. IEEE Trans Visualizat Comput Graph 18(5):689–702 May

Jönsson D (2016) Enhancing Salient features in volumetric data using illumination and transfer functions. PhD thesis, Linköping University

Jönsson D, Falk M, Ynnerman A (2016) Intuitive exploration of volumetric data using dynamic galleries. IEEE Trans Visual Comput Graph 22(1):896–905. https://doi.org/10.1109/TVCG.2015.2467294

Jönsson D, Kronander J, Ropinski T, Ynnerman A (2012) Historygrams: enabling interactive global illumination in direct volume rendering using photon mapping. IEEE Trans Visualizat Comput Graph 18(12):2364–2371

Jönsson D, Sundén E, Ynnerman A, Ropinski T (2014) A survey of volumetric illumination techniques for interactive volume rendering. Comput Graph Forum 33(1):27–51

Jönsson D, Ynnerman A (2017) Correlated photon mapping for interactive global illumination of time-varying volumetric data. IEEE Trans Visualizat Comput Graph 23(1):901–910

Kazoka D, Pilmane M (2019) 3D dissection tools in Anatomage supported interactive human anatomy teaching and learning. In: SHS Web of Conferences, vol. 68. EDP Sciences, p 02015

Kniss JM, Van Uitert R, Stephens A, Li GS, Tasdizen T, Hansen C (2005) Statistically quantitative volume visualization. In: VIS05, pp 287–294

Kroes T, Post FH, Botha CP (2012) Exposure render: an interactive photo-realistic volume rendering framework. PLOS One 7(7):1–10

Kronander J, Jönsson D, Löw J, Ljung P, Ynnerman A, Unger J (2012) Efficient visibility encoding for dynamic illumination in direct volume rendering. IEEE Trans Visualizat Comput Graph 18(3):447–462

Lafortune EP (1995) Mathematical models and monte carlo algorithms for physically based rendering. PhD thesis

Ljung P, Krüger J, Gröller E, Hadwiger M, Hansen C, Ynnerman A (2016) State of the art in transfer functions for direct volume rendering. Comput Graph Forum 35(3):669–691

Ljung P, Winskog C, Perssson A, Lundström C, Ynnerman A (2006) Full body virtual autopsies using a state-of-the-art volume rendering pipeline. IEEE Trans Visualizat Comput Graph (Proceedings Visualization/Information Visualization 2006) 12(5):869–876

Lundström C (2007) Efficient medical volume visualization an approach based on domain knowledge. PhD thesis, Linköping University

Lundström C, Ljung P, Persson A, Ynnerman A (2007) Uncertainty visualization in medical volume rendering using probabilistic animation. IEEE Trans Visual Comput Graph 13(6):1648–1655 November

Lundström C, Rydell T, Forsell C, Persson A, Ynnerman A (2011) Multi-touch table system for medical visualization: application to orthopedic surgery planning. IEEE Trans Visualizat Comput Grap 17(12):1775–1784

Lundström C, Persson A, Ross S, Ljung P, Lindholm S, Gyllensvärd F, Ynnerman A (2012) State-of-the-art of visualization in post-mortem imaging. APMIS 120: 316–26

Mahindra KA, Singel TC (2014) A comparative study of learning with "anatomage" virtual dissection table versus traditional dissection method in neuroanatomy. Ind J Clin Anat Physiol 4:177–180

Mansoori B, Seipel S, Wish-Baratz S, Herrmann KA, Gilkeson R (2013) Use of 3D visualization table as a learning tool for medical students. In: International Association of Medical Science Educators Conference, vol. 4

Marks J, Andalman B, Beardsley PA, Freeman W, Gibson S, Hodgins J, Kang T, Mirtich B, Pfister H, Ruml W, Ryall K, Seims J, Shieber S (1997) Design galleries: a general approach to setting parameters for computer graphics and animation. In: SIGGRAPH'97, pp 389–400

Martín JG, Mora CD, Henche SA (2018) Possibilities for the use of Anatomage (the anatomical real body-size table) for teaching and learning anatomy with the students. Biomed J 2:4

Max N (1995) Optical models for direct volume rendering. IEEE Trans Visualizat Comput Graph 1(2):99–108

Paladini G, Petkov K, Paulus J, Engel K (2015) Optimization techniques for cloud based interactive volumetric monte carlo path tracing. In: Eurographics/VGTC Conference on Visualization (EuroVis 2015) January 2015. Technical Talk

Preim B, Bartz D (2007) Visualization in medicine. Theory, Algorithms, and Applications. Series in Computer Graphics. Morgan Kaufmann

Preim B, Saalfeld P (2018) A survey of virtual human anatomy education systems. Comput Graph 71:132–153

Ropinski T, Meyer-Spradow J, Diepenbrock S, Mensmann J, Hinrichs Klaus H (2008) Interactive volume rendering with dynamic ambient occlusion and color bleeding. Comput Graph Forum 27(2):567–576

Soundararajan KP, Schultz T (2015) Learning probabilistic transfer functions: a comparative study of classifiers. Comput Graph Forum 34(3):111–120

Sundén E, Bock A, Jönsson D, Ynnerman A, Ropinski T (2014) Interaction techniques as a communication channel when presenting 3D visualizations. In: IEEE VIS International Workshop on 3DVis, pp 61–65

Volonte F, Robert JH, Ratib O, Triponez F (2011) A lung segmentectomy performed with 3D reconstruction images available on the operating table with an iPad. Inter Cardio Vascul Thor Sur 12(6):1066–1068

Wayne C (2013) Best of both worlds. museums, libraries, and archives in a digital age. Smithsonian Institution, Washington

Ynnerman A, Löwgren J, Tibell L (2018) Exploranation: a new science communication paradigm. IEEE Comput Graph Applicat 38(3):13–20

Ynnerman A, Rydell T, Antoine D, Hughes D, Persson A, Ljung P (2016) Interactive visualization of 3D scanned mummies at public venues. Comm ACM 59(12):72–81

Ynnerman A, Rydell T, Persson A, Ernvik A, Forsell C, Ljung P, Lundström C (2015) Multi-touch table system for medical visualization. In: Hege HC, Ropinski T (eds) EG 2015—Dirk Bartz Prize. The Eurographics Association

Yu Z, Leng S, Jorgensen SM, Li Z, Gutjahr R, Chen B, Halaweish AF, Kappler S, Yu L, Ritman EL, McCollough CH (2016) Evaluation of conventional imaging performance in a research whole-body CT system with a photon-counting detector array. Phys Med Biol 61(4):1572–1595

Chapter 13
Innovations in Microscopic Neurosurgery

Iype Cherian, Hira Burhan, Ibrahim E. Efe, Timothée Jacquesson, and Igor Lima Maldonado

Abstract Image-guidance has been the mainstay for most neurosurgical procedures to aid in accuracy and precision. Developments in visualization tools have brought into existence the current microscope and even sophisticated augmented reality devices providing a human–computer interface. The current microscope poses an ergonomic challenge particularly in scenarios like sitting position. Also, the cost associated with the present microscope hinders the accessibility of micro neurosurgery in most low-to-middle-income countries.

The Hyperscope is a modern concept of a hybrid visualization tool. This three-dimensional, ultra-high-definition camera mounted on a robotic arm is compatible with a Microsoft HoloLens. The Hyperscope is an endoscope-exoscope hybrid that can allow switching between the two using a foot control. The endoscopic image is projected alongside the exoscopic image on the same screen in ultra-high-definition rendering possible convenient and bimanual endoscope-assisted microneurosurgery. Incorporating neuronavigation and augmented reality to create a mixed image of the

I. Cherian (✉) · H. Burhan
Department of Neurosciences, Krishna Institute of Medical Sciences, "Deemed to Be University", Karad, Maharashtra, India
e-mail: drrajucherian@gmail.com

I. E. Efe
Charité - Universitätsmedizin Berlin, Corporate Member of Freie Universität Berlin, Humboldt-Universität Zu Berlin, and Berlin Institute of Health, Berlin, Germany

T. Jacquesson
Multidisciplinary Skull Base Unit, Department of Neurosurgery, Wertheimer Neurological Hospital, Lyon, France

Department of Anatomy, Rockefeller Site, University of Lyon 1, Lyon, France

Brain and Imaging Lab UMR1253, University Francois Rabelais, Tours, France

I. L. Maldonado
Inserm UMR 1253, IBrain, Université de Tours, Tours, France

CHU Tours, Tours, France

© The Author(s), under exclusive license to Springer Nature Switzerland AG 2021
J.-F. Uhl et al. (eds.), *Digital Anatomy*, Human–Computer Interaction Series,
https://doi.org/10.1007/978-3-030-61905-3_13

inputs obtained is what makes it a "hyper" scope. The ability to superimpose real over virtual images using set coordinates on the operative field can help to map the underlying structures and provide fine precision movements.

This chapter lays down the detailed concept of the Hyperscope based on current visualization tools.

13.1   The Neurosurgical Microscope from Past to Present

Neurosurgery is the medical discipline that focuses on the diagnosis and treatment of disorders affecting the brain, spinal cord, and peripheral nerves. Operative treatment often involves delicate dissection of neurovascular microstructures and thus requires the use of proper magnification and lighting (Yasargil and Krayenbuhl 1970; Uluc et al. 2009). The introduction of the surgical microscope into the neurosurgical operating theater triggered the transition towards safer and more precise neurosurgery. In the early twentieth century, many pioneers faced hardships trying to improve what had been a surgical discipline with limited therapeutic possibilities and poor clinical outcome (McFadden 2000). However, it was not until Yasargil and his group had first made use of a microscope to access skull base and cerebrovascular lesions in the 1960s that micro neurosurgery was born (Lovato et al. 2016). The earliest models were a set of lenses and prisms fixed to a simple platform and barely allowed for manipulation of position and angle. Together with Zeiss, Yasargil had designed a prototype that enabled movement of the microscope head along six axes, fitted with magnetic brakes and a mouth switch allowing for hands-free adjustments. Over the past 50 years, microscopes have gone from being simple and small tools to complex and large pieces of machinery, the present top-end models costing more than a million US dollars (McFadden 2000). Image injection, navigability of the scope, exoscope function, indocyanine green (ICG), and ancillary techniques to improve the visualization of blood flow are only some of the advances which the modern microscope has incorporated (Langer et al. nd; Zhao et al. 2019). Other improvements include the self-balancing arm to support the optics and the eyepiece with a stable magnetic lock, as well as auto zoom and auto focus features (Roehl and Buzug 2009).

Yet, many disadvantages remain today. Through the operating microscope, the surgeon cannot see around corners. At the same time, optics are primarily lens and prism based requiring a large arm to balance and stabilize the heavy counterweight. Viewing the surgical field from different angles therefore necessitates frequent two-handed repositioning of the scope. For this, the surgeon must repeatedly interrupt his or her surgical maneuvers. Besides, when using a traditional binocular microscope, the surgeon has to look through an eyepiece which commonly provides only little adjustability. This causes both operator and assistants to often work in uncomfortable positions, resulting in fatigue and lack of concentration over time. Further, neurosurgery depends on thorough preoperative planning and the use of intraoperative navigation. Current navigation technologies, however, only provide 2D images that are projected on a separate screen. Having to constantly switch from the 3D surgical field to a 2D monitor and back perturbs the surgeon's workflow and may increase the likelihood of surgical inaccuracy. Hence, the need for more ergonomic solutions capable of integrating multiple functions in one device has to be addressed.

Here, we provide an overview of recent advances in microscopic neurosurgery with a focus on augmented reality and how it can help the surgeon navigate through the patient's anatomy. We further propose a novel concept for a neurosurgical digital visualization tool fitted with exoscope and endoscope functions and integrated navigation to allow for safe, efficient, and convenient high precision microsurgery.

13.2 Robotic Applications for Neurosurgical Microscopes

Human–robot collaboration has emerged as a hot topic in health care. Hence, the industry has been responding to the needs and stringent standards of the medical sector with a rapidly expanding range of robot-based assistance technologies (Bauer et al. 2008). Initial efforts to robotize operating microscopes, however, date back to 1993 when Zeiss introduced its MKM robotic arm which could hold a microscope head, though with limited working radius (Willems et al. 2003). In the years to follow, Giorgi and colleagues mated a microscope with a robotic arm used for industrial purposes and proposed a mouth-operated joystick to direct the microscope without having to interrupt the surgical workflow (Giorgi et al. 1995; Giorgi et al. 2000). Other solutions include footpad switches, voice command, and a remote control fixed directly to a surgical instrument (Wapler et al. 1999; Kantelhardt et al. 2013). The latter was implemented by Kantelhardt et al., who assumed that hand-operated control would feel more intuitive and thus gain wider acceptance within the neurosurgical community. They fitted a conventional operating microscope with motors.

The manual control mode still intact, their model could be positioned with the aid of a joystick mounted on the aspirator or a similar instrument. The microscope was also capable of self- navigating through a three-dimensional coordinate system to reach a predefined target which it would then focus on automatically. An integrated optical coherence tomography camera allowed for three-dimensional surface scanning at micrometer-resolution (Kantelhardt et al. 2013).

Another robotic approach to next-generation microneurosurgery is the ROVOT-m. Gonen et al. have recently shared their initial experience with this robot-operated telescopic video microscope. The ROVOT-m provides an exoscopic view of the operative field. Unlike common exoscope models, it is not fixed to a static platform but instead coupled to a robotic arm. In this case, an integrated neuronavigation system cannot only track surgical instruments introduced to the operating field but also the payload held by the robot. Using a computer–machine interface, the ROVOT-m can reposition itself with no need for manual adjustment. Yet, the authors advise potential adopters of this technology to anticipate a steep learning curve (Gonen et al. 2017).

It is estimated that up to 40% of the operative time is spent on modulating micro-scope position, angle, and focus (Yaşargil 1984). One of the latest commercially avail-able robotic visualization platforms to address this issue is the ZEISS KINEVO 900, featuring position memory and point lock functions. Regardless of the pathology, the surgeon must frequently switch back and forth between different perspectives during the procedure. Position memory allows the surgeon to save certain microscope trajec-tories that he or she can go back to at a later stage of the surgery. If, for instance, a tumor requires the surgeon to look around it or approach a different portion of the lesion, the current view can be bookmarked and precisely recalled at the touch of a button. Point lock refers to the system's ability to fix a target point while the surgeon moves the microscope. In the point lock mode, the microscope, therefore, moves along a circular path around the target, allowing a view from different angles at constant focus. Both functions can help reducing the duration of the surgery as the operator would not have to spend time rediscovering structures and trajectories that he or she had already identified. This may be particularly useful in the case of a complication where the surgeon could promptly return to a previous more suitable view of the surgical field instead of having to manually explore it again (Bohl et al. 2016; Oppenlander et al. nd; Belykh et al. 2018).

13.3   High-Definition Exoscopes as Emerging Alternative

The conventional binocular microscope may require the surgeon to operate in non-ergonomic and fatiguing postures. In most models, the three-dimensional view is restricted to one eyepiece only with the surgeon's assistant having to follow the operation with 2D vision. Commonly massive in size, their storage, sterile covers, and maintenance create high additional expenses. This drove the invention of innovative technologies like the endoscope and later, the exoscope.

Endoscopy offers a wider visualization of the surgical field as well as the possi-bility to see around corners. Being a slim tubular instrument, surgeries can be performed in a less invasive manner. Especially in neuroendoscopy, however, surgical corridors are generally narrow and maneuvering instruments can be particularly challenging. Combining the advantages of the microscope and the endoscope, the exoscope was introduced to provide high magnification, wide focal distance, and yet allow the surgeon to conveniently operate using both hands. Smaller and lighter than a conventional microscope, the exoscope does not conflict with the surgical instruments. It is easily transportable and can be sterilized and fixed to a holder, newer models of which can be moved freehand using a foot pedal (Mamelak et al. 2010; Iyer and Chaichana 2018; Oertel and Burkhardt 2017; Rossini et al. 2017; Sack et al. 2018; Mamelak et al. 2008; Ricciardi et al. 2009). The surgical field is projected on a monitor, eliminating the need for the surgeon to frequently change his or her posture. Several studies have identified the resulting convenience as the best advantage of using an exoscope (Oertel and Burkhardt 2017; Rossini et al. 2017; Sack et al. 2018; Mamelak et al. 2008). Further, the monitor allows the whole staff in the Operating Room to appreciate the procedure from the operator's point of view, potentially facilitating teamwork and training. Recently, 3D 4 K high-definition visualization has been implemented, as well (De Virgilio et al. 2020; Kwan et al. 2019; Khalessi et al. 2019; Izumo et al. 2019).

Lastly, most exoscopes are more affordable than conventional microscopes with regard to purchase and maintenance which could possibly impact neurosurgery in the developing world.

According to a study published in 2015, approximately 5 billion people have no access to basic surgical care worldwide (Meara et al. 2015). Advanced microneurosurgical service within a range of 2 h is available to only 13% of the population in sub-Saharan Africa, 27% in Latin America and the Caribbean, 9% in East Asia and the Pacific, and 28% in South Asia (Punchak et al. 2018). Most life-saving procedures in neurosurgery depend on microneurosurgical techniques. Due to their high price, however, access to operating microscopes remains limited to only a few centers in many low-and middle-income countries, resulting in a significant unmet need for low-cost alternatives. The exoscope, therefore, has the potential to narrow the gap between the standard of neurosurgical care between low- and high-income countries.

13.4 Augmented Reality in Neurosurgery

Augmented Reality (AR) refers to a technology that superimposes computer-generated objects on the physical world, generating a semi-immersive environment. Injecting virtual content into the field of view, AR adds information and enhances the perception of the real scene (Shuhaiber 2004). The first implementations date back to the 1940s when radar information was displayed on head-up displays in aircraft. Neurosurgery, strongly depending on pre- and intraoperative imaging, was one of the first specialties to explore the potential use of AR in health care (Roberts et al. 1989; Mitha et al. 2013). Currently, neuronavigation systems show two-dimensional images, e.g., computed tomography or magnetic resonance imaging, on a monitor. This requires the surgeon to not only mentally transform a 2D image into a 3D object but also to project the object onto the patient's anatomy which can be a cumbersome and error-prone task (Pandya et al. 2005). Though still in its infancy, AR holds the promise to visualize the exact location of the pathology directly within the surgical field. In this way, the neurovascular structures or white matter tracts that surround the lesion and warrant careful dissection can be projected onto the surgical field.

Augmented reality for image-guidance comprises three elements, the first being the virtual overlay to be injected into the surgeon's view. Typically, this overlay is a computer-generated three-dimensional reconstruction of narrow CT or MRI cross-sections. They can be color- and texture-coded or displayed with varying transparency to improve depth perception. Secondly, in order for virtual and real objects to perfectly match, both environments need to be synchronized and the proper alignment needs to be frequently verified during the course of the procedure. Point- matching is a method where the surgeon calibrates the scenes with the help of stationary anatomical landmarks such as bony structures (Cabrilo et al. 2015). The third component is the visual output. The virtual scene can be overlaid onto a video feed of the real scene on an external monitor (video see-through), or directly injected into the oculars of an operating microscope (optical see-through) (Sielhorst et al. 2008). Alternatively, mixed reality smart glasses combining a miniature computer, a projector, and a prism screen in a pair of glasses can be used. The Google Glass along with the Project Tango software (Google Inc.) have been tested for safety and feasibility in spine surgery (Elmi-Terander et al. 1976; Golab et al. 2016; Yoon et al. 2017). Through three-dimensional visuospatial scanning of the surgical field, this technology projects the pathology, the neighboring vital anatomy, and the according trajectory directly to the surgeon's glasses. It also tracks the user's head position to keep virtual and real scene congruent even while moving the head. The HoloLens 2 (Microsoft Inc.) was recently developed to provide a more intuitive user-interface and comfortable experience as well as a brighter display with a wider field of view. Also, the computer guts are located in the back of the HoloLens allowing for the center of gravity to align with the center of the head. A camera authenticates the user via retinal scan to adjust the image according to pupillary distance.

To date, most AR applications in neurosurgery have been dedicated to simulation and training as they offer the possibility to create virtual environments where the user can build and refine skills at no risk (Malone et al. 2010). The Immersive Touch (the University of Illinois at Chicago) is an augmented reality platform for simulation of procedures such as aneurysm clipping and percutaneous pedicle screw placement with haptic feedback (Luciano et al. 2013). Elmi-Terander and colleagues recently reported on the first in-human prospective cohort study to assess the potential of AR navigation in pedicle screw insertion into the thoracic spine. An accuracy of over 94% was achieved. According to the authors,

AR holds the promise of minimizing the rate of complications and revision surgeries (Elmi-Terander et al. 2019). Abe et al. made use of an augmented reality system with a video see-through head-mounted display for percutaneous vertebroplasties on five patients. In this minimally invasive procedure, bone cement is injected into fractured vertebrae with the aim to stabilize and support the spine. To determine the exact trajectory, the surgeon makes extensive use of C-arm fluoroscopy. The AR system, however, can visualize the trajectory for the vertebroplasty needle, guiding the operator to choose the entry point as well as angle, potentially eliminating the need for radiation exposure. The authors reported on image alignment problems and thus could not afford to fully cast aside fluoroscopy (Elmi-Terander et al. 2019). A similar observation was made by other groups who were unable to accurately determine the trajectory with AR alone (Fritz J and Iordachita 2014; Wu et al. 2014) Nevertheless, the group found AR-aided percutaneous vertebroplasty to be superior to conventional surgery (Abe et al. 2013).

In skull base and tumor surgery, overlay of planned resection margins and surrounding vascular anatomy can facilitate maximal and safe resection. Superimposition of three-dimensional CT and MRI data reduced intensive care unit and hospital stay by 40 to 50% in a series of 74 patients (Lovato et al. 2016). The AR display of superficial and deep venous structures was shown to be of particular assistance in the excision of parafalcine and parasagittal meningiomas (Low et al. 2010). Mascitelli et al. integrated the Brainlab navigation software with the Zeiss Pentero 900 operating microscope for an optical see-through approach to AR-aided surgery. A total of 84 neoplastic and vascular pathologies were operated on. In aneurysm surgery, AR was used to project virtual bony anatomy onto the real skin, vessels onto the cranium, and finally the aneurysm on to the arachnoid to guide a precise dissection with minimal skin incision, craniotomy, and dural exposure. Similarly, in extra- to intracranial bypass surgery, the superficial temporal artery, being the donor artery, could be visualized on the skin prior to the initial incision. Yet, the authors called attention to the potential risk of false reliance on the technology. Inaccurate navigation, faulty reconstruction of virtual objects as well as brain shift may render the overlaid information useless and at worst harmful. Especially, low-grade gliomas appearing like physiological brain or deep-seated intra-axial tumors with no anatomical landmarks are prone to such inaccuracy. On the other hand, these lesions are most likely to benefit the most from the increasing precision of the AR technologies currently under development (Mascitelli et al. 2018). Cabrilo and colleagues predict future software to address and correct inaccuracy through automatic mismatch recognition. Further, flat-panel detectors that are used to evaluate blood flow may soon be integrated into AR systems to visualize the flow dynamics of highly vascularized lesions intraoperatively (Cabrilo et al. 2015).

13.5 *Hyperscope*–Conceptualizing a Modern Visualization Tool for Neurosurgeons Integrating Exoscope, Endoscope, and Navigation in One System

Leveraging the recent advances, we propose the *Hyperscope*—a novel concept and a direction for innovation to the current visualization tools in microneurosurgery and a direction to take for future surgical visualization endeavors. The cost of the present scopes and the sheer mass as well as the limited ergonomics limit the surgeons to a huge extent, especially in scenarios like the sitting position. The Hyperscope also proposes a mixed reality environment where the actual surgical field and the virtual reality image from the scans are superimposed using principles of navigation.

The Hyperscope is primarily a concept of a 3D ultra-high-definition visualization tool integrating the benefits of the exoscope, endoscope, and neuronavigation in one system. The *Hyperscope* would overcome the weaknesses of the conventional operating microscope and navigation and would help push the boundaries of digital microscopy and augmented reality technologies in neurosurgery.

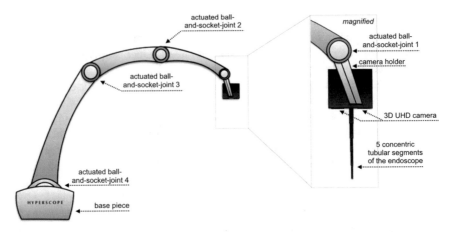

Fig. 13.1 A simplified illustration of the *Hyperscope*. The robotic arm holds the 3D ultra-high-definition camera. Ball-and-socket-joints allow multidirectional movement. An endoscope is integrated into the camera body and can be pulled out and activated on demand

The model incorporates three key innovations detailed below.

- **Robot-Aided holding arm for 3D Ultra-High-Definition camera**

Based on human–robot collaboration, we envisioned a digital visualization tool integrated into a self-balanced lightweight robotic arm (Fig. 13.1). Four ball-and-socket joints subdivide the arm into three sections and enable multidirectional movement and rotation up to 15 degrees of freedom. The proximal section communicates with the base piece whereas the distal section is linked to a camera holder with an overall payload of 4 kg. The *Hyperscope* is proprioceptive, meaning it can precisely approach any position in an XYZ coordinate system from any angle in an automated fashion. For operative convenience and patient safety, the surgeon can always over ride and manually adjust the position of the exoscope. This is ensured by back drivable actuators allowing motion with low friction and force. The *Hyperscope* arm is also fitted with pressure sensors that immediately stop movement upon contact as an additional safety measure. For free-handed robotic alignment, the system relies on accelerometer-based controlling. The surgeon wears an electronic wristband that communicates directly with the robotic arm. The band can be easily activated via a foot pedal. Upon activation, the surgeon's forearm motion is registered and then translated into the movement of the robotic arm. Would the surgeon's arm move in the horizontal or vertical axis, so would the exoscope in the XYZ axes. Similarly, rotating or tilting the forearm would manipulate the angle with no movement in the XYZ direction. The zoom can also be adjusted in this way. For additional lighting, instruments tips such as suction and bipolar may be provided with lighted tips with fiber optic technology.

A digital microscope detects an image via a camera and projects it on a screen in real time with no need for an eyepiece. At its distal end, the *Hyperscope*'s robotic arm holds a 3D ultra-high- definition camera. The high resolution and stereoscopic image of the operating field can be viewed using either a 3D ultra-high-definition monitor or the Microsoft HoloLens 2 as visual output.

- **Exoscope-Endoscope Hybridization**

The endoscope allows minimally invasive surgery on deep-seated pathologies such as tumors of the skull base or ventricles. It can be inserted via a small incision or natural opening like a nostril to minimize brain retraction and visualize regions otherwise blind to a microscope or exoscope. However, most endoscopes are purchased, set up, and used as stand-alone devices. We feel it desirable for neurosurgeons to be able to seamlessly switch back and forth between microscopic and endoscopic perspectives depending on the intraoperative situation. Instead, many procedures rely solely on either an endoscopic or a microscopic open approach. Recent efforts, therefore, aim at integrating the endoscope into the operating microscope. We believe that the hybridization of the two technologies would further encourage surgeons that are not familiar with neuroendoscopy to explore the benefits of endoscope-assisted microneurosurgery.

ZEISS introduced the QEVO Micro-Inspection Tool, a modern endoscope that can be plugged into the ZEISS KINEVO 900 digital operating microscope whenever endoscopic vision is needed. Pressing a button on the microscope's handgrip, the surgeon can easily activate the QEVO which then projects its view on a monitor alongside the exoscopic view of the KINEVO 900. Placing the device back to its tray will switch it off automatically. It also allows the operator to adjust the brightness and rotate the endoscopic image directly via the microscope's handles. Despite the ease of use, it is a hand-held device with no working channel. While using the QEVO, the surgeon is therefore restricted to performing single-handed maneuvers only (Schebesch et al. 2019). Endoscope holders were developed to enable bimanual procedures. The first robotic endoscope holder recently underwent preclinical testing on training models. Unique features included foot pedal control and a position memory function similar to that of the KINEVO 900 platform. The users perceived the robot assisted and thus bimanual technique to be faster and less stressful (Zappa et al. 2019).

While the QEVO is a separate plug-in device, the *Hyperscope* is the first concept to fully integrate an endoscope into the exoscope device. Its endoscope consists of five concentric tubular segments that can be slid over one another. When not activated, all five sections are hidden inside the exoscope. When switching from exoscope to endoscope vision via a pad on the foot pedal, the surgeon can manually extend the endoscope into the cavity where it is rigidly fixed. The depth of the endoscope can be individually tailored based on what length each section is slid out of the outer one. The endoscopic image is projected alongside the exoscopic image on the same screen in ultra-high-definition rendering possible convenient and bimanual endoscope-assisted microneurosurgery.

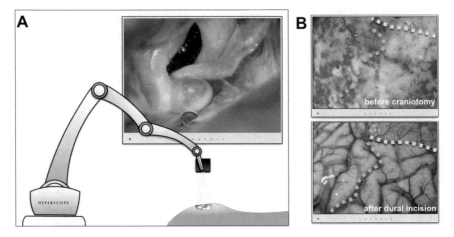

Fig. 13.2 The *Hyperscope* incorporates an Augmented-Reality-based neuronavigation feature. Using preoperative MRI scans, anatomical structures are 3D reconstructed. During operation (**A**), virtual objects are superimposed on the corresponding anatomy on the 3D screen. An internal carotid bifurcation aneurysm (green), the internal carotid artery with the anterior cerebral artery (blue), and the optic nerve (yellow) can be seen through this corridor. Projecting virtual landmarks onto yet unexposed real correlates (**B**), the *Hyperscope* can guide skin flap, craniotomy, and dural incision. The top screen shows the course of the central sulcus (blue) and the lateral sulcus (yellow) before dural opening—allowing the surgeon to know where to open the dura. The bottom screen is after the dural opening shows how virtually mapped sulci perfectly align with the real image after dural incisions on the mapped pathways

- **Real-Virtual Image Fusion**

The *Hyperscope* can guide the surgeon through augmented reality-based neuronavigation (Fig. 13.2). The real image of the surgical field and the high-definition virtual image generated from thin preoperative MRI and CT scans can be fused and displayed on the same screen. Both real and virtual scenes are visualized in 3D providing a stereoscopic see-through view of the patient in real time. As mentioned above, the robotic arm moves within an XYZ coordinate system. Through point-matching real anatomy and virtual anatomy, the patient is registered within its 3D grid. Extensive tissue resection or loss of cerebrospinal fluid may cause the brain and hence the pathology and surrounding anatomy to move in relation to the preoperative scans. Therefore, the real and the virtual scene should be regularly refreshed at a frequency of at least 120 Hz, for correct alignment during the operation.

Matching not only stationary landmarks but also surface structures such as blood vessels or gyri and sulci would allow the software to calculate brain shift and correct the overlaid information accordingly. An intraoperative CT or ultrasound may be used to further assist this.

Not having to frequently switch between the operating field and a navigation display may result in a higher level of surgeon's comfort and an unimpeded workflow. Further, being able to visualize the course of a blood vessel running below the skin or behind a bony structure without exposing it could help minimize the extent and duration of the procedure as well as complication rate.

13.6   Conclusion

The present exoscopes are a good progress in enabling a better visualization in microneurosurgery. However, the integration of navigation to the current exoscope would fulfill the ergonomic and the economical setbacks of the present tools. Having said that, the concept of the Hyperscope presented in this chapter is a novel idea that amalgamates the exoscope, endoscope, and navigation to lay down a direction for future scopes. This concept is still in its planning phase and steps need to be taken to produce the prototype for further testing and evaluation.

References

Abe Y, Sato S, Kato K, Hyakumachi T, Yanagibashi Y, Ito M, Abumi K (2013) A novel 3D guidance system using augmented reality for percutaneous vertebroplasty: technical note. J Neurosurg Spine. 19:492–501. https://doi.org/10.3171/2013.7.spine12917

Bauer A, Wollherr D, Buss M (2008) Human–robot collaboration: a survey. Int J Humanoid Robot 05:47–66. https://doi.org/10.1142/s0219843608001303

Belykh EG, Zhao X, Cavallo C, Bohl MA, Yagmurlu K, Aklinski JL, Byvaltsev VA, SanaiN, Spetzler RF, Lawton MT, Nakaji P, Preul MC (208) Laboratory evaluation of a robotic operative microscope - visualization platform for neurosurgery. Cureus 10:e3072. https://doi.org/10.7759/cureus.3072

Bohl MA, Oppenlander ME, Spetzler R (2016) A prospective cohort evaluation of a robotic, auto-navigating operating microscope. Cureus. 8:e662. https://doi.org/10.7759/cureus.662

Cabrilo I, Schaller K, Bijlenga P (2015) Augmented reality-assisted bypass surgery: embracing minimal invasiveness. World Neurosurg 83:596–602. https://doi.org/10.1016/j.wneu.2014.12.020

De Virgilio A, Mercante G, Gaino F, Yiu P, Mondello T, Malvezzi L, Colombo G, Pellini R, Spriano G (2020) Preliminary clinical experience with the 4 K3-dimensional microvideoscope (VITOM 3D) system for free flap head and neck reconstruction. Head Neck 42:138–140. https://doi.org/10.1002/hed.25979

Elmi-Terander A, Skulason H, Soderman M, Racadio J, Homan R, Babic D, van der Vaart N, Nachabe R (1976) Surgical navigation technology based on augmented reality and integrated 3D intraoperative imaging: a spine cadaveric feasibility and accuracy study. Spine (Phila Pa 1976) 41:E1303–E1311. https://doi.org/10.1097/brs.0000000000001830

Elmi-Terander A, Burstrom G, Nachabe R, Skulason H, Pedersen K, Fagerlund M, Stahl F, Charalampidis A, Soderman M, Holmin S, Babic D, Jenniskens I, Edstrom E, Gerdhem P (2019) Pedicle screw placement using augmented reality surgical navigation with intraoperative 3D imaging: a first in-human prospective cohort study. Spine (Phila Pa 1976) 44:517–525. https://doi.org/10.1097/brs.0000000000002876

Fritz J, P UT, Ungi T, Flammang AJ, Kathuria S, Fichtinger G, Iordachita, II, Carrino JA (2014) MR-guided vertebroplasty with augmented reality image overlay navigation. Cardiovasc Intervent Radiol 37:1589–1596. https://doi.org/10.1007/s00270-014-0885-2

Gildenberg PL, Labuz J (2006) Use of a volumetric target for image-guided surgery. Neurosurgery 59:651–659; discussion 651–659. https://doi.org/10.1227/01.neu.0000227474.21048.f1

Giorgi C, Eisenberg H, Costi G, Gallo E, Garibotto G, Casolino DS (1995) Robot-assisted microscope for neurosurgery. J Image Guid Surg. 1:158–163. https://pubmed.ncbi.nlm.nih.gov/907 9441/. Accessed May 2021

Giorgi C, Sala R, Riva D, Cossu A, Eisenberg H (2000) Robotics in child neurosurgery. Childs Nerv Syst 16:832–834. https://doi.org/10.1007/s003810000394

Golab MR, Breedon PJ, Vloeberghs M (2016) A wearable headset for monitoring electromyography responses within spinal surgery. Eur Spine J 25:3214–3219. https://doi.org/10.1007/s00586-016-4626-x

Gonen L, Chakravarthi SS, Monroy-Sosa A, Celix JM, Kojis N, Singh M, Jennings J, Fukui MB, Rovin RA, Kassam AB (2017) Initial experience with a robotically operated video optical telescopic-microscope in cranial neurosurgery: feasibility, safety, and clinical applications. Neurosurg Focus. 42:E9. https://doi.org/10.3171/2017.3.focus1712

Iyer R, Chaichana KL (2018) Minimally invasive resection of deep-seated high-grade gliomas using tubular retractors and Exoscopic visualization. J Neurol Surg A Cent Eur Neurosurg 79:330–336. https://doi.org/10.1055/s-0038-1641738

Izumo T, Ujifuku K, Baba S, Morofuji Y, Horie N, Matsuo T (2019) Initial Experience of ORBEYE surgical microscope for carotid endarterectomy. Asian J Neurosurg 14:839–842. https://doi.org/10.4103/ajns.AJNS_242_18

Kantelhardt SR, Finke M, Schweikard A, Giese A (2013) Evaluation of a completelyrobotized neurosurgical operating microscope. Neurosurgery 72(Suppl 1):19–26. https://doi.org/10.1227/NEU.0b013e31827235f8

Khalessi AA, Rahme R, Rennert RC, Borgas P, Steinberg JA, White TG, Santiago-Dieppa DR, Boockvar JA, Hatefi D, Pannell JS, Levy M, Langer DJ (2019) First-in-man clinical experience using a high-definition 3-dimensional exoscope system for microneurosurgery. Oper Neurosurg (Hagerstown) 16:717–725. https://doi.org/10.1093/ons/opy320

Kwan K, Schneider JR, Du V, Falting L, Boockvar JA, Oren J, Levine M, Langer DJ (2019) Lessons learned using a high-definition 3-dimensional exoscope for spinal surgery. Oper Neurosurg (Hagerstown) 16:619–625. https://doi.org/10.1093/ons/opy196

Langer DJ, White TG, Schulder M, Boockvar JA, Labib M, Lawton MT (2020) Advances in intraoperative optics: a brief review of current Exoscope Platforms. OperNeurosurg (Hagerstown). https://doi.org/10.1093/ons/opz276

Lovato RM, VitorinoAraujo JL, dePaulaGuirado VM, Veiga JC (2016) The Legacy of Yasargil: the father of modern neurosurgery. Indian J Surg 78:77–78. https://doi.org/10.1007/s12262-015-1421-6

Low D, Lee CK, Dip LL, Ng WH, Ang BT, Ng I (2010) Augmented realityneurosurgicalplanning and navigation for surgical excision of parasagittal, falcine and convexity meningiomas. Br J Neurosurg 24:69–74. https://doi.org/10.3109/02688690903506093

Luciano CJ, Banerjee PP, Sorenson JM, Foley KT, Ansari SA, Rizzi S, Germanwala AV, Kranzler L, Chittiboina P, Roitberg BZ (2013) Percutaneous spinal fixation simulation with virtual reality and haptics. Neurosurgery. 2013, 72 Suppl 1:89–96. https://doi.org/10.1227/NEU.0b013e3182750a8d

Malone HR, Syed ON, Downes MS, D'Ambrosio AL, Quest DO, Kaiser MG (2010) Simulationin neurosurgery: a review of computer-based simulation environments and their surgical applications. Neurosurgery 67:1105–1116. https://doi.org/10.1227/neu.0b013e3181ee46d0

Mamelak AN, Danielpour M, Black KL, Hagike M, Berci G (2008) A high-definition exoscope systemforneurosurgeryandothermicrosurgicaldisciplines:preliminaryreport. SurgInnov 15:38–46. https://doi.org/10.1177/1553350608315954

Mamelak AN, Nobuto T, Berci G (2010) Initial clinical experience with a high-definition exoscope system for microneurosurgery. Neurosurgery 67:476–483. https://doi.org/10.1227/01.NEU.0000372204.85227.BF

Mascitelli JR, Schlachter L, Chartrain AG, Oemke H, Gilligan J, Costa AB, Shrivastava RK, Bederson JB (2018) Navigation-linked heads-up display in intracranial surgery: early experience. Oper Neurosurg (Hagerstown) 15:184–193. https://doi.org/10.1093/ons/opx205

McFadden JT (2000) History of the operating microscope: from magnifying glass to microneurosurgery. Neurosurgery. 46:511. https://doi.org/10.1097/00006123-200002000-00059

Meara JG, Leather AJ, Hagander L, Alkire BC, Alonso N, Ameh EA, Bickler SW, Conteh L, Dare AJ, Davies J, Merisier ED, El-Halabi S, Farmer PE, Gawande A, Gillies R, Greenberg SL, Grimes CE, Gruen RL, Ismail EA, Kamara TB, Lavy C, Lundeg G, Mkandawire NC, Raykar NP, Riesel JN, Rodas E, Rose J, Roy N, Shrime MG, Sullivan R, Verguet S, Watters D, Weiser TG, Wilson IH, Yamey G, Yip W (2015) Global Surgery 2030: evidence and solutions for achieving health, welfare, and economic development. Lancet 386:569–624. https://doi.org/10.1016/S0140-6736(15)60160-X

Mitha AP, Almekhlafi MA, Janjua MJ, Albuquerque FC, McDougall CG (2013) Simulation and augmented reality in endovascular neurosurgery: lessons from aviation. Neurosurgery 72(Suppl 1):107–114. https://doi.org/10.1227/neu.0b013e31827981fd

Oertel JM, Burkhardt BW (2017) Vitom-3D for exoscopic neurosurgery: initial experience in cranial and spinal procedures. World Neurosurg 105:153–162. https://doi.org/10.1016/j.wneu.2017.05.109

Oppenlander ME, Chowdhry SA, Merkl B, Hattendorf GM, Nakaji P, Spetzler RF (2014) Robotic autopositioning of the operating microscope. Neurosurgery 10 Suppl 2:214–219; discussion 219. https://doi.org/10.1227/neu.0000000000000276

Pandya A, Siadat MR, Auner G (2005) Design, implementation and accuracy of a prototype for medical augmented reality. Comput Aided Surg. 10:23–35. https://doi.org/10.3109/10929080500221626

Punchak M, Mukhopadhyay S, Sachdev S, Hung YC, Peeters S, Rattani A, Dewan M, Johnson WD, Park KB (2018) Neurosurgical care: availability and access in low-income and middle- income countries. World Neurosurg 112:e240–e254. https://doi.org/10.1016/j.wneu.2018.01.029

Ricciardi L, Chaichana KL, Cardia A, Stifano V, Rossini Z, Olivi A, Sturiale CL (2009) The exoscope in neurosurgery: an innovative "point of view". A systematic review of the technical, surgical and educational aspects. World Neurosurg. https://doi.org/10.1016/j.wneu.2018.12.202

Roberts DW, Strohbehn JW, Hatch JF, Murray W, Kettenberger H (1989) A frameless stereotaxic integration of computerized tomographic imaging and the operating microscope. J Neurosurg 65:545–549. https://doi.org/10.3171/jns.1986.65.4.0545

Roehl E, Buzug TM (2009) Surgical microscope with automated focus adaptation. Springer, Berlin Heidelberg, Berlin, Heidelberg, pp 110–113

Rossini Z, Cardia A, Milani D, Lasio GB, Fornari M, D'Angelo V (2017) VITOM 3D: preliminary experience in cranial surgery. World Neurosurg 107:663–668. https://doi.org/10.1016/j.wneu.2017.08.083

Sack J, Steinberg JA, Rennert RC, Hatefi D, Pannell JS, Levy M, Khalessi AA (2018) Initial Experience using a high-definition 3-dimensional exoscope system for microneurosurgery. Oper Neurosurg (Hagerstown) 14:395–401. https://doi.org/10.1093/ons/opx145

Schebesch KM, Brawanski A, Tamm ER, Kuhnel TS, Hohne J (2019) QEVO((R))—a new digital endoscopic microinspection tool-a cadaveric study and first clinical experiences (case series). Surg Neurol Int 10:46. https://doi.org/10.25259/sni-45-2019

Shuhaiber JH (2004) Augmented reality in surgery. Arch Surg 139:170–174. https://doi.org/10.1001/archsurg.139.2.170

Sielhorst T, Feuerstein M, Navab N (2008) Advanced medical displays: a literature review of augmented reality. J Display Technol 4:451–467

Uluc K, Kujoth GC, Baskaya MK (2009) Operating microscopes: past, present, and future Neurosurg Focus 27:E4. https://doi.org/10.3171/2009.6.focus09120

Wapler M, Braucker M, Durr M, Hiller A, Stallkamp J, Urban V (1999) A voice-controlled robotic assistant for neuroendoscopy. Stud Health Technol Inform 62:384–387

Willems PWA, Noordmans HJ, Ramos LMP et al (2003) Clinical evaluation of stereotactic brain biopsies with an MKM-mounted instrument holder. Acta Neurochir 145(10):889–897

Wu JR, Wang ML, Liu KC, Hu MH, Lee PY (2014) Real-time advanced spinal surgery via visible patient model and augmented reality system. Comput Methods Programs Biomed 113:869–881. https://doi.org/10.1016/j.cmpb.2013.12.021

Yaşargil M (1984) Microneurosurgery Volume 1, Chapter 5 pp 301–303. Georg Thieme Verlag, Stuttgart

Yasargil MG, Krayenbuhl H (1970) The use of the binocular microscope in neurosurgery. Bibl Ophthalmol 81:62–65

Yoon JW, Chen RE, Han PK, Si P, Freeman WD, Pirris SM (2017) Technical feasibility andsafety of an intraoperative head-up display device during spine instrumentation. Int J Med Robot 13. https://doi.org/10.1002/rcs.1770

Zappa F, Mattavelli D, Madoglio A, Rampinelli V, Ferrari M, Tampalini F, FontanellaM, Nicolai P, Doglietto F, Group PR (2019) Hybrid robotics for endoscopic skull base surgery: preclinical evaluation and surgeon first impression. World Neurosurg. https://doi.org/10.1016/j.wneu.2019.10.142

Zhao X, Belykh E, Cavallo C, Valli D, Gandhi S, Preul MC, Vajkoczy P, Lawton MT, Nakaji P (2019) application of fluorescein fluorescence in vascular neurosurgery. Front Surg 6:52. https://doi.org/10.3389/fsurg.2019.00052

Chapter 14
Simulating Cataracts in Virtual Reality

Katharina Krösl

Abstract Vision impairments, such as cataracts, affect the visual perception of numerous people worldwide, but are hardly ever considered in architectural or lighting design, due to a lack of suitable tools. In this chapter, we address this issue by presenting a method to simulate vision impairments, in particular cataracts, graphically in virtual reality (VR), for people with normal sight. Such simulations can help train medical personnel, allow relatives of people with vision impairments to better understand the challenges they face in their everyday lives and also help architects and lighting designers to test their designs for accessibility. There have been different approaches and devices used for such simulations in the past. The boom of VR and augmented reality (AR) devices, following the release of the Oculus Rift and HTC Vive headsets, has provided new opportunities to create more immersive and more realistic simulations than ever before. However, the development of a vision impairment simulation is dependent on multiple factors: the designated application area, the impacts of the used hardware on a user's vision, and of course the impact of the respective vision impairment on different aspects of the human visual system. We will discuss these factors in this chapter and also introduce some basic knowledge on human vision and how to measure it. Then we will illustrate how to simulate vision impairments in VR on the example of cataracts and explain how the presented methodology can be used to calibrate simulated symptoms to the same level of severity for different users. This methodology allows for the first time to conduct quantitative user studies to investigate the impact of certain vision impairments on perception and gain insight that can inform the design process of architects and lighting designers in the future.

K. Krösl (✉)
VRVis Zentrum fuer Virtual Reality und Visualisierung Forschungs-GmbH,
Donau-City-Strasse 11, 1220 Vienna, Austria

TU Wien, Institute of Visual Computing & Human-Centered Technology,
Research Unit of Computer Graphics, E193-02, Favoritenstraße 9-11, 1040 Vienna, Austria
e-mail: kroesl@vrvis.at; kkroesl@cg.tuwien.ac.at

257

14.1 Introduction

In 2019, at least 2.2 billion people were affected by vision impairments, according to a report, published by the *World Health Organization* (WHO) (World report on vision 2019). The WHO expects this number to rise, due to increasing urbanization, behavioral and lifestyle changes, or aging of the population. The *National Eye Institute* (NEI) (NEI Office of Science Communications, Public Liaison, and Education 2018) predicts that the number of people with vision impairments will rise until it will have approximately doubled from 2010 to 2050. The data on vision impairments show a higher prevalence of eye diseases in age groups of 40 years and older, especially for conditions such as presbyopia, cataract, glaucoma, and age-related macular degeneration (World report on vision 2019). The leading cause for blindness and, with 51%, one of the major causes for vision impairments are cataracts (Pascolini and Mariotti 2012).

With an increasing number of people being affected by vision impairments, inclusive architecture and lighting design is more pressing than ever. We need to create a common understanding of the effects of vision impairments for affected people and their environment and give architects and designers suitable tools to evaluate their designs for accessibility.

Accurate vision impairment simulations are critical for facilitating inclusive design of everyday objects, architectural planning, or the development of standards and norms to increase accessibility of public spaces. Currently, experts in these fields have to rely on their professional experience and the few guidelines that are available. There is a lack of data to show precisely how to consider people with vision impairments in the design process, as well as a lack of tools to evaluate a design for accessibility. User studies with affected people could help to gain insights. However, conducting user studies to gain data for the revision of standards and norms, or to evaluate the accessibility of architectural or lighting designs, can be extremely difficult, since such studies currently require participation by many people with the same form of vision impairment to allow for statistical analysis of sufficient power.

Wood et al. (2010) used modified goggles to conduct a user study on the effects of vision impairments on nighttime driving. Unfortunately, to be able to perform a reliable statistical analyses, such studies need a high number of participants with the exact same vision capabilities, since the vision of the participants might influence the results. Getting a large enough sample size can be a challenge, especially for studies involving eye diseases such as cataracts, diabetic retinopathy, glaucoma, or macular degeneration, since these vision impairments can cause symptoms that are hard to assess precisely. Even if eye exams show a similar extent of a certain eye disease for two participants, the severity of a symptom can be experienced differently and could therefore also be described differently by affected people. Hence, real-world studies are infeasible to determine the exact effects of eye diseases such as cataracts on perception.

Conducting real-world studies with affected patients is difficult, costly, and may even be dangerous for participants. A safe and inexpensive alternative is to conduct user studies in VR with healthy participants, which is what we focus on in this chapter. We present

- *A simulation of nuclear cataracts, cortical cataracts, and subcapsular cataracts in VR, using eye tracking for gaze-dependent effects.* Multiple effects, simulating different symptoms caused by cataracts, are combined in an effects pipeline to create a simulation of various disease patterns of cataracts. In contrast to previous work, which mostly provided simple approximations of cataract vision, we present a complex, plausible, medically informed simulation of cataract vision in VR, featuring the most common cataract symptoms.
- *A Simulation of the effects of different lighting conditions on vision with cataracts.* We simulate pupil dilation and contraction, caused by different levels of illumination in a scene and its effects on cataract vision. These changes in pupil size result in more or less exposure of the opacities formed in a lens affected by cataracts, and consequently the perception of more or less disturbing shadows in the field of view of a person. Furthermore, we simulate straylight effects caused by the increased scattering of light of a clouded lens.
- *A novel methodology as an alternative to real-world user studies, when finding a sufficiently large sample group of participants with the same form of vision impairment is not possible.* We show how to conduct studies with participants with normal sight and graphically simulated vision impairments and how to calibrate these impairments to the same level of severity for every participant, regardless of their actual vision capabilities or the influence of hardware limitations of VR headsets on perception.

To understand vision impairment simulations and their implementation, it is useful to first understand the basics of human vision. In Sect. 14.2 we discuss the concept of *visual acuity* and what factors might influence the vision of a person wearing a VR headset. We also provide background on cataracts, different types of this eye disease, and how it affects a person's vision. In Sect. 14.3 we give a brief overview of different approaches to simulate vision impairments, including different display modalities. We expand on this in Sect. 14.4, and describe the structure of effects pipelines required to simulate complex eye disease patterns and discuss eye tracking for gaze-dependent effects. Important applications for vision impairment simulations include accessibility evaluations and measurements, involving user studies. Conducting user studies in VR with people with normal sight and simulated vision impairments is not trivial, if we want to obtain quantitative data, which can be used for statistical analyses. We present our methodology to calibrate symptoms of vision impairments to the same level for each user, so they experience the same amount of degraded vision, in Sect. 14.5.

14.2 Human Vision and Impairments

In this Section we briefly discuss how human vision is measured and classified, and what factors can influence someone's visual perception in a VR headset. We also give some medical background information and explain how cataracts impact a person's vision, so we can then use cataracts as an example when we look into building simulations of vision impairments.

14.2.1 Understanding Visual Acuity

Visual acuity (VA) describes in a quantifiable way a person's ability to recognize small details. It is measured by showing the subject different optotypes (standardized symbols used in medical eyesight tests, such as the Landolt C, shown in Fig. 14.1) of different sizes at a predefined distance, and determining which size can be recognized and which cannot. A person has normal vision when they can recognize a detail that spans 1 arc minute (1/60 of a degree), which would be a size of $\sim 1.75mm$ at 6 meters distance (see Fig. 14.2). This can theoretically be tested at any viewing distance as long as the detail in question is appropriately scaled in relation to the distance. Shortsighted people can see very close objects well and only have a reduced VA at a certain distance. Therefore the test distance should not be too short (e.g., not under 1 m). A common test distance is 6 meters or 20 feet.

Normal sight is defined as 1.0 *decimal acuity* or 6/6 vision (in the metric system) or 20/20 vision (using foot as unit), or better.

Fig. 14.1 *Landolt C*, or *Landolt ring*, with a gap at one of eight possible positions: top, bottom, left, right, or 45° in between

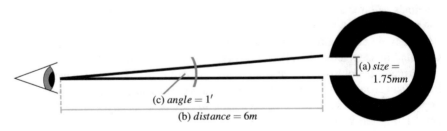

Fig. 14.2 A person with normal sight can recognize a detail, such as **a** the gap in the Landolt C, of size $\sim 1.75mm$ at **b** 6m distance. **c** The respective viewing angle corresponds to 1 arc minute (1/60 of a degree)

In 6/6 or 20/20 vision, the numerator specifies the viewing distance (6 meters or 20 feet) at which a person who is tested can recognize the same size of optotypes as a person with normal sight (1.0 *decimal acuity*, 6/6, or 20/20 vision) can from the distance given by the denominator. The result of this fraction is the decimal acuity value of the tested person.

14.2.1.1   Example

A person's VA is examined using a vision test where optotypes are displayed at a distance of $6m$. The smallest details the person recognizes at $6m$ viewing distance are of size $\sim$3.5 mm (2 arc minutes), which is double the size (or double the angle) of what a person with normal sight would be able to recognize. In other words, the tested person recognizes details of a certain size that a person with normal sight would already be able to recognize at double the viewing distance. The tested person, therefore, has a VA of 6/12 or 20/40, which is equivalent to 0.5 decimal acuity.

14.2.2   Impacts on Vision in VR Headsets

There are a number of factors that influence the perception of users and how they experience a simulation of a vision impairment in VR:

- Visual capabilities of participants (with normal sight or corrected sight).
- Resolution of the VR headset.
- Fixed focal distance of the VR headset.
- Possible misplacement of the VR headset.

When conducting user studies in VR with simulated vision impairments, it is important to recruit participants with normal sight to avoid degrading a user's vision more than intended by the simulated impairments. However, even people with normal sight have varying vision capabilities that need to be accounted for. Furthermore, the resolution of VR headsets is lower than that of the human eye. Therefore, users already experience a mild form of vision impairment when wearing a VR headset. An HMD also has a fixed focal distance to the eyes of the user. This can create a mismatch between vergence and accommodation (Kramida 2015) of the eyes of a user, leading to visual fatigue after a certain time and consequently to a further reduced VA. The lenses that are built into a VR headset focus the light in a specific area of the retina. Similar to a misplacement of glasses, an HMD that does not sit correctly on a user's head can cause images to be perceived as less sharp, resulting in an additional reduction of VA.

We have to be aware of these factors if we want to achieve a similar visual experience of a simulated vision impairment for all our users. Participants should have normal or corrected sight (wearing lenses), experiences in VR should be limited to short periods of time to avoid visual fatigue, and we have to make sure the headset is worn correctly. The low resolution of the display is a factor we can take into account in our simulation (see Sect. 14.5.2 for more details). For a realistic simulation of vision impairments, such as cataracts, we also need to understand their impact on vision, which we will cover in the following section.

14.2.3   Cataracts

The leading causes of vision impairment worldwide, as identified by the WHO, are uncorrected refractive errors and cataracts. VA is usually expressed relative to 6/6, the Snellen fraction for the test distance of *6m*, or 20/20 in feet, or the decimal value of these fractions. The WHO distinguishes between mild (VA worse than 6/12 or 20/40), moderate (VA worse than 6/18 or 20/60), and severe (VA worse than 6/60 or 20/200) vision impairment, and blindness (VA worse than 3/60 or 20/400). The global estimates of the number of visually impaired people, given in the WHO report on global data on visual impairments Pascolini and Mariotti (2012) show that about 14% (186.203 million people) of the world population over the age of 50 (1340.80 million) have a moderate to severe vision impairment (154.043 million) or are blind (32.16 million), with cataracts being the major cause of blindness.

14.2.3.1   Cataract Types

Cataracts are opacities in the lens of the eye, that occlude parts of the visual field and can also lead to vision loss, when left untreated. Depending on their characteristics and the region of the lens that is affected, cataracts are categorized as *nuclear, cortical,* or *posterior subcapsular* cataracts (Michael and Bron 2011). VR simulations of these types of cataract vision are shown in Fig. 14.3.

Nuclear cataracts are experienced as yellow tinting (or clouding) in one's vision, accompanied by increased straylight. This is due to an accumulation of yellow-brown pigment or protein in the nucleus (central area) of the lens, which results in a homogeneous clouding of the lens. Consequently, incoming light is scattered much more than in a healthy eye (Michael and Bron 2011), which results in blurry vision.

Cortical cataracts develop when fibers are damaged or protein aggregates in the lens cortex. As a result, people with cortical cataracts see peripheral radial opacities, dot-like shadows, or spoked opacities in the periphery of their visual field. Cortical cataracts with spoked opacities near the periphery are the most common form of cataracts (Michael and Bron 2011).

Posterior subcapsular cataracts are the least common type of cataract. Defective fiber production in the lens can cause opacities to form at the posterior pole of the lens.

Fig. 14.3 Simulation of **a** nuclear cataract, **b** cortical cataract and **c** posterior subcapsular cataracts when trying to read an escape-route sign, simulated in VR

These opacities are typically experienced as dark shadows in the center of the visual field and therefore have a very severe effect on vision a person's vision (Michael and Bron 2011).

14.2.3.2 Impact of Cataracts on Vision

Depending on the type of cataract and severity of symptoms, tasks like finding your way out of a building in case of an emergency can become difficult (see Fig. 14.3). According to the NEI (NEI Office of Science Communications, Public Liaison, and Education 2019) and reports from ophthalmologist and patients, the following symptoms can be caused by cataracts:

- Cloudy or blurry vision (reduced VA)
- Faded colors (reduced contrast)
- tinted vision (color shift)
- Troubles seeing at night (bloom/glare)
- Increased sensitivity to light (bloom/glare)
- Halos around lights
- Double vision

The effect of lens opacities on vision depends on their location and on pupil size. In daylight, when the pupil diameter is small, only opacities within the pupillary zone are likely to affect vision. If ambient light is further reduced and the pupil diameter becomes larger, vision is further affected as an increasing amount of straylight (light that is scattered by parts of the eye, due to optical imperfections like particles in a clouded lens Van den Berg (1986)) falls on the retina (Michael et al. 2009).

14.3 Related Work on Simulating Vision Impairments

There has been some research on simulating visual impairments across different display modalities and for different purposes like educational purposes, raising awareness, accessibility inspection, design aids, or user studies. Note that there is also a lot of work on assistive technology (Aydin et al. 2020; Billah 2019; Chakravarthula et al. 2018; Guo et al. 2016; Langlotz et al. 2018; Reichinger et al. 2018; Stearns et al. 2018; Sutton et al. 2019), including virtual reality (VR) / augmented reality (AR) simulations that compensate for low vision (Zhao et al. 2019a, b, 2020, 2015), or software specifically designed for people with vision impairments (Albouys-Perrois et al. 2018; Thévin and Machulla 2020; Wedoff et al. 2019; Zhao et al. 2018), which is beyond the scope of this chapter. In this chapter, we focus on the simulation of impaired vision and will now give a brief overview of different approaches in this area.

14.3.1 Goggles

Physical (non-VR) goggles with special lenses have been used to recreate the effects of eye diseases and educate people about how these impairments affect perception. Wood et al. (2010) used modified goggles in a study to investigate the potential effects of simulated visual impairments on nighttime driving performance and pedestrian recognition under real road conditions. Similarily, Zagar and Baggarly (2010) developed individual sets of (non-VR) goggles to simulate glaucoma, cataracts, macular degeneration, diabetic retinopathy, and retinitis pigmentosa. Physical goggles (Fig. 14.4) designed to simulate the decreased visual acuity (VA) and increased glare of generic cataracts are available commercially (Vision Rehabilitation Services LLC 2019), but with the express disclaimer that they are not intended to replicate a specific user's visual impairment.

Fig. 14.4 Commercially available vision simulator goggles from Vision Rehabilitation Services LLC. Image taken from Vision Rehabilitation Services LLC (2019). Image courtesy of Marshall Flax, Fork in the Road Vision Rehabilitation Services, LLC

Although real goggles might be suitable for educational purposes, they limit the experiment environment of user studies to the real world, where environmental changes like fire or smoke are hard to simulate safely. Furthermore, each set of goggles only simulates one particular vision impairment. They are not adjustable to the vision capabilities of users or to simulate different levels of severity of a vision impairment and have limited field of view and immersion.

14.3.2 2D Images

A widely used approach to convey the effects of vision impairments is to modify 2D images. Banks and Crindle (2008), for example, attempted to recreate the visual effects of several ocular diseases, such as glaucoma or age-related macular degeneration (AMD) by combining different image-processing effects and creating overlays and filters for 2D images or rendered images from a 3D scene (viewed on a desktop display). Hogervorst and Van Damme Hogervorst and van Damme (2006) also modified 2D images to give unimpaired persons insight into the problems people with vision impairments face every day. They conducted a user study to evaluate the relationship between blurred imagery and VA. When measuring the VA with eyesight tests using the *Landolt C*, the authors found a linear correlation between VA and a just-recognizable threshold for blurring an image. Building upon these findings, the calibration procedure presented in this chapter (see Sect. 14.5) uses a blur filter to adapt a user's vision to a certain level of VA. Very well-known depictions of vision impairments are provided by the NEI (National Eye Institute, National Institutes of Health (NEI/NIH) 2018). These images inform a lot of research work in the area of vision impairment simulations (e.g., Ates et al. 2015). However, it should be noted that these images show simplified versions of the respective vision impairments and can lead to misconceptions about vision impairments Thévin and Machulla (2020).

While 2D images are a cheap and easy way to visualize impaired vision, these static images do not allow calibrating for individual users, reacting to eye movements, or providing an immersive experience.

14.3.3 3D Simulations

Simulations in 3D environments (like in 3D computer games) offer more possibilities than 2D images to investigate and understand vision impairments, as they provide a higher level of immersion. Lewis et al. (2011), for example, used the Unreal Engine 3 to apply post-processing effects to simulate common eye diseases in a 3D game or explorable environment on a desktop screen. A later version of this system (Lewis et al. 2012), using the Microsoft XNA framework to simulate AMD, glaucoma, hyperopia, myopia, and cataracts, provides means to adjust the simulated vision impairments, but does not take vision capabilities into account.

14.3.4   Virtual Reality Simulations

With the advent of modern VR and eye-tracking technology, it is now possible to graphically simulate vision impairments in VR. Jin et al. (2005) developed a vision impairment simulation, using a complex eye anatomy model and a scotoma texture, created from perimetry exam data from real patients, to define regions of degraded vision. This texture is the same for every user and does not account for a user's vision capabilities. The goal of Väyrynen et al. (2016) was to provide a simulation that could help architectural designers to better understand the visual perception of people with vision impairments and the challenges they face in their everyday lives because of their limited vision capabilities. The authors used standard effects provided by the game engine Unity3D to approximate different vision impairments. Their cataract simulation consists only of a lens-flare component for each scene light source and a flare-layer in each virtual camera.

14.3.5   AR Simulations

Using the Oculus Rift head-worn display (HWD) and a PlayStation 4 camera as AR setup, Ates et al. (2015) conducted a user study with focus on accessibility inspection of user interfaces. Their simulation of vision impairments is based on photos of the NEI (NEI Office of Science Communications, Public Liaison, and Education 2018) and implemented through a VR media player that can render stereoscopic video files. The level of intensity of the simulated impairments can be adjusted via keyboard, but the existing VA of the user is not taken into account, and the implemented impairments are simplified approximations and do not attempt to recreate the impaired vision of specific persons. The cataract simulation, for example, is restricted to a Gaussian blur and does not include other symptoms. Eye tracking is not supported either. Werfel et al. (2016) developed an AR and VR system for empathizing with people with audiovisual sense impairments, which also includes a cataract module. They modeled audiovisual sensory impairments using real-time audio and visual filters experienced in a video–see-through AR HWD. Visual impairments, such as macular degeneration, diabetic retinopathy, and retinitis pigmentosa, were modeled according to information and illustrations from the German Association for the Blind and Visually Handicapped (DBSV). Their cataract simulation was realized with a blur, decreased saturation, and modified contrast and brightness, but without eye tracking for gaze-dependent effects. The work by Jones and Ometto (2018) aims not only at creating a teaching or empathy aid, but also a tool for accessibility evaluations. Their VR/AR simulation of different visual impairment symptoms allows adjusting symptoms, integrates eye-tracking data, and achieves near real-time rendering. In more recent work Jones et al. (2020) used their developed software *OpenVisSim* to simulate glaucoma in VR and AR. Their simulation does not take vision capabilities or hardware constraints into account, and currently only consist of a gaze-dependent region of variable blur, which is a non-negligible simplification of this vision impairment. However, the used *OpenVisSim* software also offers simulations of some other effects that could be used to improve the glaucoma simulation in future work. There are also commercial smartphone applications available that simulate vision impairments. For example, the Novartis ViaOpta Simulator (Novartis Pharma AG 2019) for Android and iOS processes the live smartphone camera feed to address a broad set of impairments, including vitreomacular traction syndrome, diabetic macular edema, glaucoma, and cataract. However, the provided cataract simulation affects only VA and color vision, can be adjusted only in severity and not per symptom, and supports just one generic cataract type. In addition, while smartphones are ubiquitous and thus can reach a broad audience, they have a far smaller field of view than current VR head-worn displays when held at a comfortable distance, are monoscopic, and do not support eye tracking for simulating gaze-dependent effects.

> Most existing approaches are targeted at educational or demonstrative purposes and do not take the user's actual vision capabilities or hardware limitations of the VR headsets into account. Hence, they are not feasible for user studies and often only provide very simplified simulations of eye diseases.

The effects pipeline presented in this chapter can be used to simulate different eye diseases and vision impairments through an appropriate combination of individual effects, the severity of symptoms can be interactively modified, and the simulation can be calibrated to vision capabilities of users and reacts to eye tracking. This allows a realistic simulation of visual impairment in diverse immersive settings. We developed our calibration methodology originally to investigate the maximum recognition distances (MRDs) of escape-route signs in buildings Krösl et al. (2018). The presented effects pipeline was first built to simulate cataracts and was evaluated in a user study (Krösl et al. 2019). We also outlined our research approach and described how our pipeline could be extended to AR (Krösl 2019). In our recent work, we also describe how our effects pipeline can be adapted to simulate different eye diseases in VR, 360° images and video–see-through AR (Krösl et al. 2020a).

14.4 Building an Effects Pipeline for Complex Eye Disease Patterns

In this chapter, we present our methodology to simulate different symptoms of eye diseases separately and then combine them for a simulation of the whole disease pattern. This approach can be applied to several eye diseases that cause different distinct symptoms.

As example, we will now discuss our effects pipeline to combine the most common cataract symptoms (see Sect. 14.2.3.2): blurred vision, contrast loss, tinted vision,

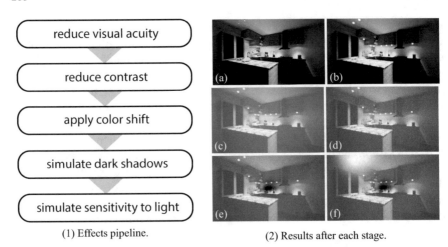

(1) Effects pipeline.   (2) Results after each stage.

Fig. 14.5 To combine all effects for a simulation of cataract vision, we take **a** the original image and first **b** reduce the VA, and **c** the contrast of the image and then **d** apply a color shift. Next we use a texture to simulate **e** the dark shadows that people with cortical or posterior subcapsular cataracts (as shown in this figure) typically experience. We modify this effect according to the brightness of the virtual environment the user is currently viewing and add a **f** bloom or glare effect to simulate straylight and sensitivity to light. Each stage in this effects pipeline simulates one symptom. To create simulations of other eye disease patterns, stages can be added, removed, or changed

clouded lens, and increased sensitivity to light. For each frame, the image that is to be displayed on the VR headset is modified in several ways by applying different effects in sequence. Figure 14.5 shows this effects pipeline and the resulting image of each stage. Note that these effects could be applied in a different order. However, the order of effects can influence the final result of the simulation, which we will discuss in more detail in Sect. 14.5.5.

14.4.1 Reduce Visual Acuity

The most common symptom present in vision impairments is the reduction of VA. We used two different approaches to simulate reduced VA in our studies (Krösl et al. 2018, 2019), a Gaussian blur, based findings by Hogervorst et al. (2006), and a depth-of-field effect, provided by Unreal Engine. Nearsighted people can sharply see objects that are very close to their eyes, while everything in the distance appears blurred. A simple uniform Gaussian blur might be sufficient to simulate the reduced VA caused by cataracts. However, since cataracts are an age-related vision impairment, many people with this eye disease often also have a refractive error (nearsightedness, farsightedness, astigmatism, or presbyopia). Therefore, an effect that is able to also simulate reduced VA dependent on viewing distance can help to create a more realistic simulation of the vision of people with cataracts. We can simulate this distance-

Fig. 14.6 **a** Original image.
b Reduced VA

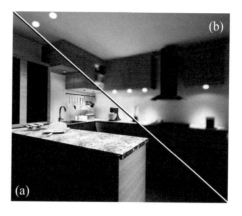

dependent reduction of VA by using a depth-of-field effect. By adjusting the sigma parameter of the effect, we can change the severity of this simulated symptom. Unreal Engine has a built-in depth-of-field effect that we can leverage for this purpose. Figure 14.6 shows the image before and after VA reduction. In the next step, the VA-reduced image C_{rVA} is further modified by reducing its contrast.

14.4.2 Reduce Contrast

A loss of contrast is often experienced as faded colors (see Fig. 14.7), which may be implemented in a number of different ways in VR. Using an approach that shrinks the histogram of a frame by using min and max values of the image is not feasible, because intensity changes from one frame to the next could change the color and intensity distribution in the image. This could yield very sudden changes in the histogram and introduce flickering artifacts. Instead, we need a way to reduce contrast that is consistent over multiple frames. When contrast sensitivity in the human eye weakens, colors become faded and harder to distinguish (see Fig. 14.7). This symptom could be simulated with a simple per-frame histogram compression, using min and max intensity values in the image. However, significant changes in intensity from one frame to the next (changing min or max intensity for the current frame) could lead to flickering artifacts. Therefore, it is essential to use a contrast reduction that does not depend on color and intensity distribution in the image. Furthermore, our simulation needs to run in real time, which means we need to avoid expensive calculations. A simple way to reduce the contrast is to interpolate between the current image C_{rVA} (with already reduced VA) and a uniformly gray image (represented by the linear RGB color value (0.5, 0.5, 0.5) in Eq. 14.1), weighted by a constant c. The following calculation is done per color channel:

$$C_{rContrast} = C_{rVA}c + 0.5(1 - c). \tag{14.1}$$

Fig. 14.7 **a** Reduced VA. **b**
Reduced contrast

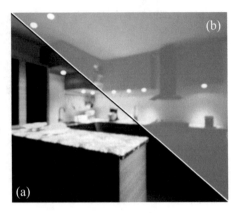

The parameter c (between 0 and 1) determines the amount of contrast reduction. This operation can also be interpreted as histogram remapping: The color values are scaled by c, which essentially shrinks the histogram of intensities in the image. Intensities are reduced by $(1 - c)$ percent, which means the image becomes a lot darker. To preserve the average intensity in the image, all values are shifted back into the middle of the available range by adding 50% of the amount the histogram was reduced, $0.5(1 - c)$. A more advanced version of this histogram remapping would operate on luminance values and not color values. Furthermore, tone mapping algorithms could also be modified to reduce contrast.

14.4.3 Apply Color Shift

In this next step, we apply a color shift to the image (see Fig. 14.8), to simulate tinted vision, which is a common symptom of cataracts (as mentioned in Sect. 14.2.3.2). There are multiple different ways to perform a color shift and different color spaces to choose from, for this operation. We will discuss one simple type of color shift and how to improve it. Because the yellowish/brownish color shift, which is often experienced by people with cataracts, results from a physical phenomenon—the absorption of parts of the incident light falling onto the retina, caused by opacities in the lens—we chose to do our calculations in the linear RGB color space instead of a perceptual color space. Other eye diseases that cause changes in color vision might require other color spaces for the respective color shift calculations, depending on what causes the color shift. To add tinted vision as symptom to our cataracts simulation, we apply a color shift after the contrast reduction step. Analogous to the contrast reduction step, we interpolate all color values with a predefined target color C_{target}:

$$C_{colorShift} = C_{rContrast}t + C_{target}(1 - t). \tag{14.2}$$

Fig. 14.8 a Reduced
Contrast. **b** Color shift,
calculated with Eq. (14.2)

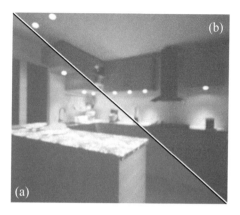

The target color C_{target} and the parameter t can be modified at runtime to change
the color for the tint and the amount of color shift that should be applied. Figure 14.8
shows the result of this operation with target color $C_{target} = (1.0, 0.718461,$
$0.177084)$ and parameter $t = 0.8$. Besides adding a tint to the image, this type of
color shift also reduces the contrast in the image to a certain extent, since this is
equivalent to an interpolation between the current image $C_{rContrast}$ (with already
reduced contrast) and the target color. To avoid reducing the contrast when applying
a color shift, we can also perform a color shift by simulating a filter that reduces the
amount of light in parts of the visible spectrum, that is absorbed or blocked by the
cataract opacities. For each color channel we calculate the color shift as

$$C_{colorShift} = C_{rContrast} - C_{rContrast}C_{filtered}, \qquad (14.3)$$

or

$$C_{colorShift} = C_{rContrast}(1 - C_{filtered}), \qquad (14.4)$$

where $C_{filtered}$ represents the amount of filtered light per color channel that does not
reach the retina.

Cataract opacities scatter, absorb, and reflect different amounts of each wavelength
of the visible light, depending on the opacity. The exact amount of transmitted light
per wavelength is hard to determine. It is easier to get a description of the experienced color tint from people with cataracts. Since the components of the color that
creates this tint, represent the amount of transmitted light per wavelength (C_t) and
are therefore complementary to the amount of light that is reflected or absorbed by
the cataract opacities ($C_{filtered}$) we can simulate the color shift as

$$C_{colorShift} = C_{rContrast}C_t. \qquad (14.5)$$

This calculation is done per color channel with the respective components of the
pixel colors $C_{rContrast}$ and transmitted light C_t.

14.4.4 Simulate Dark Shadows

Cataracts lead to a clouding of the eye lens. While for nuclear cataracts, this clouding is uniform over the whole lens, cortical cataracts also produce dark shadows in the periphery of the lens, and posterior subcapsular cataracts create a dark shadow in the center of the lens. We can simulate these shadows with an alpha texture (see Fig. 14.9) that we use to darken the image, either in the periphery (for cortical cataracts) or in the center (for posterior subcapsular cataracts), by linearly interpolating between the image color $C_{colorShift}$ of the image after the color shift and a shadow color C_{shadow}:

$$C = C_{colorShift}\alpha + C_{shadow}(1 - \alpha), \qquad (14.6)$$

where α has values between 0 and 1.

Different illumination levels cause the pupil of the human eye to get wider or narrower, allowing more or less light to enter the eye. This also affects the area of the lens that is exposed to light entering the eye. For some forms of cataracts, like cortical or posterior subcapsular cataracts, which exhibit a nonuniform clouding of the lens, the area of the pupil that is exposed to light affects the way vision is impaired. The dilation and contraction of the pupil when looking at dark areas or into bright lights determine how much of the areas of the lens, which create dark shadows, are exposed to light. Consequently, dark shadows can appear very small or affect almost the whole visual field, depending on the illumination and therefore the intensity in a virtual scene.

Posterior subcapsular cataracts cause opacities to form at the posterior pole of the lens, which creates visible shadows in the center of the field of view, while often not affecting peripheral vision (see Fig. 14.10). In bright sunlight, the pupil is very small, exposing only a small central area to light, which is largely affected by these opacities. Therefore the shadow can appear huge and cloud the whole visual field. In very dark environments, when the pupil is dilated, the affected area of the lens is only a small percentage of the lens area that is exposed to light. Light can also enter through the surrounding unaffected area. Thus only a small dark shadow might appear in the center of the field of view (Fig. 14.10a, d).

To simulate this effect, we can scale the texture that we use to create these shadows, according to the average light intensity of the image area, the user is looking at (Fig.

Fig. 14.9 Textures used to create shadows for **a** cortical cataracts and **b** posterior subcapsular cataracts, by scaling the image values with the alpha value (between 0 and 1) of this texture

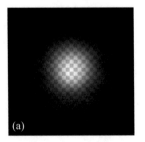
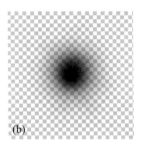

(a)   (b)

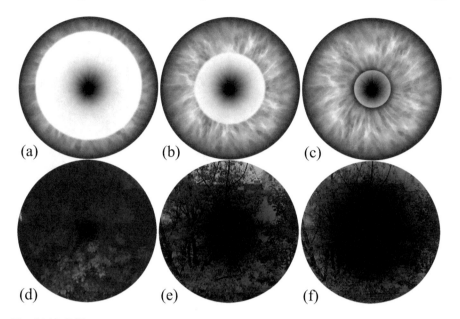

Fig. 14.10 Different pupil sizes **a, b, c** affect the severity of dark shadows, experienced **d, e, f** with posterior subcapsular cataracts. For demonstrative purposes other effects were omitted in this image. **a, d**: Vision of a very dark scene with large pupil. **b, e**: Vision of a scene with moderate brightness with smaller pupil. **c, f**: Vision of a bright scene with very small pupil

Fig. 14.11 a Color shift. **b** Dark shadows for cortical cataracts in the center of the visual field

14.11b). For very bright areas the texture and consequently the shadow becomes larger. When looking at very dark areas it becomes smaller and less disturbing. In contrast to posterior subcapsular cataracts, cortical cataracts affect predominantly the peripheral vision. To simulate dark shadows created by cortical cataracts, we use a texture with shadows in the periphery. When scaling it larger for bright areas, the shadows in the periphery can almost disappear (by leaving the field of view) and only become more disturbing in darker scenes.

Fig. 14.12 **a** Dark shadows for cortical cataracts. **b** Sensitivity to light, experienced as bloom effect of a larger light source in the upper left corner

14.4.5 Simulate Sensitivity to Light

The second way in which light affects the vision of people with cataracts is the clouded lens that scatters light in many directions onto the retina. Images become blurred and bright lights become especially problematic, because they create intense glare. We can simulate this by post-processing the image to apply a bloom or glare effect. The threshold for the bloom is set to a value below the intensity of the light sources in the scene but above the rest of the geometry. This avoids the blooming of white walls or other white objects that are not light sources.

The intensity and width of such an effect can be adjusted. The effect can also be made view-dependent, using the gaze-position from an eye tracker (see Sect. 14.4.6 for details). This simple bloom effect (see Fig. 14.12b) can already give a good impression of glare effects caused by cataracts, but is not a perceptually perfectly accurate depiction of these effects. More advanced approaches of creating glare effects can involve complex simulations of particles in the eye and dynamic effects, like taking the oscillation of the pupil into account, as described in the work of Ritschel et al. (2009), but are challenging to use in very performance-intensive real-time applications like VR simulations.

14.4.6 Gaze-Dependent Effects

Some symptoms such as the above described dark shadows (Sect. 14.4.4) or sensitivity to light (Sect. 14.4.5) that can degrade a person's vision are gaze dependent. To correctly simulate vision affected by gaze-dependent symptoms, we need to track the gaze of the user and adjust effects that should just appear in a certain area of the visual field of a person. We can, for example, move the texture, that is used to simulate the dark shadows produced by cortical or posterior subcapsular cataracts, according to

Fig. 14.13 Eye tracking
with the Pupil Labs (Pupil
Labs GmbH 2019) eye
tracker

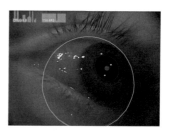

the gaze of the user. This can be done with different eye trackers. Figure 14.13 shows an example of the eye-tracking software from Pupil Labs (Pupil Labs GmbH 2019).

Furthermore, different illumination levels cause the pupil of the human eye to get wider or narrower, allowing more or less light to enter the eye. This also affects the area of the lens that is exposed to light entering the eye. For some forms of cataracts, like cortical or posterior subcapsular cataracts, that exhibit a nonuniform clouding of the lens, the area of the pupil that is exposed to light affects the way vision is impaired. Since this effect depends on the brightness of the area a person is looking at, we can make use of eye-tracking data to constantly adjust the effect, by calculating a gaze-dependent brightness to scale the shadow textures (used to simulate dark shadows) or bloom effects (simulating sensitivity to light). We can just take a cutout of the rendered image, centered around the current gaze point and calculate the average intensity or luminance in this window. To avoid sudden changes of effect sizes, a Gaussian distribution can be used as weights to give pixels at or near the gaze point a higher importance than pixels farther away.

> To avoid any noticeable delay of the effects, the eye tracker needs to be fast enough to recognize saccades (quick eye movements from one fixation to another). Therefore, the eye tracker should at least have 200 Hz cameras.

Slower cameras might cause a noticeable delay of the movement of the effects. However, even with a certain delay, gaze-adjusted effects are perceived as more realistic as without eye tracking.

There are some things to consider when using eye tracking: The performance and accuracy of the tracker might decrease for people wearing glasses or it might not work for them at all. Everything that can create reflections could potentially disturb the eye tracker, for example, eye lashes with mascara on them.

14.5   Calibrating Effects for Different Users and Hardware Devices

In order to use vision impairment simulations, like the cataract simulation described in Sect. 14.4, to evaluate accessibility, measure recognition distances or readability of signage, we need to take vision capabilities of users into account and use a methodology that allows us to calibrate simulated symptoms of vision impairments to the same level of severity for different users. In this section, we discuss vision capabilities of users, present a suitable calibration methodology, and illustrate this methodology on two examples (calibrating reduced VA and reduced contrast).

14.5.1   Vision Capabilities of Users

Let's assume, we want to conduct a user study to measure recognition distances of signage or evaluate other accessibility aspects under a certain form of vision impairment, by conducting a user study with participants with normal vision and simulated vision impairment. We have to be able to create the same visual impression for every user study participant. Only then is it possible to statistically analyze and generalize findings from a user study. Independent variables need to be controlled. For vision impairment simulations, such variables are the actual vision capabilities of a user study participants and the hardware constraints imposed by the VR headset. As explained in Sect. 14.1, finding participants with the exact same type and severity of vision impairment can be difficult or even impossible, which means, these variables can't be controlled through careful selection of participants. To overcome this problem, we can restrict our participant pool to people with normal sight. However, even people with normal sight can be expected to have different levels of VA and contrast sensitivity, since *normal sight* is not defined by one distinct value, but by a certain range. Hence, it is difficult to control these variables. However, we can take them into account when simulating a vision impairment, to create the same baseline for every user study participant. For our presented cataract simulation, we can do this, for example, by calibrating the reduced VA and the reduced contrast to the same levels for all participants.

We do not necessarily need to calibrate other effects like dark shadows, color shift, or an increased sensitivity to light, if we assume our users to have normal sight, because we would only expect significantly different perception of these effects from users that already have an eye disease.

14.5.2 Calibration Methodology

In order to calibrate simulated symptoms of vision impairments to the same level of severity, taking different vision capabilities of users into account, we impose the following restrictions:

- All users have to have normal sight (or corrected sight, wearing glasses or lenses) and no conditions that influence their vision.
- The level of severity we calibrate to, has to be worse than the vision of each participant.
- The level of severity we calibrate to, has to be worse than the vision impairment induced by the VR hardware.

These restrictions allow us to calibrate vision capabilities, such as VA or contrast sensitivity to a certain reduced level, that is perceived similar by every user study participant, by conducting vision tests in VR.

14.5.2.1 Vision Tests in VR

The basic concept is to define a level of reduced VA or contrast, that is known to be a worse vision impairment as the mild vision impairment induced by the VR headset. For this chosen level of impairment, we know what vision tests people with such vision impairment still pass and at what distance, size, or contrast of the optotype they fail the test. So we create an eyesight test in VR that shows optotypes where people with this vision impairment are supposed to fail. Each user takes this eyesight test. During the test, the simulated impairment is increased as long as the user is able to pass the test. At the time a user fails the test, we know exactly which level of simulated impairment we need to calibrate the vision of this particular user to the predefined level. Some users, with very good vision might require a higher severity of simulated impairment to fail the test at the same stage, as others.

Following this methodology yields parameter values (severity levels of simulated impairments) per user that can be used to create the same perceived level of vision impairment for every user. This methodology is illustrated in the following two Sections on the examples of visual acuity and contrast sensitivity.

14.5.3 Calibrating Reduced Visual Acuity

According to the methodology described in above, we can use an eyesight test in VR to calibrate the simulation for all our users to a specific level of reduced VA. There are different eyesight tests that can be adapted for the use in VR. One of these tests is described by the international standard *ISO 8596:2017* (International Organization for Standardization 2017). A set of five Landolt rings (see Fig. 14.1) is displayed at a predefined distance.

The test advances to the next set of smaller Landolt rings as long at the user correctly recognizes the position of the gap in the ring for at least 60% (3 out of 5) of the rings in a set. According to the test protocol, the VA of a user is determined by the last correct set. The virtual equivalent to the real-world visual acuity test setting is to place a user in a virtual room with Landolt chart (International Organization for Standardization 2017) lines on the wall at a specific test distance. An alternative, inspired by the Freiburg Vision Test (FrACT) (Bach 1996) is to show five Landolt rings of the same size at the same fixed distance in sequence and not simultaneously. We know which visual angel a person with a certain visual acuity, e.g., 6/38 or 20/125 or 0.16 decimal, can still recognize. So we chose the distance and size of the optotypes according to what a person with this 0.16 VA should be able to still recognize in the real world (6.31 arcminutes). In order to simulate this VA for a normal sighted person inside a HWD, we blur the image and stepwise increase the amount of blur (while keeping distance and size of the optotypes unchanged), with each new set of 5 Landolt rings, until the user is not able to recognize the optotypes anymore. The blur amount used in the last correctly recognized set of Landolt rings then represents the simulation of this particular reduced VA.

The value of 6/38 or 0.16 decimal, as used in the example above, represents a moderate vision impairment (VA between 6/18 and 6/60) as defined by the WHO (Pascolini and Mariotti 2012), which is well beyond the VA limit of 0.5 decimal for driving, as prescribed by most international standards (Bron et al. 2010).

14.5.4 Calibrating Reduced Contrast

Based on the methodology we used for calibrating reduced VA, we can also calibrate reduced contrast. The *Pelli–Robson contrast sensitivity test* (Pelli et al. 1988) allows us to test the contrast vision of user. Optotypes are displayed in groups of three at a fixed distance and a size that corresponds to 6/18 or 20/60 acuity, which should be easy to read for any normal sighted person. Each group further down the chart is reduced in contrast. A person has to recognize two out of three optotypes correctly to advance to the next group. The last correctly recognized group determines the contrast sensitivity (CS) of a person and is recorded as *log CS* value.

A person with reduced CS only reaches a certain *log CS* value. To calibrate to a specific level of contrast loss in VR we select a *log CS* value as target CS we want to calibrate to. During contrast calibration, one group of optotypes is displayed at a time. Starting at the selected level of reduced contrast, contrast reduction is increased after each correctly recognized group until the optotypes cannot be recognized correctly anymore. The following formula is used to reduce contrast:

$$C = C_{original}c + (1 - c). \tag{14.7}$$

In this equation, c is a constant specifying the amount of contrast reduction and $C_{original}$ the linear RGB color values. The Pelli–Robson contrast sensitivity test (Pelli et al. 1988) uses log contrast sensitivity and reduces the contrast after each group of three letters or optotypes by a factor of 0.15 log units. We can set c to a value representing a reduction of 0.15 log unit or any other amount, depending on how much contrast reduction we want to have per three optotypes. Depending on the used contrast simulation method, $C_{original}$ represents the linear RGB color values, or the luminance values in the CIELAB space for a compression of luminance values. To reduce the contrast relative to a fixed background color, usually white, we add $(1 - c)$, preserving the maximum intensity in the image. This results in a similar appearance of optotypes as on the Pelli–Robson contrast sensitivity chart (Pelli et al. 1988). Keeping the background white is also important to preserve the overall brightness in the scene, since contrast sensitivity is influenced by the illumination of the background and ambient light in the scene (Karatepe et al. 2017). Other formulas for contrast reduction (like adaptions of tonemapping algorithms) could be used as well.

The calibration ends when a group of optotypes cannot be recognized anymore. The value c, used for the last correctly recognized group can then be used to simulate the selected target CS. This value might be different for different users, depending on their vision capabilities, but allows us to calibrate the vision of every user to the same predefined level of reduced contrast, as part of the simulated cataract vision (see Sect. 14.4.2).

14.5.5 Order of Effects

With the presented calibration methodology, we are able to calibrate different simulated symptoms, but the calibration is only done for one symptom at a time. We have to keep in mind that one symptom can influence the perception of another symptom. If we, for example, do a VA test or calibrate VA with already reduced contrast, we will get a different result than when testing or calibrating VA in isolation. Every simulated symptom degrades the vision of a person to some extent. If we combine multiple symptoms and then test a user's vision with the whole simulation, we can get different results for our eyesight tests than when just testing individual symptoms. Consequently, if we reduce the VA of a user to, e.g., 50% and then reduce contrast, add a color shift, simulate dark shadows and sensitivity to light and then perform a VA test in the end, the user might just be left with, e.g., 20% VA. If we change the order of effects and perform the VA reduction last, we can already take into account the amount of VA reduction that happens due to other effects and just reduce the VA further to the predefined value. However, if we start with a contrast reduction to, e.g., 80%, then add other symptoms such as dark shadows or reduced VA, and then perform a contrast test with the whole simulation, we might get a different overall contrast sensitivity value than the one we initially calibrated the contrast vision to. Since many symptoms can influence each other, we cannot easily simulate a complex eye disease pattern that is specified beforehand by parameters like VA, contrast sensitivity, or the amount of tinted vision, since we don't know which symptom affects which vision capability to which amount. We can, however, use this methodology to simulate an eye disease pattern that is perceived similarly by different users, by defining values for individual effects as opposed to values for the whole simulated condition.

14.6 Summary

In architectural and lighting design, the large amount of people with vision impairments in our society are hardly ever considered. In order to test designs for accessibility, we need tools and methods to simulate vision impairments. In order to gain insight into the effects of vision impairments on perception we further need to calibrate simulated symptoms to the same level for different users, so we can conduct quantitative user studies.

In this chapter, we have discussed how vision impairments, such as cataracts, can be simulated in VR and what we need to do to achieve a realistic simulation. Different factors, like the visual capabilities of participants, the resolution and fixed focal distance or possible misplacement of the VR headset could influence the perception of people in VR and need to be taken into account. Visual capabilities of users in VR can be measured with eyesight tests, determining their visual acuity (VA) value.

There are different approaches to simulating vision impairments, using goggles, modified 2D images, VR or AR simulations. At the time of writing this chapter, most existing approaches did not take hardware constraints or vision capabilities of users into account. Therefore, they are not feasible for user studies where we want to take exact measurements and statistically evaluate results.

In this chapter, we have shown how an effects pipeline can be built to simulate complex eye diseases, on the example of cataracts. In order to achieve gaze-dependent effects, we discussed how eye tracking can be used to calculate brightness values at the gaze point, to adjust different effects, or move effects relative with the gaze of the user. The calibration methodology, presented in this chapter enables us to calibrate different effects, that simulate symptoms of vision impairments, to the same level for every user, by adapting medical eyesight tests and using them in the VR simulation.

The methodology presented in this chapter was evaluated in a study simulating cataract vision (Krösl et al. 2019), but can also be adapted and used for other simulations of other vision impairments (Krösl et al. 2020a), like age-related macular degeneration, cornea disease, diabetic retinopathy or glaucoma, to name just a few examples, and is also applicable to AR simulations (Krösl et al. 2020b).

Acknowledgements This research was enabled by the Doctoral College Computational Design (DCCD) of the Center for Geometry and Computational Design (GCD) at TU Wien and by the Competence Center VRVis. VRVis is funded by BMVIT, BMDW, Styria, SFG, and Vienna Business Agency in the scope of COMET—Competence Centers for Excellent Technologies (854174) which is managed by FFG.

References

World Report on Vision (2019) World Health Organization, Geneva. Licence: CC BY-NC-SA 3.0 IGO

Albouys-Perrois J, Laviole J, Briant C, Brock AM (2018) Towards a multisensory augmented reality map for blind and low vision people: a participatory design approach. In: Proceedings of the 2018 CHI conference on human factors in computing systems, pp 1–14

Ates HC, Fiannaca A, Folmer E (2015) Immersive simulation of visual impairments using a wearable see-through display. In: Proceedings of the ninth international conference on tangible, embedded, and embodied interaction. ACM, pp 225–228

Aydin AS, Feiz S, Ashok V, Ramakrishnan I (2020) Towards making videos accessible for low vision screen magnifier users. In: Proceedings of the 25th international conference on intelligent user interfaces, pp 10–21

Bach M et al (1996) The freiburg visual acuity test-automatic measurement of visual acuity. Optom Vis Sci 73(1):49–53

Banks D, McCrindle R (2008) Visual eye disease simulator. In: Proceedings of 7th ICDVRAT with ArtAbilitation, Maia, Portugal

Billah SM (2019) Transforming assistive technologies from the ground up for people with vision impairments. PhD thesis, State University of New York at Stony Brook

Bron AM, Viswanathan AC, Thelen U, de Natale R, Ferreras A, Gundgaard J, Schwartz G, Buchholz P (2010) International vision requirements for driver licensing and disability pensions: using a milestone approach in characterization of progressive eye disease. Clin Ophthalmol (Auckland, NZ) 4:1361

Chakravarthula P, Dunn D, Akşit K, Fuchs H (2018) FocusAR: auto-focus augmented reality eye-glasses for both real world and virtual imagery. IEEE Trans Vis Comput Graph 24(11):2906–2916

Guo A, Chen X, Qi H, White S, Ghosh S, Asakawa C, Bigham JP (2016) Vizlens: a robust and interactive screen reader for interfaces in the real world. In: Proceedings of the 29th annual symposium on user interface software and technology, pp 651–664

Hogervorst M, van Damme W (2006) Visualizing visual impairments. Gerontechnology 5(4):208–221

International Organization for Standardization (2017) ISO 8596:2017(en) Ophthalmic optics–Visual acuity testing–Standard and clinical optotypes and their presentation

Jin B, Ai Z, Rasmussen M (2005) Simulation of eye disease in virtual reality. In: 27th annual international conference of the engineering in medicine and biology society, 2005. IEEE-EMBS 2005. IEEE, pp 5128–5131

Jones PR, Ometto G (2018) Degraded reality: using vr/ar to simulate visual impairments. In: 2018 IEEE workshop on augmented and virtual realities for good (VAR4Good) (March 2018), pp 1–4

Jones PR, Somoskeöy T, Chow-Wing-Bom H, Crabb DP (2020) Seeing other perspectives: evaluating the use of virtual and augmented reality to simulate visual impairments (openvissim). NPJ Digit Med 3(1):1–9

Karatepe AS, Köse S, Eğrilmez S (2017) Factors affecting contrast sensitivity in healthy individuals: a pilot study. Turk J Ophthalmol 47(2):80

Kramida G (2015) Resolving the vergence-accommodation conflict in head-mounted displays. IEEE Trans Vis Comput Graph 22(7):1912–1931

Krösl K, Bauer D, Schwärzler M, Fuchs H, Suter G, Wimmer M (2018) A vr-based user study on the effects of vision impairments on recognition distances of escape-route signs in buildings. Vis Comput 34(6–8):911–923

Krösl K, Elvezio C, Wimmer M, Hürbe M, Feiner S, Karst S (2019) Icthroughvr: illuminating cataracts through virtual reality. In: 2019 IEEE conference on virtual reality and 3D user interfaces (VR), IEEE, pp 655–663

Krösl K (2019) Simulating vision impairments in vr and ar. ACM Siggraph thesis Fast Forward 2019

Krösl K, Elvezio C, Hürbe M, Karst S, Feiner S, Wimmer M (2020) XREye: simulating visual impairments in eye-tracked XR. In: 2020 IEEE conference on virtual reality and 3D user interfaces abstracts and workshops (VRW), pp 830–831

Krösl K, Elvezio C, Luidolt LR, Hürbe M, Karst S, Feiner S, Wimmer M (2020) CatARact: simulating cataracts in augmented reality. In: IEEE international symposium on mixed and augmented reality (ISMAR), IEEE

Langlotz T, Sutton J, Zollmann S, Itoh Y, Regenbrecht H (2018) Chromaglasses: computational glasses for compensating colour blindness. In: Proceedings of the 2018 CHI conference on human factors in computing systems, CHI'18. ACM, New York, NY, USA, pp 390:1–390:12

Lewis J, Brown D, Cranton W, Mason R (2011) Simulating visual impairments using the unreal engine 3 game engine. In: 2011 IEEE 1st international conference on serious games and applications for health (SeGAH), IEEE, pp 1–8

Lewis J, Shires L, Brown D (2012) Development of a visual impairment simulator using the microsoft XNA framework. In: Proceedings 9th international conference disability, virtual reality & associated technologies, Laval, France

Michael R, Bron A (2011) The ageing lens and cataract: a model of normal and pathological ageing. Philos Trans R Soc Lond B Biol Sci 366(1568):1278–1292

Michael R, Van Rijn LJ, Van Den Berg TJ, Barraquer RI, Grabner G, Wilhelm H, Coeckelbergh T, Emesz M, Marvan P, Nischler C (2009) Association of lens opacities, intraocular straylight, contrast sensitivity and visual acuity in European drivers. Acta Ophthalmol 87(6):666–671

National Eye Institute (2018) National Institutes of Health (NEI/NIH). NEI photos and images. https://nei.nih.gov/photo. Accessed 29 Nov 2018

NEI Office of Science Communications, Public Liaison, and Education (2019) At a glance: Cataracts. https://www.nei.nih.gov/learn-about-eye-health/eye-conditions-and-diseases/cataracts. Accessed 24 Oct 2019

NEI Office of Science Communications, Public Liaison, and Education (2018) Prevalence of adult vision impairment and age-related eye diseases in America. https://nei.nih.gov/eyedata/adultvision_usa. Accessed 09 Nov 2018

Novartis Pharma AG (2018) ViaOpta simulator. https://www.itcares.it/portfolio/viaoptanav/. Accessed 1 May 2021

Pascolini D, Mariotti SP (2012) Global estimates of visual impairment: 2010. Br J Ophthalmol 96(5):614–618

Pelli D, Robson J, Wilkins A (1988) The design of a new letter chart for measuring contrast sensitivity. In: Clinical Vision Sciences, Citeseer

Pupil Labs GmbH (2019) VR/AR. https://pupil-labs.com/products/vr-ar/. Accessed 24 Oct 2019

Reichinger A, Carrizosa HG, Wood J, Schröder S, Löw C, Luidolt LR, Schimkowitsch M, Fuhrmann A, Maierhofer S, Purgathofer W (2018) Pictures in your mind: using interactive gesture-controlled reliefs to explore art. ACM Trans Access Comput (TACCESS) 11(1):1–39

Ritschel T, Ihrke M, Frisvad JR, Coppens J, Myszkowski K, Seidel H-P (2009) Temporal glare: real-time dynamic simulation of the scattering in the human eye. In: Computer graphics forum, vol 28. Wiley Online Library, pp 183–192

Stearns L, Findlater L, Froehlich JE (2018) Design of an augmented reality magnification aid for low vision users. In: Proceedings of the 20th international ACM SIGACCESS conference on computers and accessibility, pp 28–39

Sutton J, Langlotz T, Itoh Y (2019) Computational glasses: vision augmentations using computational near-eye optics and displays. In: 2019 IEEE international symposium on mixed and augmented reality adjunct (ISMAR-Adjunct), IEEE, pp 438–442

Thévin L, Machulla T (2020) Three common misconceptions about visual impairments

Van den Berg T (1986) Importance of pathological intraocular light scatter for visual disability. Doc Ophthalmol 61(3–4):327–333

Väyrynen J, Colley A, Häkkilä J (2016) Head mounted display design tool for simulating visual disabilities. In: Proceedings of the 15th international conference on mobile and ubiquitous multimedia, ACM, pp 69–73

Vision Rehabilitation Services LLC (2019) Cataract simulators. https://www.lowvisionsimulators. com/products/cataract-simulators. Accessed 19 Sept 2019

Wedoff R, Ball L, Wang A, Khoo YX, Lieberman L, Rector K (2019) Virtual showdown: an accessible virtual reality game with scaffolds for youth with visual impairments. In: Proceedings of the 2019 CHI conference on human factors in computing systems, pp 1–15

Werfel F, Wiche R, Feitsch J, Geiger C (2016) Empathizing audiovisual sense impairments: interactive real-time illustration of diminished sense perception. In: Proceedings of the 7th augmented human international conference 2016, ACM, p 15

Wood J, Chaparro A, Carberry T, Chu BS (2010) Effect of simulated visual impairment on nighttime driving performance. Optom Vis Sci 87(6):379–386

Zagar M, Baggarly S (2010) Low vision simulator goggles in pharmacy education. Am J Pharm Educ 74(5):83

Zhao Y, Bennett CL, Benko H, Cutrell E, Holz C, Morris MR, Sinclair M (2018) Enabling people with visual impairments to navigate virtual reality with a haptic and auditory cane simulation. In: Proceedings of the 2018 CHI conference on human factors in computing systems, pp 1–14

Zhao Y, Cutrell E, Holz C, Morris MR, Ofek E, Wilson AD (2019) Seeingvr: a set of tools to make virtual reality more accessible to people with low vision. In: Proceedings of the 2019 CHI conference on human factors in computing systems, pp 1–14

Zhao Y, Kupferstein E, Castro BV, Feiner S, Azenkot S (2019) Designing ar visualizations to facilitate stair navigation for people with low vision. In: Proceedings of the 32nd annual ACM symposium on user interface software and technology, pp 387–402

Zhao Y, Kupferstein E, Rojnirun H, Findlater L, Azenkot S (2020) The effectiveness of visual and audio wayfinding guidance on smartglasses for people with low vision. In: Proceedings of the 2020 CHI conference on human factors in computing systems, pp 1–14

Zhao Y, Szpiro S, Azenkot S (2015) Foresee: a customizable head-mounted vision enhancement system for people with low vision. In: Proceedings of the 17th international ACM SIGACCESS conference on computers & accessibility, pp 239–249

Chapter 15
Patient-Specific Anatomy: The New Area of Anatomy Based on 3D Modelling

Luc Soler, Didier Mutter, and Jacques Marescaux

Abstract 3D anatomical medical imaging (CT-scan or MRI) can provide a vision of patient anatomy and pathology. But for any human, even experts, these images have two drawbacks: each voxel density is visualized in grey levels, which are totally inadequate for human eye cones' perception, and the volume is cut in slices, making any 3D mental representation of the real 3D anatomy of the patient highly complex. Usually, the limits of human perception are overcome by human knowledge. In anatomy, this knowledge is a mix between the average anatomy definition and anatomical variations. But how to understand an anatomical variation from slices in grey levels? In routine, such a difficulty can sometimes be so important that it creates errors. Fortunately, these mistakes can be overcome through 3D patient-specific surgical anatomy. New computer-based medical image analysis and associated 3D modelling provide a highly efficient solution by allowing a patient-specific virtual copy of their anatomic reconstruction. Some articles reporting clinical studies show that up to one-third of initial planning is modified using 3D modelling, and that this modification is always validated efficiently intraoperatively. They also demonstrate that major errors can thus be avoided. In this chapter, we propose to illustrate the 3D modelling process and the associated benefit on a set of patient clinical cases in three main domains: liver surgery, thoracic surgery, and kidney surgery. In each case, we will present the limits of usual medical image analysis due to an average anatomical definition and limited human perception. 3D modelling is provided by the Visible Patient online service and the surgeon then plans his/her surgery on a simple PC using the Visible Patient Planning software. We will then compare the result obtained from a normal anatomical analysis with the result obtained from the 3D modelling and associated preoperative planning. These examples illustrate the great benefit of using patient-specific 3D modelling and preoperative virtual planning in comparison with the usual

L. Soler (✉)
Visible Patient, 8 rue Gustave Adolphe Hirn, 67000 Strasbourg, France
e-mail: luc.soler@visiblepatient.fr

L. Soler · D. Mutter
Medical Faculty of the Strasbourg University, Strasbourg, France

D. Mutter · J. Marescaux
IRCAD, Strasbourg, France

© The Author(s), under exclusive license to Springer Nature Switzerland AG 2021   285
J.-F. Uhl et al. (eds.), *Digital Anatomy*, Human–Computer Interaction Series,
https://doi.org/10.1007/978-3-030-61905-3_15

anatomical definition using only medical image slices. These examples also confirm that this new patient-specific anatomy corrects many mistakes created by the current standard definition, increased by physician interpretation that can vary from one person to another.

15.1   Introduction

It is impossible to imagine any medical care without a minimum of anatomical knowledge. In the case of surgery, In surgery, a full understanding is mandatory. "Modern anatomical description", introduced by André Vésale in the sixteenth century, is based on a description of human anatomy from "alive human or having lived" represented by an average and standardized anatomy. But patients being different, this average anatomy has been completed by variations or exceptions. Since André Vésale, anatomy has been progressively improved thanks to new techniques and technologies, increasing variations but making the average anatomy more precise. Such an anatomy definition has the main benefit: it allows physicians to use standardized names and labels. Surgical procedures have since been more easily explained and described for better knowledge sharing. It also has a drawback: it can lead to an error when applying normalized anatomy on a patient with an anatomical variation. Fortunately, during the past century, medical imaging paved the way for a new revolution: internal anatomy of a patient could be seen without any invasive techniques. This medical imaging can thus provide patient-specific anatomical data including geometry, topology and also function of organs. But some limits linked to these new technologies appeared in parallel. Firstly, interpretation of the medical image information and of the visible anatomical variation is totally dependent on the physician's knowledge and can vary from one to another. Secondly, there are more and more anatomical variations discovered during medical image analysis or intraoperative exploration, all patients being different. These differences can induce errors in the patient anatomical description and sometimes also in therapy selection.

The liver is here a perfect illustration of such limits. Surgery offers the foremost success rates against liver tumour (more than 50% 5-year survival rate). Regretfully, less than 20% of patients are eligible for surgery due to anatomical limitations. Indeed, eligibility is based on multiple criteria and rules. It is thus established that two adjacent **liver segments** can be separated with an adequate vascular inflow and outflow as well as biliary drainage and that the standardized Future Liver Remnant (*Standardized FLR = remnant liver volume/liver volume*) must be over 20% for patients with an otherwise normal liver, 30% for patients who have received extensive preoperative systemic chemotherapy, and 40% for patients with existing chronic liver diseases such as hepatitis, fibrosis or cirrhosis. Precise knowledge of patient liver anatomy is thus a key point for any surgical procedure, including resection of liver tumours or living donor transplant, surgical eligibility being linked to the definition of liver segments. But defined preoperatively, patient liver anatomy remains a challenge. Indeed, there exist today four main anatomical definitions used routinely worldwide: The Takasaki segments definition (Takasaki 1998), essentially used in Asia, the Goldsmith and Woodburn sectors (Goldsmith and Woodburne 1957) definition, essentially used in North America, the corrected Bismuth sectors (Bismuth 1982) definition, essentially used in Europe, and the Couinaud segment (Couinaud 1999) definition used worldwide.

These definitions are based on labelling the portal tree distribution in the liver following essentially geometrical criteria on relative location in the liver: right, middle, left, anterior, posterior, lateral, median, and caudal. We can also notice that hepatic veins define separating limits between main sectors in Goldsmith and Woodburn and Bismuth definitions. This general overview also clearly illustrates that Couinaud segmentation is the most precise one. All other segmentations can be obtained by a grouping of Couinaud segments in different sets. But Couinaud segmentation contains major errors. Platzer and Maurer (Platzer and Maurer 1966) surely were the first ones to show in 1966 that the variability of segment contours was too important for any general scheme to be viable. Many researches (Nelson et al. 1990; Soyer et al. 1991; Fasel et al. 1988; Rieker et al. 2000; Fischer et al. 2002; Strunk et al. 2003; Fasel 2008) have subsequently completed that first study by providing quantifiable results thanks to 3D medical imaging. Couinaud himself (Couinaud 2002) described topographic anomalies in 2002. In 34 cases out of 111 (i.e. 30,63% of cases), he demonstrated that the real anatomical anterior sector of the liver (segment V + segment VIII) was different from his own definition. This may have surgical consequences. Thus, by clamping the right paramedian vein, portal branches which are topologically considered as being in segment VI took in fact their origin on the right paramedian branch, and were topologically in the anterior sector of the liver. Couinaud concluded that there was incoherence between vascular topology and the topography of the segments that could be corrected by using our 3D modelling and segmentation software (Soler et al. 2001) that we have clinically validated.

All patients have different anatomy that is all the more difficult to predict since some pathologies can modify vascularization. Moreover, even if the medical image contains all the information needed to extract the patient-specific anatomical configuration, it is too complex for any physician to mentally reconstruct this anatomy in 3D from the set of 2D slices provided by the medical imaging system. To overcome this human limit, we have developed a computer-assisted surgery system over the past 20 years and this system has been clinically validated (Bégin et al. 2014 Dec; Soler et al. 2014 Mutter et al. 2010, 2009) to obtain CE marking and FDA approval for clinical use. The resulting computer-assisted surgery system is composed of several steps: 3D patient modelling, preoperative surgical planning and intraoperative use of this virtual reality assistance. In this chapter, we will describe the technique and illustrate these different elements and results in three main areas: liver, thorax and kidney surgery. We will then conclude the real need for such patient-specific anatomy based on 3D modelling.

15.2   Materials and Methods

The first expected benefit of a computer-assisted system applied to patient-specific anatomy is to provide a fast, efficient and easy way to implement a view of the patient's anatomy. Any software meeting these needs should allow for the reading of images recorded during a clinical routine in DICOM, the international standard format. Moreover, such software should provide at least two types of immediate rendering: a 2D view of image slices and a 3D view. Currently, many of the available software applications for the visualization consoles of radiology departments must be paid for or can be freely downloaded from the Internet. Osirix is the most notorious and used software essentially used by radiologists. Although it is very complete, it presents two drawbacks: it only works on macOS and its user interface is not particularly intuitive for surgeons as it is too similar to the software of post-processing consoles used in radiology. Whether free or commercial, we have noticed that surgeons scarcely use these applications due to their complexity: the user interface is submerged with complicated options and lengthy training is sometimes required to properly use the software.

To overcome this recurring drawback, we have developed a software, Visible Patient Planning™ (©Visible Patient 2014 https://www.visiblepatient.com/en/pro ducts/software), that is free of charge as Osirix™. But the free version of Osirix is not certified, unlike Visible Patient Planning software that is CE-Marked and FDA approved (510 k). Moreover, Visible Patient Planning works on macOS and Windows. Whatever you use, workstation, Osirix or Visible Patient Planning, the first advantage for surgeons is direct volume rendering which is automatically computed by the software from the CT or MRI slices of the DICOM image (Fig. 15.1). This free technique provides adequate 3D visualization of anatomical and pathological structures and can thus be a useful preoperative planning tool. In order to see internal structures, the initial voxel grey level is replaced by an associated voxel colour and transparency. This transparency allows to distinguish more contrasted anatomical or pathological structures, even when they are not delineated in reality. That volume can also be cut along the three main axes (axial, frontal or sagittal) or with an oblique mouse-controlled plane. In clinical routine, direct volume rendering can be of considerable preoperative interest. This is the case for all malformation pathologies, in particular vascular or bone malformation, but also for thoracic and digestive pathologies.

Direct volume rendering is thus a very useful tool as it is freely accessible without any pre-processing; however, it does have some limitations. It cannot provide the volume of organs nor their dimensions since these organs are not delineated. For the same reason, it is not possible to provide a volume after resection, or to cut a section of these structures without cutting neighbouring structures. To overcome this limit, each anatomical and pathological structure in the medical image has to be delineated. This task, named "segmentation", can be performed through a specific workstation available on the market (Myrian™ from Intrasense, Synapse™ from Fuji) or through a distant online service that can be compared to a medical analysis

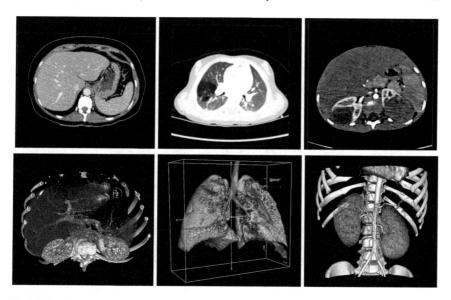

Fig. 15.1 Direct volume rendering (second line) of three different clinical cases from their DICOM image (first line), here from CT-scan of liver (left), lung (middle) and kidney (right) using Visible Patient Planning^TM software

laboratory (Visible Patient Service). In the first solution, hospitals acquire a workstation and then physicians can use it by themselves to perform 3D modelling. In the second solution, hospitals purchase the image analysis for each patient just like they would pay for a biological analysis. Moreover, this last pay per case solution is now covered by some private insurance companies (more than 45% of French citizen will be thus covered in France in September 2020), making it more easily accessible. Each solution (onsite or online) allows for a 3D surface rendering of organs as well as a volume computation of delineated structures. In that set of solutions, Visible Patient is today the only service available for any part of the body and for any pathology or organ ranging from infant to adult. The result of the 3D modelling process can be visualized from the free Visible Patient Planning software through surface rendering but can also be fused with the volume rendering (Fig. 15.2).

This surface rendering of delineated structures provides a more advanced anatomical view of the patient, but it remains insufficient for several surgeries such as partial resection that needs preoperative evaluation of future volumes remaining after resection. More advanced solutions give so the opportunity to simulate virtual resection and to obtain preoperatively the resulting volume. Some software, such as Myrian^TM, offer this possibility from virtual cutting planes. Some other, such as Synapse^TM or Visible Patient Planning^TM, have a more anatomy-oriented approach based on vascular territory simulation and virtual clip applying with interactive placement (Fig. 15.3).

By defining the vascular territories, the resulting patient-specific anatomy is therefore not only a geometrical patient-specific anatomy but also a functional anatomy

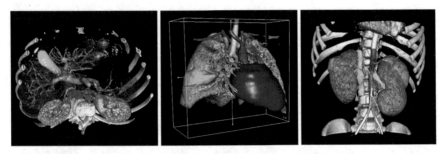

Fig. 15.2 Visible Patient Planning™ Fusion between Direct volume rendering and surface rendering of organs provided by the Visible Patient online service of the same Fig. 15.1 patients

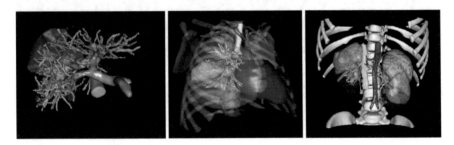

Fig. 15.3 Virtual clip applying and resulting devascularized territories simulated by Visible Patient Planning™ from the same Fig. 15.1 patients

that can be used preoperatively to define more accurately a surgical procedure. It can also be used intraoperatively to guide the surgeons thanks to the development of intraoperative tools. For instance, the Visible Patient Planning™ software can be brought inside the OP-room to directly visualize the result on the laptop or smartphone, or indirectly by plugging it on an OP-Room display or on a surgical robot such as the Da-Vinci Robot from Intuitive Surgical (Fig. 15.4).

To illustrate the clinical benefits of such patient-specific computer-assisted anatomy and surgical planning system, for several years, we have applied it to a wide

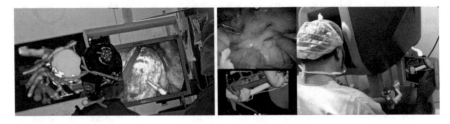

Fig. 15.4 Intraoperative use of Visible Patient Planning™ plugged on Operative Display (left) or on robotic display (right)

range of surgical procedures in digestive, thoracic, urological, endocrine, and paediatric procedures. We will here limit our description of clinical benefits to the same three patients illustrated in Figs. 15.1, 15.2 and 15.3. Then we will provide some other article references illustrating the clinical benefits of such computer-assisted patient-specific anatomy before concluding.

15.3   Results

The first patient was a 55-year old patient diagnosed with hepatocarcinoma. From the CT image only, a single tumour (yellow circle in Fig. 15.5 the left) was detected on the back of the right hepatic vein, i.e. the most right-hand part of the liver. The Visible Patient Online 3D modelling provided a 3D modelling of an additional nodule (red circle in Figs. 15.5 and 15.6) validated by the radiologist after a second medical image analysis as a potential hepatic tumour. This second tumour was located between the median hepatic vein and the right hepatic vein, i.e. again in the right liver. Additional diagnosis from the radiological department was a right liver volume of about 60% of the global liver volume, but this volume was computed without the Visible Patient Planning software. By observing this radiological analysis and using standard anatomical landmarks, the surgical team considered a right liver resection, the second tumour (red circle) position being too deep and in contact with too large vessels around to be resected or burnt through thermal ablation.

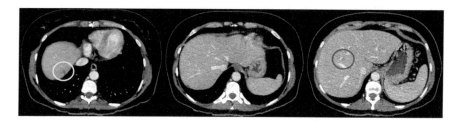

Fig. 15.5   Three axial slices of the first patient with two small hepatic tumours (circles)

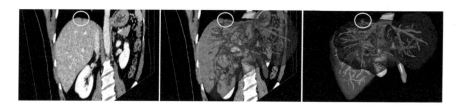

Fig. 15.6   3D modelling provided online by Visible Patient Service and allowing to visualize two hepatic tumours (circles). From virtual clip applying performed with Visible Patient Planning software, the 3D view of the left liver (blue) and the right liver (orange) shows the location of both tumours in the left liver and not in the right liver

But by using the Visible Patient Planning software, surgeons simulated this right liver resection by virtually clipping the right branch. The orange territory on Fig. 15.6 show this right liver, but none of the tumours were located in this real portal vein territory representing only 45% of the liver volume. On the opposite, by virtually clipping the left portal branch (blue territory on Fig. 15.6), both tumours appeared in this devascularized left liver representing 55% of the liver volume. Finally, by analysing more accurately the reason for this error, it appears easily in 3D that this patient had no right paramedian branches providing blood flow to segments 5 and 8. A new unusual branch coming from the left portal branch was providing the blood supply of a territory located in the usual segment 8 area, but vascularised from the left portal branch. Virtual clip applying on this branch provided thus the smallest territory to resect (13,3% of the liver volume, blue area in left image of Fig. 15.3).

This example illustrates the well-known problem of hepatic anatomical variation already described by Couinaud (Couinaud 2002) and that we pointed out in a recent article (Soler et al. 2015). In fact, studies (Soler et al. 2014 Wang et al. 2017 Jun) show that in more than one-third of hepatic surgical interventions, 3D modelling and preoperative simulation allow to correct imperfections or errors of the initial surgical planning. Another study (He et al. 2015 Sep 21) shows that for some interventions it can reduce operating time by 25% and reduce complications by more than one-third. These impressive results in link with the Couinaud analysis illustrate the huge anatomical variation of liver and its potential consequences on treatment choice. They also illustrate the great benefit of using computer-assisted patient-specific 3D modelling preoperatively to avoid such medical image interpretation errors in liver surgery.

But what about lung surgery? Today, surgeons operate lung cancers like the liver by resecting the sick part of the bronchial tree, like cutting unhealthy branches of a tree to preserve the healthy part. Like in liver surgery, the difficulty will lie in defining the territories of that bronchial tree, which are very hard to define and localize from a simple CT image. It is all the more complex as the pathology, cancerous or not, can locally modify the anatomy or mask landmarks that are however present in the image. The second example illustrated in Figs. 15.1, 15.2 and 15.3 is a 6-month old child with a lung cyst due to a cystic adenomatoid lung disease, a pathology of airways that requires the sick part of the bronchial tree to be resected, just like for a tumorous pathology. From the CT medical image of the child, the cyst (black part in the centre of the left image in Fig. 15.7) seemed located in the upper lobe of the right lung. This was the diagnosis performed by the radiological team and validated by the surgical team. But after the online 3D modelling delivered by the Visible Patient service, the virtual clip performed by the surgeon showed that the cyst (in green in the right image of Fig. 15.7) was not in the upper right lobe (in yellow). Thanks to this 3D modelling, the surgery was modified and performed flawlessly, validating the preoperative simulation's efficiency.

In fact, 3D modelling provides the same benefits in lung tumour resection as for liver tumours. Applying preoperative virtual clips allows thus to simulate bronchial territories as well as to simulate portal vein territories of the liver. Moreover, the 3D modelling of internal lung structures is not limited to the bronchial system but

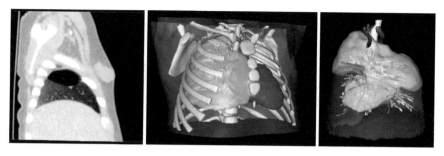

Fig. 15.7 Pulmonary adenomatoid cyst detected in the right upper lobe from a CT-scan (left) of a 6-month old patient. 3D modelling (centre) provided by Visible Patient online service and the clip applying simulation of the right upper lobe (yellow) showed that the cyst was not in this territory and avoided the error preoperatively

includes lung arteries and veins. This technique allows to avoid errors in pathological territory definition and to improve the surgical therapy planning as validated in several recent articles and clinical studies (Gossot et al. 2016; Le Moal et al. 2018; Gossot and Seguin-Givelet 2018a, b).

These two first application examples illustrated the benefits of computer-assisted patient-specific 3D anatomy in surgical procedures using the functional anatomy definition. Liver and lung have anatomically defined vascular territories that anatomical variations can disturb and thus create errors in the choice of treatment. Finally, we propose to end with kidneys, which do not have a functional anatomical segmentation because their vascularisation variations are too important.

The third example, illustrated in Figs. 15.1, 15.2 and 15.3, was a 5-year-old patient diagnosed with a double nephroblastoma. Even though this cancer is frequent in adults (over 13.000 new patients each year), only very few children are concerned each year in France (130 new patients per year). From the CT-scan only (on the left in Fig. 15.8), the expert team for this kind of surgery proposes to resect half of the right kidney and full resection of the left kidney, where, according to the images, tumour invasion is too important. This surgery will necessarily induce renal insufficiency, which will translate into maximum 6 months of dialysis. These 6 months will allow

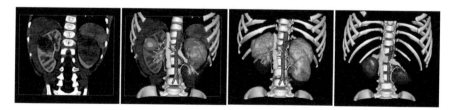

Fig. 15.8 Double nephroblastoma detected on both kidneys of a 5-year-old child. Thanks to 3D modelling and preoperative simulation of clip applying using Visible Patient service and planning software, a partial resection of each kidney was validated and realized. As illustrated on the right, 1 year later the patient seems to be cured

to check that no new tumour regrowth appears in the lower part of the remaining right kidney. After these 6 months and if no tumour has appeared, left kidney transplant will be proposed. Transplant will increase the child's life expectancy by more than 50 years but will also induce lifelong anti-rejection treatment. This therapeutic proposition is submitted to a second team, also expert in that kind of pathology, for a second medical opinion. The second team fully validates this choice of treatment.

From the modelling and by using the VISIBLE PATIENT Planning™ software, the surgeon simulates applying surgical clip (clipping vessels) and validates the possibility of resecting only half of the right kidney, 50,9% of the right kidney remaining functional after surgery. But the surprise comes from the left kidney, as when simulating surgery, the surgeon sees that he can preserve one-third of the kidney function. Yet volume computation provided by the software shows that both remaining kidney parts on the left and on the right after surgery will have a slightly greater volume than the volume of one kidney in a child of that age and size. Therefore, the surgical choice is modified and preservation surgery of the functional part of the left kidney is proposed and validated by a second team. Surgery is done in two steps, in January 2018, and by the end of January no renal insufficiency has been noted after resection of the diseased part of both the child's kidneys. The child went back to school in February. One year later, the control image shows no tumour regrowth and the child is in perfect health condition (right image).

As illustrated by this example, the same benefit observed in liver and lung is obtained on kidney by using virtual clipping on the computer-assisted 3D modelling of the patient. Such 3D modelling can be used for any other urological surgery and pathologies such as renal pelvis dilatation (Fig. 15.9 left), crossed or crossed fused renal ectopia (Fig. 15.9 middle), or kidney transplant (Fig. 15.9 right). In any case, the precise 3D visualization of vascular structures, ureters and surrounding organs provides a major benefit in surgical procedure planning (Soler 2016a, b).

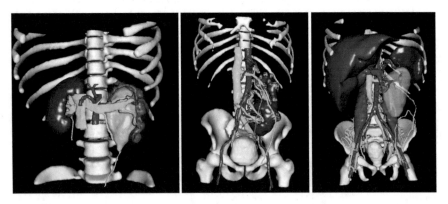

Fig. 15.9 Left renal pelvis dilatation (left), left crossed fused renal ectopia (middle) and simulation of kidney transplant (in blue on the right)

15.4 Discussion and Conclusion

These 3 examples illustrate the high benefit of using patient-specific 3D modelling and preoperative virtual planning in comparison with the usual anatomical definition using medical image slices alone. It also confirms that this new patient-specific anatomy corrects many potential mistakes created by the current standard definition, increased by the physician's interpretation that can vary from one person to another. Such patient-specific anatomy based on 3D modelling from medical imaging of the patient should now be used to propose new anatomical definition.

Thus, in a recent article[19], we have proposed a new anatomical segmentation of the liver based on 4 main rules to apply in order to correct topological errors of the four main standard segmentations. In the past, the only way to correct usual anatomical mistakes was to clamp vessels during surgery, associated vascular territories then appearing clearly. By applying these rules and using preoperative 3D modelling, we can now obtain the same results preoperatively, these rules being based on the surgical logic of vascular territory clamping and using Virtual Reality technologies. Moreover, more recent software can simulate in the same way virtual clipping on a vessel and thus provide virtually the vascular territory in real time. These rules should so be applied on any organ to optimize and personalize their functional anatomical definition.

In the past, such 3D modelling was limited and not frequently used due to processing time, complexity and cost of such systems. The fact that private health insurances cover the cost of such online modelling should generalize now the use of patient-specific anatomy based on 3D modelling. This increase of use should allow further numerous studies showing the benefit of such new anatomy in routine practice. The next step will be to use it intraoperatively through Augmented Reality (AR). AR merges virtual patient-specific anatomy onto the real surgical view. The patient will then become virtually transparent, as illustrated by our work in that domain (Soler et al. 2017). This 3D patient-specific anatomy can be indeed considered as a patient map, the next step will thus be to develop the surgeon's GPS.

Acknowledgements These research works contain a part of results obtained in the FP7 E-health project PASSPORT, funded by the European Commission's ICT program and in the French PSPC project 3D-Surg supported by BPI France.

References

Bégin A, Martel G, Lapointe R, Belblidia A, Lepanto L, Soler L, Mutter D, Marescaux J, Vandenbroucke-Menu F (2014) Accuracy of preoperative automatic measurement of the liver volume by CT-scan combined to a 3D virtual surgical planning software. Surg Endosc 28(12):3408–3412
Bismuth H (1982) Surgical anatomy and anatomical surgery of the liver. World J Surg 6:3–9

Couinaud C (1999) Liver anatomy: portal (and suprahepatic) or biliary segmentation. Digestive Surgery 1999(16):459–467

Couinaud C (2002) Erreur dans le diagnostic topographique des lésions hépatiques. Ann Chir 127:418–430

Fasel J (2008) Portal venous territories within the human liver: an anatomical reappraisal. Anat Rec 291:636–642

Fasel JHD, Selle D, Evertsz CJC, Terrier F, Peitgen HO, Gailloud P (1988) Segmental anatomy of the liver: poor correlation with CT1. Radiology 206:151–156

Fischer L, Cardenas C, Thorn M, Benner A, Grenacher L, Vetter MA, Lehnert T, Klar E, Meinzer H-P, Lamade W (2002) Limits of Couinaud's liver segment classification: a quantitative computer-based three-dimensional analysis. J Comput Assist Tomogr 26(6):962–967

Goldsmith NA, Woodburne RT (1957) Surgical anatomy pertaining to liver resection. Surg Gynecol Obstet 105:310

Gossot D, Seguin-Givelet A (2018a) Thoracoscopic right S9 + 10 segmentectomy. J Vis Surg 4:181

Gossot D, Seguin-Givelet A (2018b) Anatomical variations and pitfalls to know during thoraco-scopic segmentectomies. J Thorac Dis. 10(Suppl 10):S1134–S1144

Gossot D, Lutz J, Grigoroiu M, Brian E, Seguin-Givelet A (2016) Thoracoscopic anatomic segmentectomies for lung cancer: technical aspects. J Viz Surg 2(171):1–8

He Y-B, Bai L, Aji T, Jiang Y, Zhao J-M, Zhang J-H, Shao Y-M, Liu W-Y, Wen H (2015) Appli-cation of 3D reconstruction for surgical treatment of hepatic alveolar echinococcosis. World J Gastroenterol 21(35):10200–10207

Le Moal J, Peillon C, Dacher JN, Baste JM (2018) Three-dimensional computed tomography reconstruction for operative planning in robotic segmentectomy: a pilot study. J Thorax Dis 10(1):196–201

Mutter D, Dallemagne B, Bailey C, Soler L, Marescaux J (2009) 3D virtual reality and selective vascular control for laparoscopic left hepatic lobectomy. Surg Endosc 23:432–435

Mutter D, Soler L, Marescaux J (2010) Recent advances in liver imaging. Expert Rev Gastroenterol Hepatol 4(5):613–621(9)

Nelson RC, Chezmar JL, Sugarbaker PH, Murray DR, Bernardino ME (1990) Preoperative local-ization of focal liver lesions to specific liver segments: utility of CT during arterial portography. Radiology 176:89–94

Platzer W, Maurer H (1966) Zur Segmenteinteilung der Leber. Acta Anat 63:8–31

Rieker O, Mildenberger P, Hintze C, Schunk K, Otto G, Thelen M (2000) Segmental anatomy of the liver in computed tomography: do we localize the lesion accurately? Röfo 172(2):147–152

Soler L (2006) Surgical Simulation and Training applied to urology, Annual meeting of the British Association of Urological Surgeons, Manchester (Royaume-Uni), 30th June 2006

Soler L (2016) La technologie 3D dans la chirurgie urologique d'aujourd'hui et de demain. Congrès Urofrance Paris

Soler L, Delingette H, Malandain G, Montagnat J, Ayache N, Koehl C, Dourthe O, Malassagne B, Smith M, Mutter D, Marescaux J (2001) Fully automatic anatomical, pathological, and functional segmentation from CT scans for hepatic surgery. Comput Aided Surg 6(3):131–142

Soler L, Nicolau S, Pessaux P, Mutter D, Marescaux J (2014) Real time 3D image reconstruction guidance in liver resection surgery. Hepatob Surg Nutrit 3(2):73–81

Soler L, Mutter D, Pessaux P, Marescaux J (2015) Patient-specific anatomy: the new area of anatomy based on computer science illustrated on liver. J Vis Surg 1:21

Soler L, Nicolau S, Pessaux P, Mutter D, Marescaux J (2017) Augmented reality in minimally invasive digestive surgery, Pediatric Digestive Surgery 2017, Mario Lima Editor, Springer, 421–432

Soyer P, Roche A, Gad M et al (1991) Preoperative segmental localization of hepatic metastese: ulility of three-dimensional CT during arterial portography. Radiology 180:653–658

Strunk H, Stuckmann G, Textor J, Willinek W (2003) Limitations and pitfalls of Couinaud's segmentation of the liver in transaxial imaging. Eur Radiol 13:2472–2482

Takasaki K (1998) Glissonean pedicle transsection method for hepaticesection. A new concept of liver segmentation. J of Hepato-Biliary-Pancreatic Surg 5(3):286–291
Wang X-D, Wang H-G, Shi J, Duan W-D, Luo Y, Ji W-B, Zhang N, Dong J-H (2017) Traditional surgical planning of liver surgery is modified by 3D interactive quantitative surgical planning approach: a single-center experience with 305 patients. Hepatobiliary Pancreat Dis Int 16(3):271–278

Chapter 16
Virtual and Augmented Reality for Educational Anatomy

Bernhard Preim, Patrick Saalfeld, and Christian Hansen

Abstract Recent progress in VR and AR hardware enables a wide range of educational applications. Anatomy education, where the complex spatial relations of the human anatomy need to be *imagined*, may benefit from the immersive experience. Also the integration of virtual information and real information, e.g., muscles and bone overlaid on the user's body, are beneficial for imaging the interplay of various anatomical structures. VR and AR systems for anatomy education compete with other media to support anatomy teaching, such as interactive 3D visualization and anatomy textbooks. We discuss the constraints that must be considered when designing VR and AR systems that enable efficient knowledge transfer.

16.1 Introduction

Anatomy education aims at providing medical students with an in-depth understanding of the morphology and function of anatomical structures, their position, and spatial relations, e.g., connectivity and innervation. Students should be able to locate anatomical structures, which is an essential prerequisite for radiological and surgical interventions. They should also be aware of the variability of the morphology and location, e.g., of branching patterns of vascular structures. Furthermore, they should understand functional aspects, e.g., the range of motion possible with a joint or the function of the beating heart. Recently, anatomy education is often combined with radiological image data, e.g., CT and MRI data, to convey the appearance of anatomy in cross-sectional images, as physicians encounter it in clinical practice. The clinical

B. Preim (✉) · P. Saalfeld · C. Hansen
Department of Simulation and Graphics, University of Magdeburg, Magdeburg, Germany
e-mail: bernhard@isg.cs.uni-magdeburg.de

P. Saalfeld
e-mail: saalfeld@isg.cs.uni-magdeburg.de

C. Hansen
e-mail: hansen@isg.cs.uni-magdeburg.de

299

reference demonstrated by describing possible diseases, patients, and examination methods also plays an important role as a motivator to students.

Traditional methods for anatomy education involve lectures, the use of text books, and atlases as well as cadaver dissections (Preim and Saalfeld 2018). Cadaver dissection is essential because it is an active type of learning, involving the training of manual dexterity and communication skills (Brenton et al. 2007). The realism of these dissections, however, is limited, because the color and texture of anatomical structures in cadavers differ strongly from living patients. Moreover, cadaver dissection is an expensive type of training and cadavers can only be used for a short period. Textbook presentations provide valuable information, but the 2D nature does not well support the recognition of real three-dimensional structures, e.g., in a cadaver.

Anatomy education is not only a component of the education in medicine, but also essential in other health-related disciplines, e.g., physiotherapy and sport, where an in-depth understanding of muscles is essential. For students in other disciplines, cadaver dissection is not available at all. Interactive 3D visualizations are another active type of learning and enable learners to explore the spatial relations themselves, e.g., by peeling off structures, rotating a geometric model, and clipping parts of it. This computer-based training, like training in other areas, may be particularly effective when it is combined with specific learning tasks and integrated with other more traditional forms of learning. Anatomy education was one of the driving applications of medical visualization with the VOXEL-MAN as the outstanding example (Pommert et al. 2001). Like other anatomy education systems, it was based on the Visible Human Dataset from the National Library of Medicine (Spitzer et al. 1996). These high-quality 3D datasets originate from two bodies that were given to science, frozen, and digitized into horizontally spaced slices. In recent years, also special hardware was developed dedicated to anatomy education. Anatomy tables providing virtual dissection facilities on a table with a similar size as the OR table were introduced. As an example, the SECTRA table provides a touch-based interface and high-quality volume rendering for anatomy education (Lundstrom et al. 2011). The Anatomage table experienced broad acceptance in anatomy education. Its specific value for learning anatomy was assessed in many studies, e.g., by Custer and Michael (2015).

VR and AR may strongly contribute to an *imagination* of complex anatomical regions. As an example, hand anatomy (Boonbrahm et al. 2018) or anatomy of the ear region (Nicholson et al. 2006) involve a high density of delicate interwoven anatomical structures, such as bones, tendons, and nerves. Interactive 3D visualizations or even immersive VR solutions have the potential to enable learners to develop mental models of these regions. AR provides an overlay of virtual information on real-world objects, such as a physical model. This overlay should be displayed in a natural and seamless manner to correctly convey the spatial relations.

The advent of web-based graphics standards, in particular the introduction of WebGL, and the progress of affordable VR and AR glasses triggered the development of new systems in research and practice. The high degree of immersion in VR may enable new and strong learning experiences, e.g., users may virtually move inside the skull, and walk along the foramina—small holes that enable vascular structures and nerves to enter another part of the skull. AR enables to add guiding information, e.g., arrows and labels, on the real world. In particular, mobile AR has a strong potential for wide use. VR and AR are not restricted to the visual sense. It may involve audio feedback, e.g., to indicate how close the user is to a target structure, and tactile feedback, e.g., to convey the stiffness of different tissue types. Besides technically motivated developments, progress was also made w.r.t. motivational design partially inspired by serious games and w.r.t. a proper balance between self-directed learning and guidance. The essential questions from an application point of view are, of course, how the learning is affected and which students will benefit from the introduction of VR and AR in anatomy education.

Organization. In this chapter, we first discuss fundamentals related to the educational use of VR and AR (Sect. 16.2). We continue with an overview of AR solutions for anatomy education (Sect. 16.3), which involves also evaluations related to the learning effect. The following section is dedicated to VR solutions for anatomy education (Sect. 16.4). It includes an overview of semi-immersive and immersive systems and describes immersive systems in more detail.

16.2   Fundamentals of VR and AR in Medical Education

In this section, we discuss fundamentals that are essential for both VR and AR solutions. We start with a discussion of cognition and learning theory and continue with some remarks on the generation of 3D models that are the basis for any type of interactive exploration in VR and AR.

16.2.1   Learning Theoretical Foundations

Anatomy education may be discussed from a *teacher perspective* and from a *learner perspective*. The teacher perspective involves the tools necessary to *author* interactive 3D visualizations and relate them to the symbolic information (anatomic names, categories, relations) as well as self-assessment tools to directly support learning. These tools comprise segmentation and model reconstruction. In this survey, however, we focus on the *learners' perspective*, i.e., the character and quality of the available information and the techniques to explore this information and acquire knowledge.

For the students, the usage of current VR and AR technologies has several advantages (Martín-Gutiérrez et al. 2017):

- Their motivation and engagement is increased by being immersed.
- VR and AR solutions are more affordable.

- The recent development of virtual technologies makes them more accessible regarding the hardware and anatomical learning content.
- The interaction with the virtual content allows a higher degree of freedom.

The design of VR and AR environments has to consider a number of aspects related to perception, cognition, motivation, and learning theory. We will focus on learning theory as the most specific basis for computer-assisted educational tools.

According to the majority of researchers in education, *active* types of learning have the highest potential to achieve a sustainable learning effort. As a consequence, a carefully prepared video sequence for passive watching may be a good starting point to get familiar with a topic, e.g., anatomy in a particular region. But to achieve a deeper understanding, it must be complemented with other types of learning. Jang et al. (2017) highlight the additional learning effect due to interactive manipulation of 3D content compared to passive viewing of prepared video sequences showing the same content.

VR and AR can be designed to leverage the benefits of active learning. Merriënboer et al. (2002) described and tested a 4-level instruction model for guiding training applications. It supports the transfer of procedural knowledge and was also applied and tested for medical training systems (Mönch et al. 2013), which indicates its relevance for anatomy education.

Knowledge gain. Learning theory also provides means to assess the success of a teaching aids. A general strategy is to assess the *knowledge gain*, i.e., learners take part in a pre-test, get instructed how to use a teaching aid, then actually use the teaching aid, and a final test reveals how much knowledge was gained (Sakellariou et al. 2009). A variant of this strategy is to repeat the final test later to analyze whether the gained knowledge is actually remembered. Knowledge gain in anatomy may be tested by asking users to name certain anatomical structures or to describe them w.r.t. their function, course, and innervation. Probably most teaching aids will lead at least to a slight knowledge gain. Therefore, it is more interesting to compare different teaching aids in their efficiency. An example for such a comparative evaluation related to anatomy teaching is given by Ritter et al. (2002). They let users work with an interactive 3D visualization of the foot anatomy with functions related to labeling and exploded views, where distances between objects are slightly enlarged to reduce occlusion and support shape perception. One group had the additional possibility to use the system as a 3D jigsaw puzzle (see Fig. 16.1), where anatomical structures (muscles, ligaments) need to be docked to the correct parts of the skeleton (see Ritter et al. 2000 for a description of the system). In the other group, this feature was disabled. Thus, the implicit guidance of the jigsaw puzzle could be separately analyzed.

Cognitive load theory considers the limited resources of the human brain to process information. Due to the limited short term memory, information overload may interfere with learning (Sweller 2004). Cognitive load is further distinguished in *intrinsic* cognitive load that comprises the information that need to be processed, and *extrinsic* cognitive load that comprises other information that consume cognitive resources without any effect on learning. Obviously, extrinsic cognitive load, e.g.,

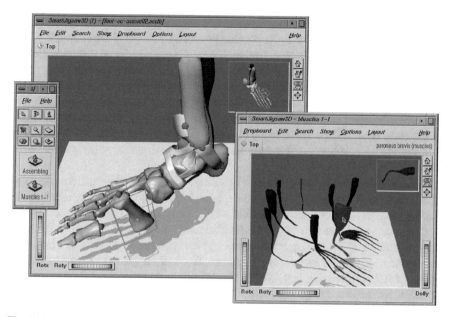

Fig. 16.1 A 3D model of the bones and tendons is correctly assembled in the left view. In the right view, the user has models of all muscles in the foot region. Her task is to move them in the left viewer and dock them at the right position (Courtesy of Felix Ritter, University of Magdeburg)

anything that may distract, should be minimized (Sweller 2004). Thus, it is essential that the cognitive load caused by the user interface is as low as possible to avoid a reduced learning experience. Cognitive load theory is also essential for other 3D teaching aids for anatomy, e.g., computer-generated animations (Ruiz et al. 2009).

Spatial ability. Research on anatomy education clearly reveals that the efficiency of teaching aids may strongly depend on the *spatial ability* of learners. A general strategy is to carry out a standardized mental rotation test with all learners and classify them as low, moderate, and high spatial ability learners (Preim and Saalfeld 2018). This classification is used to separately analyze for these groups how effective learning actually is. In a similar way, it can be expected that VR and AR solutions for anatomy education are not equally beneficial for all user groups.

Embodied cognition. The handling of 3D anatomical structures in a computer system resembles the way students explore such structures in reality, e.g., when they rotate physical models of muscles and tendons in their hands. The advantage of this similarity is due to the phenomenon of *embodied cognition*, i.e., learning benefits from the involvement of body movements. Embodied cognition considers the perceptual and motor systems as essential ingredients of cognitive processes (Chittaro and Ranon 2007). As an example, Kosslyn et al. (2001) found that physical rotation leads to a stronger activation of the motor cortex in a subsequent mental rotation task.

Blended learning. VR/AR systems will be particularly efficient if they are embedded with other forms of learning, supported by anatomy teachers explaining in which stages and how to use them as a kind of *blended learning*. As an example, Philips et al. (2013) show how a virtual anatomy system with radiological image data of a cadaver is shown at a large display in the dissection room to explain students the structures of the corresponding real cadaver. Other blended learning approaches in anatomy education employ tablet PCs with data of the cadaver in the dissection room (Murakami et al. 2014). These types of blended learning are related to the classical morphological anatomy combined with cross-sectional anatomy (Murakami et al. 2014).

Cooperation and other aspects of learning. Communication functions to share thoughts and questions with other students and teachers may further enhance the learning experience (Preim and Saalfeld 2018). Thus, the design of cooperative VR systems and an evaluation how the cooperative functions are actually used is essential. The evaluation of VR/AR systems should also consider other aspects that influence acceptance and finally technology adoption. Radu et al. (2012) also analyzed the effect of AR solutions on collaboration, motivation, attention, and distraction. VR solutions are assessed, e.g,. with respect to immersion and cybersickness.

16.2.2 Model Generation

VR and AR solutions for anatomy education are based on high-quality geometric models. The models should be realistic and provide sufficient detail for the learning tasks they should support. Therefore, often a dataset with particular high spatial resolution and high signal-to-noise ratio is employed. The Visible Human Dataset and the Visible Korean Dataset (employed by Jain et al. 2017) are examples for such datasets. Relevant structures need to be accurately segmented from these datasets. For visualization purposes, the segmented data is transformed in surface meshes and smoothed to avoid distracting details that are often artefacts from the image acquisition. Constrained-based smoothing techniques ensure that a smoothing effect can be achieved without a significant decrease in accuracy. Even in high-quality data, not all relevant anatomical structures may be completely visible. Small structures, such as nerves and also some vascular structures, may be only partially visible.

Some muscles, e.g., the two chest muscles *pectoralis major* and *pectorialis minor* are very thin and lie over each other. Even with a high spatial resolution of 0.5 mm, they are not clearly distinguishable. Thus, the information actually extracted from the image data may be added by information provided by an expert in anatomy using her knowledge on the course of anatomical structures. A final step for providing realistic models is to add textures at least to some anatomical structures where the surface texture is prominent, e.g., muscles. Texturing is mostly done manually using 3D modeling applications. This approach works for single surface models but is too time-consuming if the process from image acquisition to a textured model should be automated. To automatically apply textures to models, procedural terrain modeling techniques can be used. As an example, Saalfeld et al. (2017) applied tri-planar texture mapping to arbitrary vascular structures.

For any serious use in anatomy education, all steps of this process and the final result need to be verified by an expert in anatomy. The model generation described by Pommert et al. (2001) elaborates on these steps with appropriate illustrations.

As a source for anatomy education, commercial 3D model catalogs can be employed. Leading vendors are Digimation (formerly Viewpoint Datalabs), providing almost 600 excellent anatomic models (http://www.digimation.com/the-archive/), and TURBOSQUID, where even 9,000 anatomy models in very good quality are available (https://www.turbosquid.com/3d-model/anatomy). The Open Anatomy Browser (OABrowser) is an example of a free, web-based anatomy atlas viewer (Halle et al. 2017). It provides a growing number of atlases dedicated to different regions, e.g., the brain or inner ear.

16.3 AR in Anatomy Education

AR solutions have to address a number of tasks, including (Boonbrahm et al. 2018):

- capture the pose of real-world objects, e.g., the human hand, head or other body parts,
- map virtual information, e.g., 3D models on a physical object, such as a mannequin, a model produced with 3D printing, or the users' body,
- real-time occlusion handling, and
- update information on the pose of real-world objects in real time to adapt the overlay of virtual information at the right position.

To tackle these problems, a number of toolkits are available. AR Toolkit for marker-based tracking is a classic toolkit that is widely spread (Kato 2007). Vuforia[1] is a more recent toolkit for markerless and marker-based AR where the focus is on mobile solutions (support for various smartphones, tablets, and the HoloLens). We mention it because several AR anatomy education systems were developed based on Vuforia (Boonbrahm et al. 2018).

Before we discuss the specifics of anatomy education, we briefly discuss the use of AR in the two broader areas to which anatomy belongs: medicine and education.

16.3.1 AR in Medicine

AR in medicine is a comprehensive research topic, largely driven by use cases in minimally-invasive surgery and interventions, i.e., situations where the direct sight in the human body is not possible and thus the precise localization of a pathology

---

[1]https://www.vuforia.com/.

is difficult. The potential of AR in these areas is to provide *guidance information*, e.g., to project the position of a tumor or a larger artery on top of an organ to support the physician in locating a pathology and avoiding damage to nearby risk structures. Bajura et al. (1992) and, as a follow-up, Fuchs et al. (1998) did pioneering work in this area providing an AR system for overlaying live ultrasound data on a real (pregnant) patient. A video camera attached to a head-mounted display (HMD) captures the patient and this real data is combined with the ultrasound data acquired by a second user.

The requirements in operative medicine are high: a correct registration of virtual information, derived from (segmented) medical image data, and intraoperative information, often ultrasound or video images, are required. This is already challenging because the character of preoperative data, e.g., CT data, is strongly different from intraoperative information. Moreover, the patient moves and the tissue shifts. Therefore, it is extremely demanding to achieve this correct registration in real-time. The movement deforms anatomical structures in an irregular manner, i.e., structures are not *rigidly* translated or rotated but change their shape. Due to these difficulties, most systems only reached the stage of a proof-of-concept and were assessed based on phantom data or in animal experiments (Scheuering et al. 2003) but did not reach the stage where they serve actual interventions on patients. These examples indicate that camera tracking, computer vision, visual displays, and registration are the major technical areas relevant in medical AR (Chen et al. 2017).

Chen et al. (2017) list only very few examples of actually available AR systems in medical treatment, including a "vein viewer" (Chapman et al. 2011), a system that projects the venous network on top of an arm to support vascular access procedures. Education and rehabilitation are topics in medical AR that gain momentum according to Chen et al. (2017).

Anatomy education has reduced requirements compared to real-time support for interventional procedures. Virtual content can be projected to the user who sees the inner structures in a mirror (Blum et al. 2012) or projected on a cadaver. In both cases, movement is irrelevant.

16.3.2   AR in Education

AR systems are increasingly employed for educational purposes (Akçayır and Akçayır 2017). The number of publications in this area strongly increased after 2012. Bacca et al. (2014) report on the use of AR in various educational settings ranging from natural sciences to social sciences and health. The major motivation to employ AR is to improve student engagement, and the systems analyzed could demonstrate that this was achieved in most systems. However, 30 out of 32 studies were cross-sectional, i.e., the AR system was used and evaluated only once. Bacca et al. (2014) argue that long-term evaluations are necessary to understand whether these motivational effects are stable beyond an initial stage. Also problems and limitations were discussed: users in some evaluations reported on difficulties of interpreting superimposed virtual information and on a distracting effect of this information. As a consequence, the actual visualization technique used to display virtual information needs to be carefully considered. As an example, a tumor may be shown overlaid on an organ as a colored object, as a semi-transparent object or even less obtrusive only with its contours leaving most of the real information unaltered. The interpretation problems are often due to conflicting depth cues: any simple visualization that presents a virtual object overlaid on real data leads to the impression that the virtual object is on top, which of course is wrong in case of a deep-seated pathology. Sielhorst et al. (2006) provide an in-depth discussion of this problem, which they characterize as follows: "Even though the data is presented at the correct place, the physician often perceives the spatial position of the visualization to be closer or further because of virtual/real overlay."

Bacca et al. (2014) also report on two studies addressing an important aspect in any kind of computer-assisted learning: *personalization*, i.e., the preferences and needs of individual learners. Personalization may be supported with a flexible system that enables many adjustments to customize its behavior, but also with an adaptive system that tries to *learn* the preferences of the individual user.

With respect to technology, they observe that most AR systems employ optical markers for tracking the pose of real-world data. These markers are designed to be easily recognizable, e.g., they have a strong contrast and regular patterns. Because technology has improved considerably in the last 5 years, markerless AR might be more important in the future. Markerless AR is more demanding because the pose of the users' hands or body needs to be identified by a depth camera without the support of markers (Chen et al. 2017).

16.3.3 Mobile AR in Anatomy Education

Mobile devices are considered the ideal technical context for AR in educational settings, because AR systems on this basis are small, light, portable, and cost-effective (Akçayır and Akçayır 2017). They are increasingly used for anatomy education. Mayfield et al. (2013) demonstrate the use of tablets integrated into an anatomy dissection course and could demonstrate that the tablet use provides relevant additional information. The use of smartphones and tablets for AR is a major trend that can also be observed in medical applications (Chen et al. 2017). Because these devices are relatively cheap, mobile AR is particularly important for medical education where expensive devices, like in image-guided surgery, could not get wide acceptance. Kamphuis et al. (2014) discussed concepts for this use case (see Fig. 16.2). Jain et al. (2017) prepared a detailed model of the anatomy in the head and neck region by extracting many anatomical structures from the Visible Korean Dataset. They render these data in physical space on top of a tablet. The crucial question is the accuracy of this rendering which they investigated by rendering a simple geometry, namely, a cube. Interactions are provided to select different categories of anatomical structures (see Fig. 16.3).

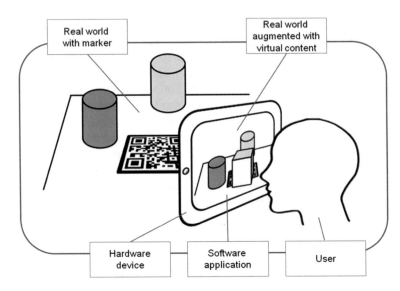

Fig. 16.2 Conceptual sketch of mobile AR. A QR code with its black-and-white pattern serves as a marker to correctly overlay the real and virtual content (From: Kamphuis et al. 2014)

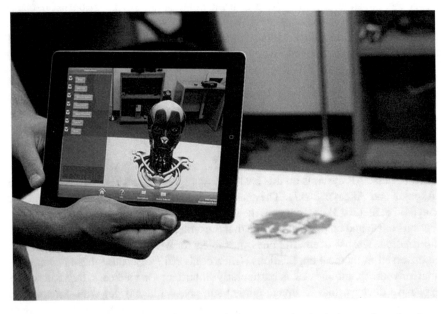

Fig. 16.3 A mobile AR system renders anatomical structures in physical space by enhancing a camera feed with surface models (From: Jain et al. 2017)

16.3.4 Interaction

Interaction is a crucial component of an AR system. Generic interaction tasks for exploring 3D models of human anatomy include rotation, the movement of clipping planes, hiding/displaying single anatomical structures or groups, such as muscles, and to emphasize selected structures. Typical interaction mechanisms from desktop computers, e.g., menu selection, are not intuitive in an AR system (Jain et al. 2017). Thus, more *natural* interaction styles are employed. Natural interaction, however, means that user input needs to be interpreted, which involves uncertainty and ambiguity, i.e., the system not always interprets gestures, for example, as intended.

Gesture-based input is frequently used and requires that the users' hands and arms are reliably detected. Combined with a (rather small) set of easy-to-distinguish gestures, such as the well-known wave, spread, and pinch gestures, commands can be issued. The *Microsoft Kinect* and the *Leap Motion* sensor are examples of depth cameras that enable gesture recognition. Both cameras convert the acquired data to a skeleton representation, e.g., a graph-based representation that represents joints as nodes and edges as the segments between them. The *Kinect* sensor captures 25 joints (Shotton et al. 2011) representing whole body movements (in a lower spatial resolution), whereas the Leap Motion sensor is restricted to the hand and arm.

Speech recognition is another *natural* interaction style that may be employed for AR. However, there are currently no systems using this interaction style for anatomy education.

16.3.5 Anatomy Education

In this subsection, we discuss specific examples of AR systems developed for anatomy education. As real component of an AR system, physical models of anatomical structures or generic anatomic mannequins may be used. Physical models can be created in a straightforward manner with rapid prototyping technology (Thomas et al. 2010). Medical image data can be used as input, segmented, and exported as an STL file that can be processed by a 3D printer. A tracking solution for indoor situations is another essential component. Compared to generic mannequins, physical models derived from medical image data are patient-specific and thus allow to study the variability of the human anatomy if a series of datasets are prepared for this purpose.

AR in anatomy with 3D printing. Thomas et al. (2010) present the BARETA system, an early full AR system based on rapid prototyping and electromagnetic tracking. The physical model is created from an MR scan of a patient and the same dataset is used to generate virtual information and enable clipping and slab rendering (see Fig. 16.4). The electromagnetic tracking was preferred over optical tracking to avoid line-of-sight problems that would easily occur if the user's hand occludes the camera. An electromagnetic sensor is attached to the physical model to support track-

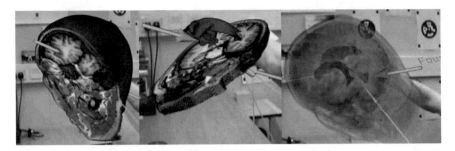

Fig. 16.4 Clipping planes and slab rendering in the BARETA system (From: Thomas et al. 2010)

ing. The virtual content is rendered with surface and volume rendering using VTK. Essential neuro-anatomical structures are labeled and emphasized with pointers.

Thomas et al. (2010) provide a number of ideas to apply their BARETA system to other aspects of the anatomy and to extend its functionality. Physical models, for example, may be associated with a pump where a liquid may be circulated that mimics blood flow.

Virtual mirror. Blum et al. (2012) introduce the compelling idea to use the users' body as real-world object where virtual information is overlaid. Organs and skeletal structures are displayed on the users' body at approximately the position where the organs actually are. This gives an intuitive sense of their location in relation to the skin and other landmarks. Because this application requires full body tracking, the *Kinect* is used as input device. The system is controlled with gestures and enables an exploration in different planes. As a consequence of user input, not only the information mapped on the user's body changes. In addition, a slice of the underlying CT or MRI dataset is displayed enabling to also learn *cross-sectional anatomy*. A large physical mirror projects the overlay of virtual data and the learners' body on a large screen (see Fig. 16.5).

While the initial system relied upon the skeleton-based representation of the Kinect, a later variant improved the accuracy considerably (Meng et al. 2013; Ma et al. 2016). This was achieved by letting the users select five landmarks of their skeletal anatomy. These landmarks were selected together with orthopedic surgeons and can be reliably touched. This additional information is used to increase the accuracy. After this improvement, the system was used at a larger scale and integrated in the regular curriculum.

Moreover, a special learning application was developed for learning muscle structures. The AR system represents muscles as connected with two bones and models movements such as contraction (Ma et al. 2016). In different learning modes, either individual muscles are projected on the users' body or muscle groups and bones involved in a particular movement. Thus, when the user moves, e.g., her arms, the muscle projection is updated accordingly. This was the only example we found that supports *functional anatomy learning*.

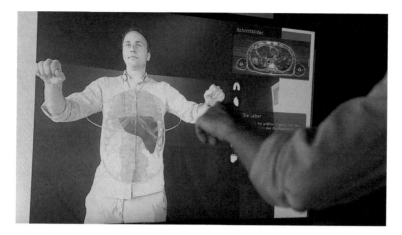

Fig. 16.5 The Mirracle system provides an overlay of internal organs (extracted from the Visible Korean Dataset) on the users' body. At the upper right, a cross-section from the photographic dataset is displayed (From: Meng et al. 2013)

Evaluation of the virtual mirror. Kugelmann et al. (2018) describe how the system was used at the Ludwig-Maximilians University (LMU) in Munich, where 880 medical students employed it. The questionnaires filled by the students indicate the potential for providing an understanding of the spatial relations.

Bork et al. (2019) recently compared the AR mirror with the Anatomage table (recall Custer and Michael 2015) and a textbook to study the specific learning effects of the AR mirror. Again, the combination of the mirrored anatomy with cross-sectional imaging was employed (recall Fig. 16.5). Anatomy students had a higher knowledge gain with the Anatomage table and the AR mirror and a subgroup analysis revealed that in particular low spatial ability learners (according to a mental rotation test) benefit most. The study has a high validity based on the careful design and the large number of participants. Minor critics from a few students relate to the long-term use that may be perceived as tiring. This highlights the importance to evaluate VR and AR systems over longer periods.

Hand anatomy. The system developed by Boonbrahm et al. (2018) is dedicated to study hand anatomy, an area where it is particularly different to understand the large number of structures that are very close to each other. Similar to the virtual mirror, the human body, in particular the human hand and arm, is used as physical object. Bones, muscles, and vascular structures may be selected and projected on the users' arm. In a self-tutorial mode, users may touch a point on their arm and see the closest bone emphasized and labeled on the second arm (see Fig. 16.6).

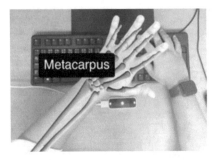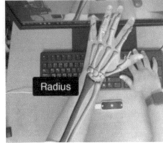

Fig. 16.6 Selected parts of the hand are labeled. Label placement is essential: the label on the left occludes part of the relevant image (From: Boonbrahm et al. 2018)

16.4   VR in Anatomy Education

In this section, which is an updated version of a similar section from Preim and Saalfeld (2018), we discuss semi-immersive and immersive VR applications for anatomy education. For immersive applications, we describe in more detail how labels are used to support education. Additionally, VR systems for selected anatomical regions are presented, including comparative user studies, which could partly show benefits for VR systems compared to traditional learning approaches.

16.4.1   Semi-Immersive VR

Semi-immersive systems use a standard desktop displays but allow the perception of additional depth cues, such as stereoscopy and motion parallax. In an extensive survey, Hackett et al. (2016) list 38 virtual anatomy systems using stereoscopic viewing (mostly with shutter glasses) or autostereoscopic viewing where the stereo rendering is perceived without glasses. The learners reported a number of perceived advantages related to the understanding of spatial structures. This was prominent for complex vascular structures where the benefit of stereo perception is particularly large (Abildgaard et al. 2010). Beside the advantages, problems were stated related to perception conflicts and artifacts that are more severe with autostereoscopic displays (Tourancheau et al. 2012). These perception problems have several causes, e.g., the vergence-accommodation conflict or ghosting, where the eye separation of the stereo images does not work accurately. Luursema et al. (2008) found that low spatial ability learners benefit stronger from stereoscopic displays. These results correlate to a reduced perceived cognitive load compared to a display without stereo. Brown et al. (2012) visualized CT data and interactive labels as 3D stereoscopic images. A study with 183 medical students showed that the 3D system-aided their understanding of anatomy.

Fig. 16.7 Image of the Dextroscope VR system used in the study of Jang et al. (2017) to present inner ear structures to medical students. The student uses the system on the left side, whereas the right side is the interface for the teacher (From: Jang et al. 2017)

Weber et al. (2016) used real-life photos of dissected brains to create stereoscopic content. These were compared with static images, interactive 3D models, and interactive stereoscopic models. Their study showed significantly higher scores for the interactive groups. However, no difference was shown for the stereoscopic and non-stereoscopic group.

The already mentioned work from Jang et al. (2017) also employed a stereoscopic system, i.e., the Dextroscope (see Fig. 16.7) to present inner ear structures to medical students. They investigated differences in directly manipulating structures against passively viewing them. The participants in the manipulation group were more successful in drawing the observed structures afterwards. Additionally, the direct manipulation was more beneficial for students with low spatial ability compared to high spatial ability students.

The zSpace is an example of a semi-immersive device that exhibits a further advantage beside stereoscopy: the passive glasses are equipped with markers that can be tracked. This allows to get the user's head position and enables motion parallax, an additional depth cue that improves the spatial perception. The VISIBLE

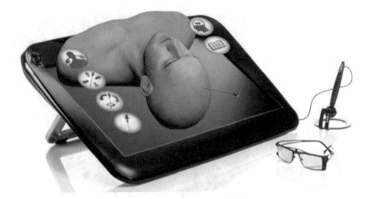

Fig. 16.8 The neurosurgery training application running on the zSpace presented by John et al. (2016) (artistic impression used in the figure to depict the 3D effect) (From: John et al. 2016)

BODY[2] system was ported to the zSpace to employ its benefits. The laboratory of the TOBB University of Economics and Technology in Turkey is equipped with several zSpaces. Here, students and residents can use them and perform "virtual dissections" before they participate in cadaveric dissections. Saalfeld et al. (2016) used the zSpace to sketch vascular structures. The interactive creation of different vascular config-urations allows teachers to illustrate complex structures and students to understand them in their spatial form.

Although primarily aiming at neurosurgery training, the system presented by John et al. (2016) provides concepts useful for anatomy education as well (see Fig. 16.8). Surprisingly, it demonstrates that stereoscopy did not add any value to the chosen tasks.

16.4.2   Immersive VR

Immersive VR systems "provide an interactive environment that reinforces the sen-sation of an immersion into computer-specified tasks" (Huang et al. 2010). The following aspects are often discussed as essential for VR learning environments as they are related to a high motivation (Huang et al. 2010):

- intuitive interaction,
- the sense of physical imagination, and
- the feeling of immersion.

The recent development of VR headsets, e.g., the latest generations of HTC Vive, has the potential for a wide adoption also for anatomy education (see Fig. 16.9). The design of VR-based systems aims at exploiting the potential of VR for an enhanced

---

[2]https://www.visiblebody.com/.

Fig. 16.9 A virtual dissection laboratory for the Institute of Medicine at Chester University based on recent VR hardware (Courtesy of Nigel John)

understanding and at avoiding motion sickness and fatigue that may occur when such systems are used for a longer time.

16.4.2.1 Labeling

Labels establish a bidirectional link between visual representations of anatomical structures and, at the most basic, their textual names. This form of link represents a foundation in anatomy education, as the long history of used anatomy atlases shows. The positioning and rendering of labels is often realized in the 2D image space, even in interactive 3D applications (Preim et al. 1997). The 3D models are rendered into a 2D image. This is the basis for a label layout calculation, followed by actually drawing the labels on or beside the anatomical structures. This approach is not feasible for immersive applications where the user is part of the 3D environment—here, labels need a 3D position to be rendered in a useful way.

This positioning can either be performed automatically or interactively. Pick et al. (2010) present an approach for the automatic placement on an immersive multi-screen display system. Their algorithm is based on the shadow volume technique. This volume is calculated by "assuming" a point light at the user's head position that casts light onto all objects. This generates a shadow volume behind each object. Because this shadow area is occluded for the user, labels are not allowed there. Occlusions during interaction are prevented with a force-based approach to re-position the labels. The *interactive positioning* of labels avoids the usage of a complex layout algorithm, but introduces challenges regarding user interaction. An example of interactively placed labels is presented by Assenmacher et al. (2006). They introduce the toolkit *IDEA* that allows the creation of generic annotations, including labels, images, and also spoken annotations. Furthermore, they show the usage of the toolkit for data analysis at the example of air flows in the human nasal cavity.

A fundamental investigation of annotation systems for multiple VR platforms was carried out by Pick et al. (2016). They presented a workflow design that facilitates the most important label and annotation operations (creation, review, modification). This design is extensible to support further annotation types and interaction techniques.

These general approaches for labeling and annotation are also applicable for anatomy education. Beside displaying the name only, additional information types such as images, videos, and animations can be shown as annotations.

16.4.2.2   3D Puzzle

The possibility to build up a more complex anatomical system out of single pieces was already mentioned in Sect. 16.2.1. In VR, this idea is even more compelling, as the investigation and placement of single puzzle pieces can be realized more naturally. From an education point of view, puzzling is related to constructivist learning theory, where learning happens through an experiential process. This process leads to the construction of knowledge that has *personal meaning* (Mota et al. 2010).

A 3D puzzle to help students learn the anatomy of various organs and systems was presented by Messier et al. (2016). Their system was piloted in a comparative study, where three display paradigms (2D monitor, stereo monitor, and Oculus Rift) and two input devices (space mouse and standard keyboard) were tested. In a master thesis, this idea was further evolved by giving the user more possibilities for manipulation and guidance (Pohlandt 2017). Here, the medical student can choose between several anatomical structures and scale them freely from their original size to large scales. Beside assembling a puzzle (see Fig. 16.10), the VR puzzle allows the disassembly in a specific order, which mimics the dissection course. The interaction is carried out with the HTC Vive controllers. Therefore, a more natural interaction is possible compared to Messier et al. (2016). The prototype supports different stages of learning by allowing the user to enable/disable several guiding features. For example, slight guidance is offered by coloring the model from red to green that shows the user if the object is oriented and positioned correctly. Another guidance is called "ghost copy", which is inspired by a real-world jigsaw puzzle where the user usually has a picture of the finished puzzle at hand. If the user takes two pieces, the goal position and orientation of the second piece is shown on the first piece. The user can then just match the piece he is holding with the copy he is seeing.

The most recent prototype in this direction was presented by Seo et al. (2018). They presented "Anatomy Builder VR", a system to learn skeletal canine anatomy. The student is able to take pieces out of a digital bone box containing 3D bone models, based on scans of real bones. After that, the student can inspect the model

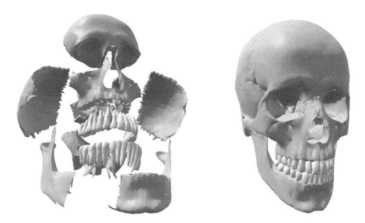

Fig. 16.10 The anatomy of the skull is learned in a VR environment by assembling the skull from an unsolved (left) to a solved (right) state (Inspired by: Pohlandt 2017)

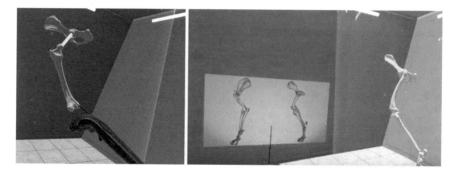

Fig. 16.11 Anatomy Builder VR allows to snap bones together (left) until a skeleton is assembled. As guidance, a reference image is shown (right) (From: Seo et al. 2018)

and place it in an "anti-gravity" field, where it is suspended in place. Positioning a bone near a connection on an already positioned bone leads to the appearance of a yellow line. If the user releases the controller trigger, the two joints snap together (see Fig. 16.11). This is repeated until the skeleton is assembled to the user's satisfaction. Seo et al. (2018) compared their system against a real bone box containing real thoracic limb bones and digital pelvic limb bones. Participants enjoyed using VR considerably more than the bone box. Additionally, they stated that they were able to manipulate bones and put them together with ease in VR. For the identification of bones, the VR environment did not show benefits compared to the bone box.

16.4.2.3   Volume Rendering

An interesting system that can be utilized for anatomy education is SpectoVR[3] that is developed within the Miracle project of the University of Basel.[4] Although the system is described as a tool for diagnosis and treatment planning, it would clearly be useful in teaching. It combines a powerful and fast volume renderer with interaction possibilities for VR. The system allows loading medical image data, adjust transfer functions, and slice through the data with a cutting plane to reveal the clipping surface. This functionality allows students and teachers to explore realistic anatomical data in an immersive environment.

A possibility to create a photorealistic rendering out of volumetric data is the Cinematic Rendering technology from Siemens Healthineers (Fellner 2016) that enables the visualization of clinical images of the human body. This technique was used in a 3D projection space for an audience of up to 150 people (Fellner et al. 2017) (see Fig. 16.12). Users wore 3D shutter glasses to get a stereoscopic impression and were able to control the visualization with a game controller. This is an impressive setting for virtual anatomy. However, the hardware requirements to render the high-resolution images efficiently are huge and this setup is not easily replicated in an auditorium of a medical faculty.

16.4.2.4   Cooperative Systems

Cooperative systems come in two flavors: *collocated cooperative VR systems* enable that different users inspect a virtual environment at the same place, whereas *remote cooperative VR systems* allow distantly located students and teachers to learn together in a "shared reality" (Brown 2000).

In collocated cooperative VR, a group of users is in a shared VR room, e.g., in a CAVE or a dome projection, seeing, and being aware of each other. As an example, Fairen et al. (2017) introduced a VR system for anatomy education that was run on a Power wall and in a CAVE. In both settings, small groups of up to ten students could follow a VR exploration of complex anatomical structures, such as the heart with ventricles, atriums, and valves. One student is tracked and guides the inspection, whereas the others are more passive. This system is quite promising. However, the hardware is very expensive and typically not available in medical faculties. Thus, the usage of such a system beyond a research project is difficult to accomplish.

There exist several possibilities for remote cooperative settings, such as group instructions or one-on-one scenarios. Also, a VR headset may be used, which is more affordable and practical to be used compared to, e.g., CAVEs. These settings differ in effectiveness of learning, applicability and costs. Moorman (2006) showed a one-on-one scenario where an anatomist in her office supports a learner via video conferencing. A master thesis uses a one-on-one tutoring system to teach the human

---

[3]https://www.diffuse.ch/.
[4]https://dbe.unibas.ch/en/research/flagship-project-miracle/.

Fig. 16.12 The "dissection theater of the future" shows a whole body data set with stereoscopic Cinematic Rendering (Fellner et al. 2017)

skull base (Schmeier 2017). Here, an asymmetric hardware setup is used. The teacher uses a zSpace to be able to see the surroundings and easily input text via a keyboard. The student uses a HTC Vive and explores a scaled-up skull from inside. For collaboration, the teacher may set landmarks and sketch 3D annotations in front of the student or re-positions the student to direct her attention to something. Additionally, the student's first person view is shared with the teacher.

If a VR headset is used, special care must be taken for the representation of the students and teachers in the virtual environment. Avatars can be used to mimic movement, interactions, and looking directions of the users. A teacher can benefit from knowing the location of all students. However, it can be cluttering and distracting for students to know where the fellow students are.

16.4.3   VR for Specific Anatomical Regions

Haley et al. (2019) presented a VR system for anatomy education of the ear. The users can interact with high-resolution 2D data and their corresponding 3D anatomical structures simultaneously. The preliminary evaluation showed the general usefulness in clinical practice and ease of use.

An example for an immersive system of the nasal cavity anatomy was presented by Marks et al. (2017). An interesting addition to the visualization of the 3D structures is the depiction of airflow that was calculated with a computational fluid dynamics simulation. Their user study was performed with mechanical engineering students as a part of their curriculum, as flow is an essential part of their lectures. Despite the fact that medical students were not involved, the qualitative feedback indicates that their educational VR application has advantages for the students in terms of engagement, understanding, and retention.

A comparative study with 42 undergraduate medical anatomy students was presented by Maresky et al. (2019). They focus on the cardiac anatomy reasoned by its complex three-dimensional nature. The study was performed with first-year undergraduate medical students who did not participate in any cardiac anatomy lecture. First, they completed a pre-intervention quiz containing conventional cardiac anatomy questions and visual-spatial questions. After that, the students were distributed into two groups, where the VR group used an immersive cardiac VR experience and the other group continued their study normally. Then, the students completed a post-intervention quiz. The students in the VR group scored 21.4 % higher in conventional cardiac anatomy questions and 26.4 % higher in visual-spatial questions.

Another comparative study regarding the learning and retention of neuroanatomy was performed by Ekstrand et al. (2018). They investigated the impact of a VR system compared to traditional paper-based methods. In their study with 60 first to second-year medical students, again, a pre-test and post-test were performed. Additionally, a test was administered 5–9 days later to assess the retention. In contrast to the study of Maresky et al. (2019), no significant differences between the VR group and the paper-based group could be found, neither for scores obtained immediately after the study nor after the retention period.

16.5   Concluding Remarks

VR and AR are beginning to enter anatomy education. While most systems described in this chapter are research prototypes that were not evaluated in large scale and over longer periods, first systems, e.g., the mirror-based system by Kugelmann et al. are integrated in an anatomy curriculum. VR and AR solutions may provide student-centered and interactive learning experiences that add to more traditional types of anatomy education. VR and AR, however, need to meet a number of requirements for exploiting this potential. User interfaces need to be simple and users should be *guided*. Light, portable, and reliable technology is necessary. Any significant delays severely hamper the user experience. Akccayir and Akccayir (Akçayır and Akçayır 2017) discuss that many AR solutions suffered from the difficult use of technology and considerable setup times. Additionally, pedagogical aspects have to be considered during the development of virtual learning environments to maximize the learning outcome (Fowler 2015). This has to be done by integrating academic staff with a background in education early in the design process (Martín-Gutiérrez et al. 2017).

While a number of systems exist that are based on verified high-quality geometric models, we have found very few VR/AR systems that support functional anatomy learning (Ma et al. 2016). Based on the H-Anim standard[5] (Jung and Behr 2008) that defines at four different levels joints and their possible movements, it would be possible to enhance geometric models with this functional information. The potential of haptics to further support anatomy learning and add it to the visual sensations is considerable. In the future, it is likely that haptics will be integrated similar to surgical simulators. While widely available force feedback systems primarily indicate stiffness/elasticity, more fine-grained tactile sensations are desirable. As a final thought for future work: There are likely considerable gender differences in the way VR is perceived. The motivational aspects and the sense of presence may be different for woman and man (Felnhofer et al. 2012). More research in this area is needed and, probably, systems need to be adjusted to the gender of the user.

Acknowledgements We thank the Virtual Anatomy team (Wolfgang D'Hanis, Lars Dornheim, Kerstin Kellermann, Alexandra Kohrmann, Philipp Pohlenz, Thomas Roskoden and Hermann-Josef Rothkötter) for many discussions about anatomy teaching, Nigel John, Benjamin Köhler, Maria Luz, Timo Ropinski, Florian Heinrich, Vuthea Chheang, and Noeska Smit for fruitful discussions on the chapter as well as Steven Birr, Konrad Mühler, Daniel Pohlandt, Felix Ritter, Aylin Albrecht and Anna Schmeier for their implementations.

References

Abildgaard A, Witwit AK, Karlsen JS, Jacobsen EA, Tennoe B, Ringstad G, Due-Tonnessen P (2010) An autostereoscopic 3D display can improve visualization of 3D models from intracranial MR angiography. Int J Comput Assist Radiol Surg 5(5):549–54

Adams H, Shinn J, Morrel WG, Noble J, Bodenheimer B (2019) Development and evaluation of an immersive virtual reality system for medical imaging of the ear. In: Proceedings of SPIE medical imaging: image-guided procedures, robotic interventions, and modeling, vol 10951, p 1095111

Akçayır M, Akçayır G (2017) Advantages and challenges associated with augmented reality for education: a systematic review of the literature. Educ Res Rev 20:1–11

Assenmacher I, Hentschel B, Ni C, Kuhlen T, Bischof C (2006) Interactive data annotation in virtual environments. In: Proceedings of Eurographics conference on virtual environments, pp 119–126

Bacca J, Baldiris S, Fabregat R, Graf S, Kinshuk (2014) Augmented reality trends in education: a systematic review of research and applications. J Educ Technol Soc 17(4):133–149

Bajura M, Fuchs H, Ohbuchi R (1992) Merging virtual objects with the real world: seeing ultrasound imagery within the patient. In: Proceedings of SIGGRAPH, pp 203–210

Blum T, Kleeberger V, Bichlmeier C, Navab N (2012) Mirracle: an augmented reality magic mirror system for anatomy education. In: Proceedings of IEEE Virtual Reality Workshops (VRW), pp 115–116

Boonbrahm P, Kaewrat C, Pengkaew P, Boonbrahm S, Meni V (2018) Study of the hand anatomy using real hand and augmented reality. Int J Interact Mobile Technol (iJIM) 12(7):181–90

Bork F, Stratmann L, Enssle S, Eck U, Navab N, Waschke J, Kugelmann D (2019) The benefits of an augmented reality magic mirror system for integrated radiology teaching in gross anatomy. In: Anatomical sciences education

---

[5]http://www.web3d.org/working-groups/humanoid-animation-h-anim.

Brenton H, Hernandez J, Bello F, Strutton P, Purkayastha S, Firth T, Darzi A (2007) Using multi-media and Web3D to enhance anatomy teaching. Comput Educ 49(1):32–53

Brown JR (2000) Enabling educational collaboration-a new shared reality. Comput Graph 24(2):289–92

Brown PM, Hamilton NM, Denison AR (2012) A novel 3D stereoscopic anatomy tutorial. Clin Teach 9(1):50–53

Chapman LL, Sullivan B, Pacheco AL, Draleau CP, Becker BM (2011) Veinviewer-assisted intravenous catheter placement in a pediatric emergency department. Acad Emerg Med 18(9):966–71

Chen C, Chen W, Peng J, Cheng B, Pan T, Kuo H (2017) A real-time markerless augmented reality framework based on slam technique. In: International symposium on pervasive systems, algorithms and networks, vol 2017, pp 127–132

Chen L, Day TW, Tang W, John NW (2017) Recent developments and future challenges in medical mixed reality. In: Proceedings of ISMAR, pp 123–35

Chittaro L, Ranon R (2007) Web3D technologies in learning, education and training: motivations, issues, opportunities. Comput Educ 49(1):3–18

Custer TM, Michael K (2015) The utilization of the anatomage virtual dissection table in the education of imaging science students. J Tomogr Simul 1

de Faria JWV, Teixeira MJ, de Moura Sousa Jr L, Otoch JP, Figueiredo EG (2016) Virtual and stereoscopic anatomy: when virtual reality meets medical education. J Neurosurg 125(5):1105–1111

Ekstrand C, Jamal A, Nguyen R, Kudryk A, Mann J, Mendez I (2018) Immersive and interactive virtual reality to improve learning and retention of neuroanatomy in medical students: a randomized controlled study. CMAJ Open 6(1):E103–9

Fairén M, Farrés M, Moyés J, Insa E (2017) Virtual Reality to teach anatomy. In: Proceedings of Eurographics-Education papers, pp 51–58

Fellner FA (2016) Introducing cinematic rendering: a novel technique for post-processing medical imaging data. J Biomed Sci Eng 10(8):170–175

Fellner FA, Engel K, Kremer C (2017) Virtual anatomy: the dissecting theatre of the future-implementation of cinematic rendering in a large 8 K high-resolution projection environment. J Biomed Sci Eng 10(8):367–75

Felnhofer A, Kothgassner OD, Beutl L, Hlavacs H, Kryspin-Exner I (2012) Is virtual reality made for men only? Exploring gender differences in the sense of presence. In: Proceedings of the international society on presence research, pp 103–112

Fowler C (2015) Virtual reality and learning: Where is the pedagogy? Br J Educ Technol 46(2):412–22

Fuchs H, Livingston MA, Raskar R, Colucci D, Keller K, State A, Crawford JR, Rademacher P, Drake SH, Meyer AA (1998) Augmented reality visualization for laparoscopic surgery. In: Proceedings of MICCAI, pp 934–943

Hackett M, Proctor M (2016) Three-dimensional display technologies for anatomical education: a literature review. J Sci Educ Technol 25:641–54

Halle M, Demeusy V, Kikinis R (2017) The open anatomy browser: a collaborative web-based viewer for interoperable anatomy atlases. Frontiers in Neuroinformatics 11

Huang H-M, Rauch U, Liaw S-S (2010) Investigating learners' attitudes toward virtual reality learning environments: based on a constructivist approach. Comput Educ 55(3):1171–82

Jain N, Youngblood P, Hasel M, Srivastava S (2017) An augmented reality tool for learning spatial anatomy on mobile devices. Clin Anat 30(6):736–41

Jang S, Vitale JM, Jyung RW, Black JB (2017) Direct manipulation is better than passive viewing for learning anatomy in a three-dimensional virtual reality environment. Comput Educ 106:150–65

John NW, Phillips NI, ap Cenydd L, Pop SR, Coope D, Kamaly-Asl I, de Souza C, Watt SJ (2016) The use of stereoscopy in a neurosurgery training virtual environment. Presence 25(4):289–298

Jung Y, Behr J (2008) Extending H-Anim and X3D for advanced animation control. In: Proceedings of the ACM symposium on 3D web technology, pp 57–65

Kamphuis C, Barsom E, Schijven M, Christoph N (2014) Augmented reality in medical education? Perspect Med Educ 3(4):300–11

Kato H (2007) Inside ARToolKit. In: Proceedings of IEEE international workshop on augmented reality toolkit

Kosslyn SM, Thompson WL, Wrage M (2001) Imaging rotation by endogeneous versus exogeneous forces: distinct neural mechanisms. NeuroReport 12(11):2519–25

Kugelmann D, Stratmann L, Nühlen N, Bork F, Hoffmann S, Samarbarksh G et al (2018) An augmented reality magic mirror as additive teaching device for gross anatomy. Ann Anat Anat Anz 215:71–77

Lundstrom C, Rydell T, Forsell C, Persson A, Ynnerman A (2011) Multi-touch table system for medical visualization: application to orthopedic surgery planning. IEEE Trans Vis Comput Graph 17(12):1775–84

Luursema J-M, Verwey WB, Kommers PA, Annema J-H (2008) The role of stereopsis in virtual anatomical learning. Interact Comput 20(4–5):455–60

Ma M, Fallavollita P, Seelbach I, Von Der Heide AM, Euler E, Waschke J, Navab N (2016) Personalized augmented reality for anatomy education. Clin Anat 29(4):446–53

Ma M, Jutzi P, Bork F, Seelbach I, von der Heide AM, Navab N, Fallavollita P (2016) Interactive mixed reality for muscle structure and function learning. In: International conference on medical imaging and augmented reality, pp 117–128

Maresky HS, Oikonomou A, Ali I, Ditkofsky N, Pakkal M, Ballyk B (2019) Virtual reality and cardiac anatomy: exploring immersive three-dimensional cardiac imaging, a pilot study in undergraduate medical anatomy education. Clin Anat 32(2):238–43

Marks S, White D, Singh M (2017) Getting up your nose: a virtual reality education tool for nasal cavity anatomy. In: SIGGRAPH Asia 2017 symposium on education, pp 1:1–1:7

Martín-Gutiérrez J, Mora CE, Aâorbe-Díaz B, González-Marrero A (2017) Virtual technologies trends in education. Eurasia J Math Sci Technol Educ 13(2):469–86

Mayfield CH, Ohara PT, O'Sullivan PS (2013) Perceptions of a mobile technology on learning strategies in the anatomy laboratory. Anat Sci Educ 6(2):81–89

Meng M, Fallavollita P, Blum T, Eck U, Sandor C, Weidert S, Waschke J, Navab N (2013) Kinect for interactive AR anatomy learning. In: Proceedings of ISMAR, pp 277–278

Messier E, Wilcox J, Dawson-Elli A, Diaz G, Linte CA (2016) An interactive 3D virtual anatomy puzzle for learning and simulation-initial demonstration and evaluation. Stud Health Technol Inform 20:233–240

Mönch J, Mühler K, Hansen C, Oldhafer K-J, Stavrou G, Hillert C, Logge C, Preim B (2013) The liversurgerytrainer: training of computer-based planning in liver resection surgery. Int J Comput Assist Radiol Surg 8(5):809–18

Moorman SJ (2006) Prof-in-a-box: using internet-videoconferencing to assist students in the gross anatomy laboratory. BMC Med Educ 6(1):55

Mota MF, da Mata FR, Aversi-Ferreira TA (2010) Constructivist pedagogic method used in the teaching of human anatomy. Int J Morphol 28(2):369–74

Murakami T, Tajika Y, Ueno H, Awata S, Hirasawa S, Sugimoto M, Kominato Y, Tsushima Y, Endo K, Yorifuji H (2014) An integrated teaching method of gross anatomy and computed tomography radiology. Anat Sci Educ 7(6):438–449

Nicholson DT, Chalk C, Funnell W, Daniel S (2006) Can virtual reality improve anatomy education? A randomised controlled study of a computer-generated three-dimensional anatomical ear model. Med Educ 40(11):1081–1087

Phillips AW, Smith S, Straus C (2013) The role of radiology in preclinical anatomy: a critical review of the past, present, and future. Acad Radiol 20(3):297–304

Pick S, Hentschel B, Tedjo-Palczynski I, Wolter M, Kuhlen T (2010) Automated positioning of annotations in immersive virtual environments. In: Proceedings Eurographics conference on virtual environments, pp 1–8

Pick S, Weyers B, Hentschel B, Kuhlen TW (2016) Design and evaluation of data annotation workflows for cave-like virtual environments. IEEE Trans Vis Comput Graph 22(4):1452–1461

Pohlandt D (2017) Supporting anatomy education with a 3D puzzle in a virtual reality environment. Master's thesis, Department of Computer Science, University of Magdeburg

Pommert A, Höhne KH, Pflesser B, Richter E, Riemer M, Schiemann T, Schubert R, Schumacher U, Tiede U (2001) Creating a high-resolution spatial/symbolic model of the inner organs based on the Visible Human. Med Image Anal 5(3):221–28

Preim B, Raab A, Strothotte T (1997) Coherent zooming of illustrations with 3D-graphics and text. In: Proceedings of graphics interface, pp 105–113

Preim B, Saalfeld P (2018) A survey of virtual human anatomy education systems. Comput Graph 71:132–53

Radu I (2012) Why should my students use ar? A comparative review of the educational impacts of augmented-reality. In: IEEE international symposium on mixed and augmented reality (ISMAR), pp 313–314

Ritter F, Berendt B, Fischer B, Richter R, Preim B (2002) Virtual 3D Jigsaw puzzles: studying the effect of exploring spatial relations with implicit guidance. In: Proceedings of Mensch & Computer, pp 363–372

Ritter F, Preim B, Deussen O, Strothotte T (2000) Using a 3D puzzle as a metaphor for learning spatial relations. In: Proceedings of graphics interface, pp 171–178

Ruiz JG, Cook DA, Levinson AJ (2009) Computer animations in medical education: a critical literature review. Med Educ 43(9):838–46

Saalfeld P, Glaßer S, Beuing O, Preim B (2017) The faust framework: free-form annotations on unfolding vascular structures for treatment planning. Comput Graph 65:12–21

Saalfeld P, Stojnic A, Preim B, Oeltze-Jafra S. (2016) Semi-immersive 3D sketching of vascular structures for medical education. In: Proceedings of VCBM, pp 123–132

Sakellariou S, Ward BM, Charissis V, Chanock D, Anderson P (2009) Design and implementation of augmented reality environment for complex anatomy training: inguinal canal case study. In: Proceedings of international conference on virtual and mixed reality, pp 605–614

Scheuering M, Schenk A, Schneider A, Preim B, Greiner G (2003) Intraoperative augmented reality for minimally invasive liver interventions. In: Proceedings of medical imaging

Schmeier A (2017) Student and teacher meet in a shared virtual environment: a VR one-on-one tutoring system to support anatomy education. Master's thesis, Department of Computer Science, University of Magdeburg

Seo JH, Smith BM, Cook M, Malone E, Pine M, Leal S, Bai Z, Suh J (2018) Anatomy builder VR: applying a constructive learning method in the virtual reality canine skeletal system. In: Andre T (ed) Advances in human factors in training, education, and learning sciences. Springer International Publishing, Cham, pp 245–252

Shotton J, Fitzgibbon A, Cook M, Sharp T, Finocchio M, Moore R, Kipman A, Blake A (2011) Real-time human pose recognition in parts from single depth images. In: Proceedings of IEEE computer vision and pattern recognition (CVPR), pp 1297–1304

Sielhorst T, Bichlmeier C, Heining SM, Navab N (2006) Depth perception–a major issue in medical ar: evaluation study by twenty surgeons. In: Proceedings of MICCAI, pp 364–372

Spitzer VM, Ackerman MJ, Scherzinger AL, Whitlock D (1996) The visible human male: a technical report. J Am Med Inform Assoc 3(2):118–30

Sweller J (2004) Instructional design consequences of an analogy between evolution by natural selection and human cognitive architecture. Instr Sci 32(1–2):9–31

Thomas RG, William John N, Delieu JM (2010) Augmented reality for anatomical education. J Vis Commun Med 33(1):6–15

Tourancheau S, Sjöström M, Olsson R, Persson A, Ericson T, Rudling J, Norén B (2012) Subjective evaluation of user experience in interactive 3D visualization in a medical context. In: Proceedings of medical imaging

Van Merriënboer JJ, Clark RE, De Croock MB (2002) Blueprints for complex learning: the 4C/ID-model. Educ Tech Res Dev 50(2):39–61

Chapter 17
The Road to Birth: Using Digital Technology to Visualise Pregnancy Anatomy

Donovan Jones⊙, Michael Hazelton⊙, Darrell J. R. Evans⊙, Vendela Pento, Zi Siang See⊙, Luka Van Leugenhaege⊙, and Shanna Fealy⊙

Abstract Pregnancy is a time of profound anatomical and physiological reorganisation. Understanding these dynamic changes is essential to providing safe and effective maternity care. Traditional teaching methods such as the use of static cadaver specimens are unable to illustrate and give appreciation to the concurrent fetal and maternal changes that occur during this time. This chapter describes the development of the Road to Birth (RtB) a collaborative, multi-modal, digital anatomy program, aimed to provide undergraduate midwifery students with a novel, visual, interactive and accessible, pregnancy education tool. The RtB digital anatomy program provides

D. Jones (✉) · S. Fealy
School of Nursing and Midwifery Indigenous Health, Faculty of Science, Charles Sturt University, Port Macquarie, NSW, Australia
e-mail: donojones@csu.edu.au

S. Fealy
e-mail: sfealy@csu.edu.au

M. Hazelton
School of Nursing and Midwifery, Faculty of Health and Medicine, University of Newcastle, Newcastle, NSW, Australia
e-mail: michael.hazelton@newcastle.edu.au

D. J. R. Evans
School of Medicine and Public Health, Faculty of Health and Medicine, University of Newcastle, Newcastle, NSW, Australia
e-mail: darrell.evans@newcastle.edu.au

V. Pento
Innovation Technology Resources Division, University of Newcastle, Newcastle, NSW, Australia
e-mail: vendela.pento@newcastle.edu.au

Z. S. See
School of Creative Industries, University of Newcastle, Newcastle, NSW, Australia
e-mail: ziziang.see@newcastle.edu.au

L. Van Leugenhaege
Lecturer-Researcher Research Coordinator ACME: AP Research Cell Midwifery Science, AP University of Applied Sciences Antwerp, Midwifery Department, Antwerp, Belgium
e-mail: luka.vanleugenhaege@ap.be

users with an internal view of pregnancy and fetal development, spanning 0–40 weeks of pregnancy, up to the immediate postpartum period. Users of the program have the opportunity to observe, interact and manipulate detailed 3D models and visualise the growth of a fetus and maternal anatomical changes simultaneously. The program is inclusive of detailed digital fetal and placental models that display both normal and pathological birth positions. The models are accompanied by written educational content that is hypothesised to support both technical and non-technical skill development. The RtB program has been deployed and tested amongst two international cohorts of undergraduate midwifery students. Findings indicate that the RtB as a mobile application and as an immersive virtual reality program have the potential to be useful pregnancy education tools with further empirical testing underway.

17.1   Introduction

Pregnancy is a dynamic human process characterised by profound ongoing anatomical and physiological reorganisation of the mother for the growth and development of the baby (Rankin 2017). As with all undergraduate health professional students, those studying midwifery require a thorough understanding of pregnancy-related anatomical sciences, forming the basis of all clinical decision-making and clinical reasoning (Yammine and Violato 2015). Midwifery educators are being challenged to move beyond traditional teaching and education methods such as lectures, use of 2d images and 3d models, and high- and low-fidelity simulation methods for technical and non-technical/cognitive skill acquisition (Cant and Cooper 2017; Fealy et al. 2019b).

The provision of anatomy education has evolved significantly over recent decades in response to a range of pedagogical developments, including challenges to traditional teaching methods, a broadening of intended learning outcomes and a growing focus on technology-enabled approaches and resources (Estai and Bunt 2016; Sugand et al. 2010). The gradual shift towards including digital technology approaches to teaching and learning in anatomy, often at the expense of using cadaveric material, is not without controversy with an ongoing debate about the effectiveness and primacy of one approach over another evident (McMenamin et al. 2018). Whilst there is no doubt that the use of anatomical dissection and prosection provides an excellent appreciation of three-dimensional anatomy and human variation (Evans et al. 2018), the incorporation of digital anatomy, particularly through interactive digital technologies such as virtual and mixed reality affords other advantages. These may include overcoming organisational limitations such as lack of access to cadaveric specimens, ethical and financial considerations of using human material and various logistical limitations (Fealy et al. 2019b, McMenamin et al. 2018). In addition, such approaches may provide enhanced learning opportunities by demonstrating anatomy that is difficult to visualise through dissection, incorporating a more functional focus and importantly illustrating anatomy undergoing developmental or physiological change, which is not possible with cadaveric material.

The teaching and learning of reproductive anatomy in midwifery is an example, where the adoption of digital technology can facilitate learning beyond what is currently possible. Typically, reproductive anatomy has relied on prosected cadaveric material and anatomical models (McMenamin et al. 2018). These have provided the appropriate context of positioning and association, and understanding of intricate structural relations, alongside detailed instruction and theory (i.e. didactic lectures). However, such an approach has not easily afforded the opportunity to focus on the dynamic and dramatic morphological changes that occur during the different stages of pregnancy and the effects on the structure and function of other body systems including the digestive, urinary, respiratory and cardiovascular systems (Rankin 2017). Such an appreciation and understanding of these processes are vital for a range of healthcare professionals, so they are able to appropriately support and care for their pregnant patients, detect complications, identify symptomology and devise treatment plans.

The use and integration of digital modalities have a unique potential to demonstrate changes within both a spatial and temporal context (Barmaki et al. 2019). The inclusion of visualisations to illustrate embryonic and fetal development, including fetal and placental positioning may enable a cognitive learning frame that assists students in gaining a holistic understanding of the anatomy of pregnancy. One of the potential advantages of developing such technologies is the ability to adapt and change digital content, so that learning tools may be used to engage students at varying knowledge levels and for deployment amongst culturally diverse audiences. Overall, the emphasis on interactive digital visualisations for midwifery and related education has the potential to enhance traditional teaching and learning methods (Downer et al. 2020; Fealy et al. 2019b) and build on the development and use of anatomical digital simulation in a range of other healthcare disciplines (Moro and Gregory 2019; Moro et al. 2017; Rousian et al. 2018; Uppot et al. 2019). The future for digital anatomy simulation in pregnancy is immense and likely to be further augmented by the development of active intervention and extended scenario functionality such as the introduction of touch and feel sensation technology (haptics). Such advancements should be anticipated with enthusiasm, but also caution is needed to ensure these advancements are scientifically driven with the focus on healthcare and education outcomes and not just for technology's sake.

The initial concept for the RtB program evolved out of the need to find an engaging, accessible and dynamic medium to teach midwifery students the concepts of fetal positioning as well as maternal anatomy and physiology. Moving away from approaches that are typically static in nature (cadaveric specimens and plastic models) to more innovative pedagogies such as digital technology integration deployed via Immersive Virtual Reality (IVR) and or Smart Phone or Tablet (SPT) devices. Both these mediums have been suggested to afford advantages such as allowing users to access educational content from their personal devices, anywhere and/or increasing access to immersive classroom learning and simulation experiences (Fealy et al. 2019b). Moreover, retention of anatomical content is often lost over time and to re-teach such sessions in cadaveric specimen laboratories requires forward planning, equipment, staff, space and associated costs (Fealy et al. 2019b; Sugand et al. 2010).

Therefore, it was important to ensure that the RtB program encompassed a self-directed, learner-driven approach that focused on user interaction, participation and contextualisation, as an active learning approach has been shown to improve knowledge retention and promote deep learning (Kerby et al. 2010; Sugand et al. 2010). In addition, as an accessible package, it was imperative that the RtB program be developed for multiple digital platforms ensuring it could be used by students in the classroom but also beyond the classroom to allow students to revisit and reinforce their learning when the need arises, such as before workplace learning or for exam revision.

17.2   The Road to Birth

The RtB program is comprised of four main digital anatomy interfaces or functions that include: (1) Base anatomy, (2) Pregnancy timeline, (3) Birth considerations, and (4) Quiz mode functions. Each function was targeted at introducing, explaining and assessing new knowledge and concepts aligned to specific midwifery student learning outcomes. The design encompassed a progressive and scaffolded approach to ensure incremental learning, i.e. novice to expert, occured (Sharma and Hannafin 2007).

In the base anatomy function, users are able to look through, explore and interact with different anatomical layers of the female body. The anatomy function allows users to manipulate individual anatomical layers by turning them on and off in order to get a better understanding of anatomical structural relationships and their functionality. The individual anatomical layers include the dermal, muscular, circulatory, respiratory, digestive, urinary, reproductive, nervous, lymphatic and skeletal layers. In the pregnancy timeline function, users are able to manipulate the weeks of pregnancy from 0 to 40 weeks of gestation. This demonstrates both fetal development, whilst simultaneously demonstrating maternal organ displacement as a result of the growing fetus. This was designed to assist with conceptualisation and understanding of common pregnancy symptoms such as backpain, pelvic pain, heartburn and breathlessness. Users are able to manipulate the various anatomical features adding and removing selected anatomical layers including the fetus, as displayed in Fig. 17.1. In addition, a floating information panel is featured on the left-hand side of the user's view. This panel includes essential information corresponding to the selected weeks of gestation as displayed in Fig. 17.2. The information panel may be tailored with any amount of text/information for use amongst broader health education audiences. This particular function of the RtB program was drawn from pedological learner defined manipulation approaches, which are suggested to increase learning motivation, learner engagement and knowledge retention (Barmaki et al. 2019; Moro and Gregory 2019; Rizzolo and William 2006; Stepan et al. 2017).

The birth considerations function is an important addition that affords the user an internal view of pregnancy, illustrating both fetal and placental positioning. This

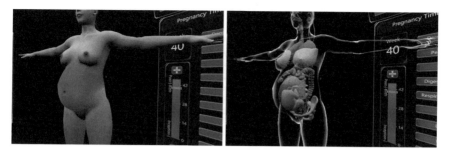

Fig. 17.1 Pregnancy timeline module with dermal layer on/off

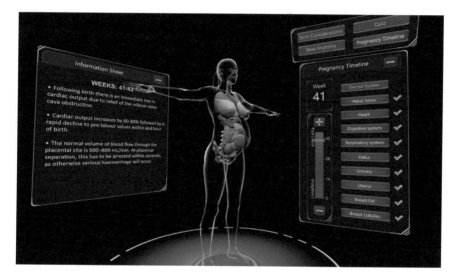

Fig. 17.2 Displays the pregnancy timeline function with associated text information

function may be used to highlight normal and pathological fetal and placental presentations that need practitioner and patient education prior to birth (Jones et al. 2019). Fetal positions include longitudinal, transverse and lateral positioning as well as common fetal presentations such as cephalic (head down), breech (bottom down), flexed and de-flexed fetal head presentations (Rankin 2017). A number of these positions are favourable for a normal vaginal birth, whilst others may make a vaginal birth more difficult and lead to complications in the birthing process such as instrumental (forceps or ventouse) or caesarean birth (Rankin 2017). These fetal positions/presentations need to be clearly understood by maternity care providers. The birth consideration function was designed to assist the novice practitioner with conceptualising the fetal positioning in utero, increasing spatial awareness. This function additionally allows for the concurrent visualisation of the fetus and the placenta as displayed in Fig. 17.3, allowing for more advanced learners to identify

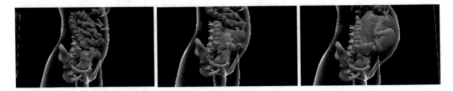

Fig. 17.3 Fetal development (visualization). Displays fetal and placental positioning in the birth considerations function

pathological placental positioning such as placenta previa, which if not identified can result in fetal and maternal morbidity and mortality (Rankin 2017).

The RtB program additionally includes a customisable quiz function. This function allows for self-assessment of the visual and text-based content within the RtB program in unison. The quiz function can be tailored to the learner's level of knowledge. The ability to embed learner-specific quizzes may facilitate and reinforce understanding of pregnancy anatomy and enhance the overall learning experience (Wiliam 2011). The advantage of this function is that it may also be used in conjunction with background analytics allowing for educator feedback such as knowledge of when students are accessing the program and access to student test score results, supporting ongoing teaching evaluation and student engagement (Sharma and Hannafin 2007; Wiliam 2011).

17.3   A Collaborative Design Approach

The RtB program was conceptualised by subject matter experts in the field of midwifery at The University of Newcastle, a large regional Australian University. A collaboration between the institution's immersive technology innovations team and midwifery experts was formed with the aim to develop a multi-platform, immersive and interactive digital anatomy prototype, for use by undergraduate midwifery students. The primary educational objective was to provide midwifery students with an interactive learning resource that assists the user to visualise the internal anatomical changes of pregnancy and fetal positioning in utero.

The application was developed in a 12-week timeframe, specifically with the goal of developing immersive anatomy content that would engage the student/users and then determining if this adds value beyond traditional teaching methods. There have been other animations and applications that show the physiology of pregnancy, but this application breaks new ground in the interactive nature of the digital visualisations. Users can turn on and off different anatomical layers and choose what they want to focus on. Additionally, users can readily work through the pregnancy timeline function to see how the fetus develops throughout pregnancy, as well as what happens to the internal organs of the mother. Users can also change the viewpoint relative to the digital visualisation and examine the visualisations from different angles and magnifications, thus being able to zoom in and out of the respective digital anatomy.

The team applied the Scrum framework for the development process. Scrum is defined as a pragmatic, iterative and incremental approach for complex product development and is based on three key principles, (1) Transparency (2) Inspection and (3) Adaption (Sutherland and Schwaber 2013).The Scrum methodology is based on the empirical process control theory or empiricism. Empiricism explains that knowledge comes from experience and decision-making based on what is known (Sutherland and Schwaber 2013). The guiding principle of transparency allows for the team to collectively and clearly define the final product outcome and development endpoint. The inspection and adaption principles are characterised by four key processes that include sprint planning, daily scrum, sprint review, and a sprint retrospective (Sutherland and Schwaber 2013). Sprints are considered "the heart" of the Scrum framework. Sprint planning is conducted with the whole team with the deliverables or outcomes of each sprint discussed and sprint duration defined (Sutherland and Schwaber 2013). The RtB program was conceptualised with the entire team (midwifery content experts, digital design, and information technology experts), over a three-month time period with the actual program development process undertaken over twelve weeks with sprints planned every two weeks until the program was completed and ready for testing.

17.4 Immersive Virtual Reality (IVR)

A sense of presence is a crucial element in providing users with compelling IVR experiences (Morie 2007, Sherman and Craig 2018). Immersive virtual reality is defined by Sherman and Craig (2018), as a technology medium composed of interactive computer simulations that sense the participant's position and actions and replaces or augments the feedback to one or more senses. This provides the user with feelings of being mentally immersed or present in the simulation (Morie 2007; Sherman and Craig 2018). Typically, IVR provides the user with a high degree of interactivity, usually using a head-mount-device (HMD) and hand controllers. Whilst the extent and application of IVR within undergraduate midwifery education is largely unknown (Fealy et al. 2019b) evidence suggests that immersive technology mediums such as IVR may be practical for use within health care, such as surgical skill acquisition, technical and non-technical skill nursing and midwifery simulation training and for the provision of educational content such as anatomy and physiology (Dey et al. 2018; Domingo and Bradley 2018; Horst and Dörner 2018; Fealy et al. 2019b; Zorzal et al. 2019). Given the potential educational and financial benefits of IVR compared to traditional anatomical cadaveric and simulation modalities, the team developed the RtB with a primary focus for use as an IVR program (Fealy et al. 2019b).

17.5 Mobile Health (SPT)

Mobile smartphone and tablet applications (apps) have become part of the contemporary student and university experience (Hashim 2018; Heflin et al. 2017). Mobile Health or Mhealth defined by the World Health Organization as a medical or public health related practice that is supported by mobile devices including mobile phones and tablets, are increasingly being tested for their potential to overcome barriers to information provision such as geographical, organisational and attitudinal barriers (Fealy et al. 2019a). In undergraduate health education, similar accessibility barriers exist, limiting knowledge development and clinical skill acquisition (Al-Emran et al. 2016; Heflin et al. 2017). The use and integration of SPT devices for educational (teaching and learning) purposes may afford learners the opportunity for more flexible learning on campus and at home, potentially encouraging and facilitating self-directed learning (Al-Emran et al. 2016; Hashim 2018; Heflin et al. 2017). Therefore, it was important for the design team to build the RtB as a smartphone application accessible on both iOS and Android platforms to facilitate on- and off-campus learning.

17.6 Interaction Design and Approach

The RtB program was created in 12 weeks. To support the rapid prototype development the team acquired re-configured pre-made 3D models of female and fetal anatomy. These models were then manipulated for interaction and use across multiple technology mediums. The RtB utilised 5 different design techniques over 7 different models: dermal, uterus, placenta, umbilical cord, fetus, breasts, digestive and respiratory system. The models were manipulated and animated in Unity 3D using C# visual programming language.

The design techniques are explained as follows:

I.   Dermal: Three models were created to represent the different key stages: pre-pregnancy, week 42 and 5 days postpartum. The dermal stretching was created by an animation that changes between the 3 models, creating a visual effect of the model in different stages of pregnancy as displayed in Fig. 17.1.

II. The Uterus and Placenta: To create the visual effect of the uterus growing and surrounding the fetus, the team used hierarchical bones. These hierarchical bones moulded the uterus into the correct size and position for each pregnancy week to accommodate the fetus and be correctly proportioned for the skin. The placenta was placed in the uterus in a similar manner, adjusting into the uterus.

III. Fetus and Umbilical cord: The fetus was created by acquiring 42 pre-made models. These models were manually scaled and positioned into the uterus to create an effect of growth while the pregnancy progresses. The cord was manually aligned with the navel in each fetus model and positioned in a natural manner according to the position of the fetus as per Fig. 17.3.

IV. Digestive and respiratory system: To allow for the uterus (and fetus) to grow, the digestive system and respiratory system were required to move to correspond with the growing fetus and uterus. This animation was created by using bones and inverse kinematic joints achieving the overall realistic effect.

V. Breast animation: The growth and changing of the breasts, specifically the mammary glands during various stages of gestation, were created by scaling the breast model at different rates according to the pregnancy timeline as displayed in Fig. 17.2.

To ensure greater access to the RtB and the best learning outcomes the application was designed to be accessed via multiple platforms. Thus, the application was made for use in Immersive Virtual Reality on Steam VR-enabled headsets, using HTC Vive and Windows Mixed Reality (MR) systems as per Fig. 17.4b. Additionally, the program was made for use on a Microsoft HoloLens as displayed in Fig. 17.4a. Figure 17.5a and 17.5b, shows the use of the IVR controller within the RtB program, with the beam pointer used for interaction with the virtual interface, such as when controlling the pregnancy timeline.

The program was additionally developed for SPT devices on both the Android and iOS platforms as displayed in Fig. 17.6a. The program was additionally made

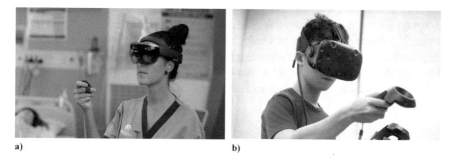

a)   b)

Fig. 17.4 a Displays the RtB program being used on Microsoft HoloLens head-mounted device) and **b** RtB being used on a Samsung Odyssey head-mounted device

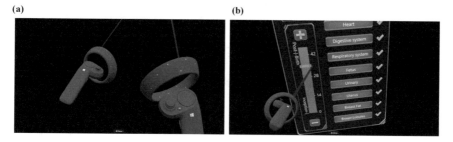

(a)   (b)

Fig. 17.5 Displays control function in the RtB IVR program **a** IVR controllers and **b** beam pointer for interacting with the user interface in VR

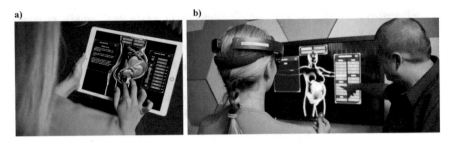

Fig. 17.6 Displays the use of the RtB program **a** RtB tablet application and **b** RtB Microsoft HoloLens projection

available for use on PC and Mac computers. The aim of the multimodal approach was to ensure broad educational application and to support student access and use. For example, the IVR medium could potentially be a good fit for classroom/lab learning, with the mobile device and computer mediums allowing for learning beyond the classroom. The advantage of the MR system is that it is highly portable, whereas the IVR experience relies on a tethered computer for computation and graphics processing. The MR option is especially convenient for providing on-demand demonstrations. One of the potential advantages of developing the RtB into a mobile application for installation onto end-user devices was the potential to collect additional user data such as a user's location, usage time and quiz outcomes. Moreover, when a user's device is connected to Wi-Fi this information can be sent to an external database for collection and analysis and reporting purposes. Additionally, when connected to Wi-Fi the RtB program checks for a "use-by date" thereby supporting research teams and or institutions to have control to disable the RtB program from the end device as necessary.

17.7   Initial Testing and Evaluation

To assess that the RtB prototype met the minimum design and educational objectives, two separate quality assurance studies were conducted. The first was conducted using a convenience sample of n = 19 Australian, second and third-year undergraduate midwifery students, using the RtB SPT application (Ethics Approval No. QA193 University of Newcastle). The second was conducted at one university in Antwerp Belgium amongst a convenience sample of n = 139 first-year undergraduate midwifery students, using the RtB IVR program.

Study 1—Australian Sample

In the Australian study, the RtB program was distributed to students via smartphone and tablet devices. Students were able to interact with the program over a 30 min time frame and were asked to provide feedback using a short anonymous paper-based survey. The following questions were asked:

- Can you please tell us what you like about the program?
- How could you see the program supporting your learning as a student?
- Demographic questions such as degree being studied and year of program.

Responses were assessed for commonalities and reported using descriptive statistics (proportions and percentages) using Microsoft® Excel v16.24. Responses were grouped into the following outcome categories: (1) identifies the program as a useful learning resource; (2) mentions the program as useful for anatomy and fetal positioning; (3) mentions the program as being useful for other populations; (4) mentions the visual aspect of the program as advantageous; (5) mentions ease of use of the program. Of the responses, all participants (n = 19) mentoined the program as being a useful learning resource. Seventy-four per cent (n = 14) of responses mentioned that the program was useful for learning pregnancy and/or fetal anatomy and fetal positioning in utero. Thirty-seven per cent (n = 7) mentioned that the program could also be useful for pregnant women; no other populations were identified within the responses. Fifty-three percent (n = 10) of responses mentioned visualisations as being advantageous with twenty-six per cent (n = 5) of responses specifically mentioning the ease of use of the program.

The results of the quality assurance testing within the Australian sample indicated that the RtB prototype provided undergraduate midwifery students with an educational resource that assisted with visualising internal anatomical changes of pregnancy and understanding of fetal positioning in utero. Further testing is required to ascertain the usefulness of the program in its other modalities (Microsoft HoloLens, Mac and PC). As suggested by a number of the quality assurance study responses, further exploration of the use of RtB as an educational resource for pregnant women is also warranted (Dey et al. 2018; Stone 2011; Fealy et al. 2019b; Jones et al. 2019).

Study 2—Belgium Sample

In the Belgium study, midwifery students were invited to interact with the RtB programme by using IVR. Students were offered the chance to voluntarily participate in a 1 hour testing session at their study institution. As these sessions were voluntary, the University of Antwerp considered this to not need formal ethical approval but rather registered these sessions as independent 'skills lab' training. Students were under no obligation to take part in the session, and the session was not linked to any formal course or assessment work.

Due to large numbers of volunteers, students were arranged into small groups with a maximum of five students in each group. At the beginning of the session, a midwifery lecturer gave basic instructions on how to use the IVR headsets and how to navigate the RtB interface. Students were able to freely explore and interact with the RtB program for as long as they wanted during the session. Feedback on the RtB program was collected via an electronic survey that was emailed to students after the session had been completed with the survey returned by n = 21 students.

The post-session survey was comprised of the System Usability Scale (SUS) originally developed by J. Brook in 1986 (Lewis 2018). The SUS is a 10-item questionnaire for the subjective assessment of the usability of software systems (Lewis 2018). The 10 questionnaire items are measured using 5-point Likert scales with responses, ranging from (1) strongly disagree to (5) strongly agree as displayed in Table 17.1. Scores for each of the 10 items are combined into a universal score of between 0 and 100 with higher scores indicating better usability. Scores of > 68 generally indicate above-average usability (Lewis 2018).

The students were additionally asked to complete a 10-item content-specific survey about the RtB program. The items were measured using a 5-point Likert

Table 17.1 System usability scale

	System usability scale	Strongly disagree			Strongly agree	
	Questionnaire items	1	2	3	4	5
1	I think that I would like to use this system frequently	–	–	–	–	–
2	I found the system unnecessarily complex	–	–	–	–	–
3	I thought the system was easy to use	–	–	–	–	–
4	I think that I would need the support of a technical person to be able to use this system	–	–	–	–	–
5	I found the various functions in this system were well integrated	–	–	–	–	–
6	I thought there was too much inconsistency in this system	–	–	–	–	–
7	I would imagine that most people would learn to use this system very quickly	–	–	–	–	–
8	I found the system very cumbersome to use	–	–	–	–	–
9	I felt very confident using the system	–	–	–	–	–
10	I needed to learn a lot of things before I could get going with this system	–	–	–	–	–

scale with responses ranging from 1 strongly disagree to 5 strongly agree as per Table 17.2. All survey scores were analysed using Microsoft® Excel v16.24.

Of the 21 surveys analysed, findings revealed that the RtB IVR program observed a mean SUS score of 80, (SD 47), with raw total survey scores ranging between 55 and 100. Findings from the SUS survey indicated that students considered the RtB to have above-average usability. The results of the content-specific survey are presented in Table 17.3. The findings suggest that the RTB improved student understanding of female reproductive anatomy and fetal positing. Responses additionally indicated that the program was fun to use with no perceived negative impacts on learning noted.

Overall the initial testing of the RtB program in the IVR and SPT application modalities was positively received by both samples of undergraduate midwifery

Table 17.2 Road to Birth content specific survey

	System usability scale	Strongly disagree			Strongly agree	
	Questionnaire items	1	2	3	4	5
1	The use of the Road to Birth programme improved my understanding of female reproductive anatomy	–	–	–	–	–
2	The use of the Road to Birth programme was more engaging than relying on reading/lecture material alone	–	–	–	–	–
3	The use of the Road to Birth programme negatively impacted the course content	–	–	–	–	–
4	The use of the Road to Birth programme improved my ability to understand fetal positions in utero	–	–	–	–	–
5	I found learning with the Road to Birth programme fun	–	–	–	–	–
6	The use of Road to Birth programme kept me motivated	–	–	–	–	–
7	The Road to Birth programme as teaching methods should be used more by lecturers in the Department of Health and Social Care	–	–	–	–	–
8	I did not engage well with learning fetal positions using the Road to Birth programme	–	–	–	–	–
9	I felt that the Road to Birth programme engaged my own learning	–	–	–	–	–
10	I found it convenient to understand and experience content in the Road to Birth programme	–	–	–	–	–

Table 17.3 Using "The Road to Birth" with virtual reality in midwifery education in Belgium (n = 21)

Question/quote	Mean ± SD	% (strongly) agree (score 4–5)
The use of "the Road to Birth" programme improved my understanding of female reproductive anatomy	4,10 ± 0,70	81
The use of "the Road to Birth" programme was more engaging than relying on reading/lecture material alone	3,81 ± 0,87	61,9
The use of "the Road to Birth" programme improved my ability to understand fetal positions in utero	4,38 ± 0,67	90,5
I found learning with "the Road to Birth" programme fun	4,76 ± 0,44	100
The use of this technology kept me motivated	4,33 ± 0,73	85,7
The Road to Birth programme as a teaching method should be used more by lecturers in the Department of Health and Social Care	4,67 ± 0,48	100
I felt that "the Road to Birth" programme engaged my own learning	4,14 ± 0,79	76,2
I found it convenient to understand and experience content in "the Road to Birth" programme	4,00 ± 0,89	71,4
		% (strongly) disagree (score 1–2)
The use of "the Road to Birth" programme negatively impacted the course content	1,24 ± 0,54	95,3
I did not engage well with learning fetal positions using "the Road to Birth" programme	1,86 ± 1,27	71,4

students. Additionally, midwifery students seemed motivated to use the RtB program with empirical testing now needed to determine the effectiveness of the program compared to traditional educational techniques.

17.8 Discussion

The development of the RtB program is a modest step forward in the design and use of digital technology for teaching pregnancy-related anatomy and physiology to undergraduate midwifery students. Anatomy education is an ever-evolving specialty (Yammine and Violato 2018). The integration of interactive 3D visualisations into midwifery education has the potential to enhance traditional teaching and learning methods (Fealy et al. 2019b). In particular, the use of interactive technologies such as virtual reality and mobile phone/tablet mediums have the potential to overcome barriers associated with traditional teaching and simulation methods such as the need for cadaveric specimens and simulation laboratories with known ethical, financial, geographical and organisational limitations (Sugand et al. 2010). Whilst there is a paucity of digital anatomy programmes focusing on pregnancy and fetal development, Fealy et al. (2019b), discuss the importance of the development of such programmes in conjunction with content specialists such as midwives and university educators, in close collaboration with computer graphic design and coding specialists. It is not just the content but how students/users engage with the content that is of particular interest. As Jang et al. (2017) have suggested, it is one thing to attempt to increase understanding of anatomy content with this also dependant on the manner in which students interact with the content. Traditional educational modes such as the use of textbooks (reading is a passive process) allow for minimal retention of content (Weyhe et al. 2018). Practical experiences as part of an active learning process may increase content recall by up to 75% (Weyhe et al. 2018). It is this *active learning process*" that suggests digital technology is an important next step in educational content delivery (McMenamin et al. 2018; Moro and Gregory 2019).

The conceptualisation of the RtB was born out of a need to support students learning by employing innovative teaching and learning approaches that challenge the traditional didactic lecture and simulation techniques and overcoming institutional barriers to education (Fealy et al. 2019b). From our early data it is postulated that the RtB program may afford firstly, better engagement with the content being delivered; Secondly, an increased understanding of anatomy and spatial awareness; Thirdly, may assist students to avoid an over-reliance on classroom/lab instruction. Importantly students indicated that the RtB program assisted in improving learning through visualising pregnancy and/or fetal positions with no adverse outcomes to learning. Whilst the initial findings are positive many other aspects regarding the advantages and disadvantages of digital anatomy for pregnancy anatomy are yet to be empirically explored. Such studies require ongoing responsive technical development and institutional IT infrastructure to support large study populations (Fealy et al. 2019b). As the use of digital technology grows in popularity and acceptability for the delivery of anatomy content, and in the case of RtB pregnancy visualisation anatomy as a medium for the delivery of pregnancy content, the development of a validated tool that targets knowledge retention and spatial relationships between anatomical structures, would be required (Yammine and Violato 2018). Additionally, developments in the use of digital anatomy visualisation in the training of health professional students, health professionals, and wider application beyond the university requires research in clinical and community settings. Future developments, for instance, could seek to directly include consumer/patient input at all stages of development, implementation and evaluation (Roche et al. 2019). Such a collaborative approach could see programmes being employed in improving the relationship between the care provider and the care recipient.

An important area of consideration is the embedding of formative tests that could be undertaken in IVR to provide immediate feedback to the user on their knowledge and understanding of the content covered. An advantage of such embedded tests would be their capacity to complement online learning systems, providing repeatable, consistent anatomical information that could be used with minimal equipment anywhere in the world without direct human instruction. Such developments would be consistent with the principles and strengths of active learning.

As interesting and enticing as they are, such possibilities emerging from the development of the RtB point to just some of the ways in which digital technologies could be used to transform the way we learn about pregnancy-related anatomy.

17.9 Conclusion

This chapter has described the development and initial testing of a digital anatomy visualisation program. Our findings suggest that the RtB deployed via IVR and SPT technology mediums are well received by undergraduate midwifery students and may assist with learning anatomy content, with further empirical testing required. The RtB is an exciting but modest step in using digital technology for pregnancy-related anatomy and physiology education.

Acknowledgements The authors would like to thank The University of Newcastle Innovations team and Professor Sally Chan Head of School, Nursing and Midwifery, Faculty of Health and Medicine, The University of Newcastle for support during this program's development.

References

Al-Emran M, Elsherif H, Shaalan K (2016) Investigating attitudes towards the use of mobile learning in higher education. Comput Human Behav 56:93–102

Barmaki R, Yu K, Pearlman R, Shingles R, Bork F, Osgood G, Navab N (2019) Enhancement of anatomical education using augmented reality: an empirical study of body painting. Anatom Sci Edu 12:599–609

Cant R, Cooper S (2017) Use of simulation-based learning in undergraduate nurse education: an umbrella systematic review. YNEDT 49:63–71

Dey A, Billinghurst M, Lindeman R, Swan E (2018) A systematic review of 10 years of augmented reality usability studies: 2005 to 2014. Front Robot 37:1–28

Domingo J, Bradley E (2018) Education student perceptions of virtual reality as a learning tool. J Edu Technol 46:329–342

Downer T, Gray M, Anderson P (2020) Three-dimensional technology: evaluating the use of visualisation in midwifery education. Clinical Simul Nurs 39:27–32

Estai M, Bunt S (2016) Best teaching practices in anatomy education: a critical review. Ann Anat 208:151–157

Evans D, Pawlina W, Lachman N (2018) Human skills for human[istic] anatomy: an emphasis on nontraditional discipline-independent skills. Anatom Sci Edu 11:221–224

Fealy S, Chan S, Wynne O, Eileen D, Ebert L, Ho R, Zhang M, Jones D (2019a) The support for new mums project: a protocol for a pilot randomised controlled trial designed to test a postnatal psychoeducation smartphone application. J Avanc Nursi 1–13

Fealy S, Jones D, Hutton A, Graham K, McNeill L, Sweet L, Hazelton M (2019b) The integration of immersive virtual reality in tertiary nursing and midwifery education_ A scoping review. YNEDT 79:14–19

Hashim H (2018) Application of technology in the digital era education. Int J Res Counsel Edu 1:1–5

Heflin H, Shewmaker J, Nguyen J (2017) Impact of mobile technology on student attitudes, engagement, and learning. Comput Edu 107:91–99

Horst R, Dörner R (2018) Opportunities for virtual and mixed reality knowledge demonstration. IEEE

Jang S, Vitale J, Jyung R, Black J (2017) Direct manipulation is better than passive viewing for learning anatomy in a three-dimensional virtual reality environment. Comput Edu 106:150–165

Jones D, Siang See Z, Billinghurst M, Goodman L, Fealy S (2019) Extended reality for midwifery learning: MR VR demonstration. VRCAI 19. ACM, New York, USA

Kerby J, Shukur Z, Shalhoub J (2010) The relationships between learning outcomes and methods of teaching anatomy as perceived by medical students. Clinical Anat 24:489–497

Lewis J (2018) The System Usability Scale: Past, Present, and Future. Int J Human Comput Inter 34:577–590

McMenamin P, McLachlan J, Wilson A, McBride J, Pickering J, Evans D, Winkelmann A (2018) Do we really need cadavers anymore to learn anatomy in undergraduate medicine? Med Teacher 40:1020–1029

Morie J (2007) Performing in (virtual) spaces: Embodiment and being in virtual environments. Int J Perform Arts Digital Media 3:123–138

Moro C, Gregory S (2019) Utilising anatomical and physiological visualisations to enhance the face-to-face student learning experience in biomedical sciences and medicine. Adv Experim Med Biol 1156:41–48

Moro C, Stromberga Z, Raikos A, Stirling A (2017) The effectiveness of virtual and augmented reality in health sciences and medical anatomy. Anatom Sci Edu 10:549–559

Rankin J (2017) Physiology in childbearing with anatomical and related biosciences. Elsevier, Sydney Australia

Rizzolo L, William S (2006) Should we continue teaching anatomy by dissection when?. Anatom Record 298B:215–218

Roche K, Liu S, Siegel S (2019) The effects of virtual reality on mental wellness: a literature review. Mental Health Family Med 14:811–818

Rousian M, Koster M, Mulders A, Koning A, Steegers T, Steegers E (2018) Virtual reality imaging techniques in the study of embryonic and early placental health. Placenta 64:S29–S35

Sharma P, Hannafin M (2007) Scaffolding in technology-enhanced learning environments. Inter Learn Environ 15:27–46

Sherman W, Craig A (2018) Understanding virtual reality: interface, application, and design

Stepan K, Zeiger J, Hanchuk S, Del Signore A, Shrivastava R, Govindaraj S, Iloreta A (2017) Immersive virtual reality as a teaching tool for neuroanatomy. Int Forum Allergy Rhinol 7:1006–1013

Stone R (2011) The (human) science of medical virtual learning environments. Philos Trans R Soc B Biol Sci 366:276–285

Sugand K, Abrahams P, Khurana A (2010) The anatomy of anatomy: a review for its modernization. Anatom Sci Edu 3:83–93

Sutherland J, Schwaber K (2013). The scrum guide TM

Uppot R, Laguna B, McCarthy C, De Novi G, Phelps A, Siegel E, Courtier J (2019) Implementing virtual and augmented reality tools for radiology education and training, communication, and clinical care. Radiology 291:570–580

Weyhe D, Uslar V, Weyhe F, Kaluschke M, Zachmann G (2018) Immersive Anatomy Atlas-Empirical Study Investigating the Usability of a Virtual Reality Environment as a Learning Tool for Anatomy. Front Surg 5:73

Wiliam D (2011) What is assessment for learning? Stud Edu Eval 37:3–14

Yammine K, Violato C (2015) A meta-analysis of the educational effectiveness of three-dimensional visualization technologies in teaching anatomy. Anatom Sci Edu 8:525–538

Yammine K, Violato C (2018) A meta-analysis of the educational effectiveness of three-dimensional visualization technologies in teaching anatomy. Anatom Sci Edu 8:525–538

Zorzal E, Sousa M, Mendes D, Kuffner dos Anjos R, Medeiros D, Paulo S, Rodrigues P, Mendes J, Delmas V, Uhl J, Mogorron J, Jorge A, Lopes D (2019) Anatomy studio: A tool for virtual dissection through augmented 3D reconstruction. Comput Graph 85:74–84

Chapter 18
Toward Constructivist Approach Using Virtual Reality in Anatomy Education

Jinsil Hwaryoung Seo, Erica Malone, Brian Beams, and Michelle Pine

Abstract We present *Anatomy Builder VR* and *Muscle Action VR* that examine how a virtual reality (VR) system can support embodied learning in anatomy education. The backbone of these projects is to pursue an alternative constructivist pedagogical model for learning human and canine anatomy. In *Anatomy Builder VR*, a user can walk around and examine anatomical models from different perspectives. Direct manipulations in the program allow learners to interact with either individual bones or groups of bones, to determine their viewing orientation and to control the pace of the content manipulation. In *Muscle Action VR*, a user learns about human muscles and their functions through moving one's own body in an immersive VR environment, as well as interacting with dynamic anatomy content. Our studies showed that participants enjoyed interactive learning within the VR programs. We suggest applying constructivist methods in VR that support active and experiential learning in anatomy.

J. H. Seo (✉)
Department of Visualization, Texas a&M University, 3137 Langford Building C, College Station, TX 77843-3137, USA
e-mail: hwaryoung@tamu.edu

E. Malone
College of Veterinary Medicine, Department of Biomedical Sciences, University of Missouri, W105 Veterinary Medicine 1600 E. Rollins St., Columbia, MO 65211, USA
e-mail: sassa-fras27@tamu.edu

B. Beams
Santa Clara University, 500 El Camino Real, Santa Clara, CA 95053, USA
e-mail: bmsmith@scu.edu

M. Pine
Texas Veterinary Medical Center, Texas a&M University, TAMU 4458, College Station, TX 77843-4458, USA
e-mail: MPine@cvm.tamu.edu

18.1 Introduction

Anatomy education is fundamental in life science and health-related education. In anatomy education, it has been traditionally believed that cadaver dissection is the optimal teaching and learning method (Winkelmann et al. 2007). Cadaver dissection definitely provides tangible knowledge of the shape and size of the organs, bones, and muscles. However, dissection offers only a subtractive and deconstructive perspective (i.e., skin to bone) of the body structure. When students start with the complexity of complete anatomical specimens, it becomes visually confusing and students may have a hard time grasping the underlying basic aspects of anatomical form (Miller 2000). Consequently, many students have difficulties with mentally visualizing the three-dimensional (3D) body from the inside out (i.e., bone to the skin), as well as how individual body parts are positioned relative to the entire body.

Unfortunately, even with the availability of 3D interactive tools including virtual reality applications (Parikh et al. 2004; Temkin et al. 2006), the issue of visualizing movement still remains to be addressed. These interactive tools mainly focus on the identification of anatomical components and passive user navigation. Students must still mentally manipulate 3D objects using information learned from 2D representations (Pedersen 2012). For students' planning for future careers such as orthopedic surgery, physical therapy, choreography, or animation, they need to know the muscle's action, how it interacts with other muscles, and which normal movements it facilitates. This level of complexity is not easily conveyed via 2D static representations (Skinder-Meredith Smith and Mathias 2010). In addition, existing educational programs don't provide a flexible learning environment that allows a student to make a mistake and then learn from it. Alternative learning materials that focus on constructivist approaches have been introduced in anatomy education: 3D printing (Li et al. 2012; Rose et al. 2015) and other physical simulation techniques (Myers et al. 2001; Waters et al. 2011). The constructivism theory of learning states that students should construct their own understanding by actively participating in their environment. However, these alternative methods also have limitations: it is difficult to create an anatomical model that makes movements, interactions are limited, and a single model can only present limited information.

Recent technical innovations, including interactive and immersive technologies, have brought new opportunities into anatomy education. Virtual reality and augmented reality technologies provide a personalized learning environment, high-quality 3D visualizations, and interactions with contents. However, the majority of anatomy education applications still focus on the identification of anatomical parts, provide simple navigations of the structure, and do not fully support 3D spatial visualizations and dynamic content manipulations (Jang et al. 2016). Even with the availability of 3D interaction tools, mentally visualizing movement remains problematic. Therefore, students still struggle to understand biomechanics to accurately determine movement caused by specific muscle contractions (Cake 2006).

We created *Anatomy Builder VR* and *Muscle Action VR*, embodied learning virtual reality applications, to promote embodied, multi-modal mental simulation of anatomical structures and functions. *Anatomy Builder VR* is a comparative anatomy lab that students can enter to examine how a virtual reality system can support embodied learning in anatomy education. The backbone of the project is to pursue an alternative constructionist pedagogical model for learning canine and human anatomy. Direct manipulations in the program allow learners to interact with either individual bones or groups of bones, to determine their viewing orientation, and to control the pace of the content manipulation. *Muscle Action VR* pursues interactive and embodied learning for functional anatomy, inspired by art practices including clay sculpting and dancing. In *Muscle Action VR*, a user learns about human muscles and their functions through moving one's own body in an immersive VR environment, as well as interacting with dynamic anatomy content.

18.2 Background and Prior Research

18.2.1 Challenges in Traditional Anatomy Education

For over 400 years, cadaveric dissection has been the mainstay for teaching gross anatomy (Azer and Eizenberg 2007). However, in recent years there has been a trend in both the medical and veterinary schools toward reducing contact hours dedicated to traditional dissection-based anatomy courses (Drake et al. 2009; Heylings 2002). This reduction is the result of many factors, both pedagogical and practical. Traditional anatomy teaching methods have emphasized learning by deconstructing anatomical structures through full-body dissection. However, there is an argument that practitioners will be treating living patients, therefore the need is to learn anatomy in that context (McLachlan 2004). Additionally, there has been a shift toward an integrated and/or system-based approach to medical education. The argument is that there are other pedagogical practices that more readily promote student engagement and active learning (Rizzolo et al. 2006; Turney 2007).

Other reasons for the reduction in contact hours include the more practical aspects of teaching anatomy utilizing cadavers, such as the costs associated with maintaining a dissection laboratory as well as the ethical considerations for legally obtaining and storing cadavers (Aziz and Ashraf 2002). Another important factor is student and instructor safety. Typical embalming solutions contain formaldehyde at levels ranging from 0.5% up to 3.7%. Formaldehyde exposure can lead to many known health hazards such as respiratory and dermal irritation (Ajao et al. 2011; Tanaka et al. 2003). Recently, formaldehyde has been designated as a carcinogen (NTP 2011). All of these factors have resulted in some medical schools teaching anatomy with either limited or even no dissection of cadavers (McLachlan 2004; Sugand et al. 2010). To date, there is an ongoing debate as to the most effective way to teach anatomy, and no individual tool has the demonstrated ability to meet all curriculum requirements (Kerby et al. 2011).

18.2.2 Prior Research

When viewing the process of anatomy learning from the students' perspective, an additional set of challenges arise. While there is more than one way to approach dissection (regional versus system based), the process of dissection is one of deconstruction. Layers of tissues are sequentially peeled away from most superficial to deepest. Additionally, muscles are cut and connective tissue and even organs are removed. This is necessary in many cases to find and view underlying, hard to access structures. However, this means that students are unable to replace the structures in their natural spatial arrangement, and any errors made during the dissection cannot be repaired.

In order to design and build student-centered embodied learning applications, we began by asking students what they would like to have. Texas A&M University offers a cadaver based undergraduate anatomy course to students in the Biomedical Sciences major. Students enrolled in this anatomy course (N = 23) were invited to participate in an anonymous online survey that asked them open-ended questions pertaining to their views on learning anatomy. We didn't collect the participants' identification information. Students were asked to list their primary study methods used to learn anatomy. Two main methods emerged from their responses: cadaver review and print notes. All of the students stated that they spent time in the anatomy laboratory reviewing using the cadaver. The second most relied upon method was to read and memorize their notes.

Students were also asked to respond to questions relating to what types of study aids they would have liked to use and what they would create as an additional study aid. One primary theme emerged from these two questions: Interactivity. Students wanted an aid that they could interact with that had diagrams, pictures, and physical elements. Here are representative student quotes from the survey:

"An interactive, virtual lab that correlated with our lab directly".

I would love a picture that you can flip up the layers of the muscles, like in the labs. That way, it is easier to visualize while we aren't with the cadavers.

Some kind of physical model that can allow students to visualize the origins/insertions and actions of the muscles.

While this was not stated directly, the implication is that they wanted a study tool that could be accessed outside of the lab and could be directly manipulated. We have performed a thematic analysis of the student feedback and laid the foundation of our creation of *Anatomy Builder VR* and *Muscle Action VR*, on the pedagogical theory of constructivism.

18.2.3 Constructivist Approaches in Anatomy Education

According to the constructivist learning theories, learning is the personal construction resulting from an experiential process. It is a dynamic process, a personal experience that leads to the construction of knowledge that has personal meaning (Mota et al. 2010). The use of physically interactive tools to aid in learning is supported by constructivist learning principles. In addition, characteristics of learning anatomy such as its visual, dynamic, 3D, and tactile nature present a unique environment for the implementation of such tools (Mione et al. 2016; Jonassen and Rohrer-Murphy 1999; Winterbottom 2017). Typically, gross anatomy courses utilize cadaver dissection to facilitate learning specific structures as well as spatial relationships of one structure to another. The deconstructive nature of dissection, however, directs students to first examine the big picture and then discover underlying details. While this approach may be useful for many, some students are more successful when they are able to build-up knowledge from the smallest details to the larger picture. Further, if students are able to successfully work through the deconstructive process of dissection, then mirror the process in a constructive way, they are more likely to have a comprehensive understanding of the location of structures and the spatial relationships between them (Malone et al. 2018).

While diagrams, drawings, and cadavers are sufficient tools for learning and recognizing structures, all of these aids have one common disadvantage—they cannot demonstrate movement (Canty et al. 2015). In addition, these tools do not allow students an opportunity to come to their own conclusions and construct their own understanding of the material. Incorporating aspects of visualization sciences that allow students to be engaged in the construction of their own knowledge could provide valuable new tools for teaching and learning anatomy (Canty et al. 2015; Malone et al. 2016a). Our interdisciplinary research, called Creative Anatomy, initially pursued this approach in the undergraduate gross anatomy classes via utilizing tangible and embodied methods. Here are our prior works that guided us toward constructivist learning using virtual reality.

18.2.3.1 Building Musculoskeletal System in Clay

We utilized the *Anatomy in Clay Learning System®* in our classes to evaluate how sculptural methods could benefit students to learn three-dimensional anatomical structures (Fig. 18.1). Jon Zahourek is a traditionally trained fine artist who created this entirely hands-on approach to learning anatomy in the late 1970s. The *Anatomy in Clay Learning System®* is now used in more than 6,000 classrooms nationwide. The system allows students to build essentially any gross anatomical structure out of modeling clay and place it on a model skeleton. In regard to learning muscles, the system is especially efficient for illustrating origins, insertions, shapes, and relationships of muscles to one another. Students are able to isolate each muscle and build

Fig. 18.1 Building muscle system using "Anatomy in Clay" at the biomedical anatomy class: Students were asked to build muscles of the arm in clay for both the dog and the human after learning about the muscles in a lecture video, via class activities, and a canine cadaveric dissection. Students were guided through this activity with a step-by-step packet as well as aided by multiple instructors trained in human and canine anatomy

them up from the base level of the skeleton to the surface of the skin (Anatomy in Clay Learning System 2019).

To evaluate how supplementing learning by dissection with a constructive analogy affects students' knowledge acquisition and application, 165 undergraduate anatomy students were asked to build pelvic limb muscles in clay following dissection of the same muscles. Prior to the clay building activity, students had completed the following class assignments: (1) watched lecture videos presenting information regarding pelvic limb muscles, (2) participated in class activities in which they were asked to apply information from the videos, and (3) completed a dissection of the pelvic limb on canine cadavers. During one lab period, students participated in a guided activity involving building muscles on a skeletal model of the dog or human (Anatomy in Clay® CANIKEN® & MANIKEN®) in order from the deepest muscles to the most superficial (Malone et al. 2018). Students' feedback from this activity was extremely positive. Some of their written feedback includes quotes listed below.

> It was really helpful to be able to see the muscles being drawn from nothing, to see where they were in relation to one another and learning about their actions.

> I think this was a really good idea because it makes learning anatomy more fun and provides a change of pace for us. I definitely remember the information from that studio session better than most dissections.

I liked this activity most as far as helping learn because it helped put into perspective each muscle in the limb from insertion to origin to everything in between.

18.2.3.2   A Kinetic Model for Learning Gross Anatomy

To address how a constructivist approach could aid students with an understanding of biomechanical concepts, the movement was simulated with a kinetic model of a canine thoracic limb (Fig. 18.2). Students in an undergraduate anatomy course were asked to interact with the model, guided by an activity designed to help them construct their own understanding of biomechanical concepts. Anatomical structures such as bones, ligaments, tendons, and muscles were simulated in order to create movement. The simulated bone was made from plastic casting resin and built to withstand pressure from all angles while maintaining a small, delicate appearance. Simulated ligaments and tendons were created with elastic bands so that they were able to withstand pressure while simultaneously giving and stretching with movement. Simulated muscles were created from a string so that they allowed attachment of tendons, stretch and contract, and smooth, continuous motion (Malone and Pine 2014).

Fig. 18.2 Kinetic model: This kinetic model represents the thoracic limb of a canine and the strings can be pulled to create the actions that muscles would create in life. Students used this model to study basic concepts in biomechanics such as flexion and extension of joints

18.2.3.3 An Interactive Simulation Model to Improve the Students' Ability to Visualize Movement

As an expansion of the previous model, our team created a physically based kinetic simulation model (Fig. 18.3) of the canine pelvic limb that provided student interaction via a computer interface (Malone et al. 2016b). Bones of the pelvic limb were molded and cast using the same technique employed for the thoracic limb model. Components such as structural support, muscle function simulation, simulated muscle structure, and simulated skeletal structure all had to be considered during model construction. Four servo motors were mounted onto the model. Two of these motors were mounted directly to the femur of the model, while the remaining two were mounted on the top of the Plexiglas stand. Screw eyes were put on the model and the stand at major origin and insertion points. White nylon string was attached to the furthest point from the center of rotation on the servo motor arm. The other end of the string was fixed onto a screw eye representing a point of insertion or origin common to a muscle group. A computer was connected to an Arduino microcontroller via a USB port and a serial monitor, or text-input window was displayed on the computer screen. The user would type the name of a muscle into the serial monitor and hit the return button on the computer. Motors that represent that muscle's action would then be activated, turning the motor arm 90–180 degrees, thus creating the action of the muscle that was typed into the program (Malone et al. 2017).

Fig. 18.3 Interactive Simulation Model being used in anatomy lab: This model utilized servo motors mounted to bones cast in plastic to create the actions of muscles in the canine pelvic limb. Students typed the name of a muscle into a computer and the simulation model would create the action of the specified muscle

Students were divided into two groups—control and experimental. Both groups completed a short quiz after the unit on pelvic limb musculoskeletal anatomy. The students in the experimental group then were provided with an opportunity to interact with the model during the lab. The average quiz score for the experimental group improved significantly from a mean of 49.83–65.26%. The average quiz score for the control group did not improve significantly (49.50–58.60%). We received positive feedback from students. 75.55% of the students found the model easy to use, and 68.4% of students agreed or strongly agreed that the model helped with at least one concept related to movement.

Even though these physical and kinetic models provided benefits in the classrooms, we have encountered numerous challenges that cannot easily be resolved. The physical systems usually have fixed structures so that it is difficult to move around and replace parts as you wish. Physical materials have limitations to visualize movements and deformation of muscles. In addition, it is very difficult to provide personalized feedback to students using physical systems. Therefore, we started looking into incorporating embodied actions and interactive computer graphics using immersive technology.

18.2.4 Immersive Applications in Anatomy Education

Immersive technologies such as virtual reality and augmented reality are becoming popular tools in instructional technology, and educators are beginning to utilize them in their classrooms. Many meaningful efforts toward immersive learning technology have been made into anatomy and medical education. First, the creation of highly detailed and accurate three-dimensional models using computed tomography (CT) scans and magnetic resonance imaging (MRI) has significantly improved (Nicholson et al. 2006; Noller et al. 2005). Educators, taking notice of the increased accessibility and accuracy of these models, have begun to rely upon virtual 3D models to illustrate structures and relay concepts that are difficult or impossible to show on a cadaver.

Methods to increase the level of student interaction have been explored across multiple different platforms for more than two decades. In 1994, researchers in Greece published an article about a computer-based veterinary anatomy tutoring system. While the program was developed before the advent of accurate and easily accessible 3D models, the researchers concluded that the unique ability of an interactive computer program to individualize the learning experience, focus and combine certain aspects of veterinary anatomy and provide immediate feedback was undoubtedly beneficial (Theodoropoulos et al. 1994). Additionally, researchers at Linkoping University in Sweden developed a web-based educational virtual reality (VR) tool to improve anatomy learning. The program was well-received by medical students studying gross anatomy and generally preferred over study with a textbook and cadaver alone (Petersson et al. 2009).

Once the ability to incorporate virtual 3D models into interactive programs and virtual environments was fully developed, the development of interactive computer-based programs began to increase. A collaboration between the Oregon Health and Science University and McGill University used MRI data from a scan of a human cadaver to create a 3D model of the inner ear. Students who had access to a web-based 3D tutorial, scored an average of 83% on a post-instructional quiz while students without access to the 3D model only scored an average of 65% (Nicholson et al. 2006). Researchers at Oregon State University are exploring the use of a virtual 3D model of the canine skull and hyoid apparatus to allow interactive virtual articulation (Viehdorfer et al. 2014). At Texas A&M University Catherine Ruoff developed a computer program that demonstrated the anatomy of the equine paranasal sinuses. She concluded that anatomical structures that are difficult to visualize can be sufficiently illustrated and understood by allowing the user to interact with 3D models by rotating them in space and choosing which aspects of the model to focus on (Ruoff 2011).

A team of anatomists and computer scientists in Munich, Germany, have created what they refer to as an "augmented reality magic mirror" (Blum et al. 2012) which they have named "Mirracle." Users stand in front of a large display which is equipped with a depth-sensing and pose-tracking camera. Different views of a virtual 3D model are displayed on the screen overlaying the image of the user based on the user's position and gestures, essentially providing a mirror that allows the user to interactively explore their own anatomy (Blum et al. 2012). With the increasing use of virtual methods for visualizing and experiencing anatomy, many educators felt that the inherently physical nature of anatomy might soon be overlooked (Preece et al. 2013), however, it was not long before interplay of haptics and virtual tools were introduced. The Ohio University Virtual Haptic Back provides both visual and haptic feedback combining the use of a virtual 3D model with haptic feedback technology. Data collected during a study of this model showed that the accuracy of identification increased and required palpatory examination time decreased (Howell et al. 2005).

At the University of Magdeburg, Germany, a survey was conducted to evaluate methods involving visualization and interaction in anatomy education and how these methods are integrated into virtual anatomy systems. The researchers cite many learning theories that support the use and design of virtual anatomy systems. These theories include constructivism and embodied cognition, problem-based learning, and blended learning. Based on their analyses, these researchers concluded that virtual anatomy systems play an essential role in allowing students to explore shapes and spatial relationships (Preim and Saalfeld 2018). More recently veterinary students teamed up with visual arts students at Virginia Tech to create an immersive dog anatomy environment. The system allowed students to explore anatomical structures beyond the confinement of surgical views. Students were even able to zoom into certain organs to view layers of tissue (Virginia Tech 2019).

With the advancement of new technologies, virtual reality systems enable users to interact directly with anatomical structures in a 3D environment. This raises a new question: does manipulating the anatomical components in a virtual space support the users' embodied learning and ability to visualize the structure mentally? Our goal is to develop virtual reality learning environments that support a constructivist learning approach and a flexible learning environment. These environments allow a student to make/manipulate a musculoskeletal system, as well as learn from any mistakes made throughout that process. The recent development of body movement tracking ability in virtual reality has allowed us to implement this idea. We investigated how virtual reality technology with hand controllers and body tracking benefit students' learning while studying human/canine anatomy.

We present two case studies (*Anatomy Builder VR and Muscle Action VR*) in this chapter. *Anatomy Builder VR* allows the user to experience different components of human/canine anatomy by physical manipulations: recognizing bones, selecting bones, and putting bones together in the 3D orientation that they would be in a live animal. *Muscle Action VR* provides embodied interactions to learn about human muscles and their functions.

18.3 Case Study One: *Anatomy Builder VR*

18.3.1 Overview of Anatomy Builder VR

Anatomy Builder VR examines how a virtual reality system can support embodied learning in anatomy education through spatial navigation and dynamic content manipulations. In the VR environment, a user can walk around and examine anatomical models from different perspectives. Direct manipulations in the program allow learners to interact with either individual bones or groups of bones in order to determine their viewing orientation and control the pace of the content manipulation. *Anatomy Builder VR* consists of four main labs: Pre-Med, Pre-Vet, Sandbox, and Game room. A user can access each lab from the main lobby (Fig. 18.4) of the application.

Fig. 18.4 Main Lobby of Anatomy Builder VR

Pre-Med and Pre-Vet Labs provide guided lessons and quizzes about directional terms (Fig. 18.5), upper/lower limbs of a human skeleton, thoracic/pelvic limbs of a canine skeleton (Fig. 18.6). Students also learn skeletal systems through 3D skeletal puzzle questions.

Sandbox (Fig. 18.7) includes major activities in *Anatomy Builder VR*. This provides an unstructured learning environment where a student can freely assemble human and canine skeletons in the "anti-gravity" field. Placing a bone in the anti-gravity field suspends it in place. Picking up another bone and placing it near a connection on the already field-bound bone will make a yellow line appear. When the user lets go of the controller trigger, the two bones snap together. The user repeats this action until the skeleton is assembled to satisfaction. Reference materials to complete the articulation of the limb are displayed on a side. Individual and grouped bones can be scaled to provide extreme details.

Game room is a playful space where a student can learn the names and shapes of bones (Fig. 18.8). A user can select certain body regions of human and canine to test their knowledge about the bones that belong to the regions. Individual bones are encapsulated in basketballs. Once the game starts, a user can shoot a ball with a bone that is displayed on the board.

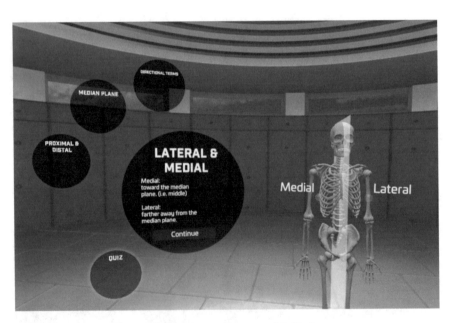

Fig. 18.5 Guided lesson about directional terms in the Pre-Med Lab

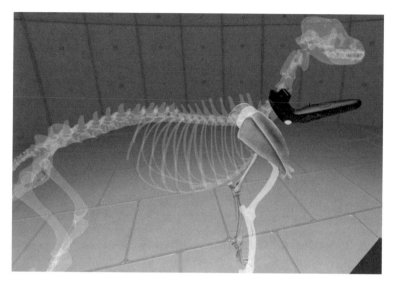

Fig. 18.6 3D puzzle quiz of a thoracic limb in the Pre-Vet Lab

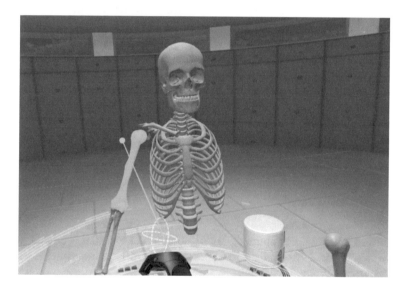

Fig. 18.7 Skeleton assembly in the anti-gravity field of the Sandbox

18.3.2 Development of Anatomy Builder VR

The Anatomy Builder VR program utilizes the HTC VIVE virtual reality platform. VIVE is a consumer-grade virtual reality hardware, primarily developed for use with video games. The platform comes with a high definition head-mounted display

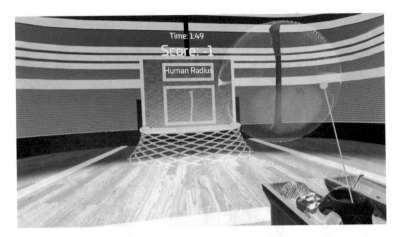

Fig. 18.8 A user is shooting a ball with a human radius in the Game Room

(HMD), two motion controllers, and two infrared tracking stations. The tracking stations, placed on opposite ends of a room, allow for room-scale virtual interactions. The project was been developed in Unity3D, a real-time development platform. All scripting is done in C#. Unity's development environment allows for easy integration of the VIVE with pre-built scripts and API. This allowed us to rapidly develop a functioning prototype and begin design on the user-specific interactions. Interaction with the virtual environment is primarily done with the VIVE controllers. The controllers have several buttons that are available to be programmed for various interactions.

18.3.3   Interaction Tasks in Anatomy Builder VR

There are multiple interaction tasks that would specifically support embodied learning of skeletal contents. All these tasks have been realized with VIVE Controllers.

Recognition of bones. For optimal identification of the bones, it is crucial that the user can view the bones from all angles. Therefore, natural head movement is required to be able to inspect individual objects.

Selection of bones. The prerequisite for 3D interaction is the selection of one of the virtual bones. Placing a bone in the anti-gravity field suspends it in place.

Transformation of bones. The transformation task includes rotating and translating the 3D bones. Since this is a task that the student is required to spend most of the time on, the success of learning the spatial relationships highly depends on the selection and interaction techniques.

Assembly of bones. Selecting and transforming a set of 3D bones is the process of assembling bones in the correct positions. When a user picks up a second bone and places one of its ends near a connection on the already field-bound bone, a yellow line appears. When the user lets go of the controller trigger, the two bone ends immediately attach together (snap) creating a joint. The user repeats this action until the skeleton is assembled to satisfaction. Assembly is entirely up to the user, allowing for incorrect bone combinations. This allows the user to make mistakes, learn, and try again.

18.3.4   User Study

In the pilot study, we investigated how a virtual reality system with direct manipulation may affect learning anatomy. The main focus of the study was to identify and assemble bones in the same orientation as they would be in a live dog, using real thoracic limb bones in a bone box and digital bones in the *Anatomy Builder VR*. For the purpose of the study, we recruited 24 undergraduate students. 66.7% of participants were females and the remaining 33.3% were males. The age range of these individuals spanned from 18 to 23, and each age was represented by roughly the same amount of people. However, a mere 8.3% of participants were 23 years old. Their majors were from departments across the university and they had never taken a college-level anatomy class before. The participants took a pre-study survey, experienced *Anatomy Builder VR*, and a post-study survey. We used a built-in processing system that recorded the duration of each participant's use and quiz scores. During the VR session, the participants were given a brief introduction to how a VR system worked and then fitted with the VIVE headset and controllers. Upon entering the *Anatomy Builder VR* program, the student was given time to become comfortable with the controls before beginning the tasks. All participants tried the Pre-Vet lab. Each student's study ended with a short interview about their learning experience.

On average, each participant spent 13.4 min in the VR system. The participants' experiences with the VR system were very positive. In the surveys, most of the participants (90%) rated as Strongly agree for the statement, "*I enjoyed using virtual reality to complete the activity.*" and 8.7% as Agree. Using the method with a constructivist focus, 63.6% of the participants responded as Agree on the statement, "*I was able to manipulate bones and put them together with ease in VR*", 27.3% responded as Strongly agree and 9.1% responded as Neutral. In the written responses, some participants expressed difficulties in certain interactions: rotating and scaling bones. However, most participants (88%) expressed positive aspects of learning the canine skeleton system using *Anatomy Builder VR*:

"This is so great. I think now anatomy students can learn things in a totally interactive world. Maybe they don't need to go to the lab" (ID 09)

"…being able to leave the bones in a specific orientation in VR was a good compromise for me mentally because I didn't have to continually revisit each bone or use energy holding them in the right place two at a time." (ID 10)

"It actually made it easier because I was able to better manipulate the bones because they were held up "in space". Also, it made more sense when the bones "connected" to each other." (ID 11).

18.4   Case Study TWO: *Muscle Action VR*

Muscle Action VR pursues interactive and embodied learning for functional anatomy, inspired by art practices including clay sculpting and dancing. In *Muscle Action VR*, a user learns about human muscles and their functions through moving one's own body in an immersive VR environment, as well as interacting with dynamic anatomy content. This allows learners to interact with either individual muscles or groups of muscles, to identify parts of the muscular system and control the pace of the content manipulation. *Muscle Action VR* utilizes the HTC VIVE virtual reality platform including VIVE trackers (Fig. 18.9).

18.4.1   Overview of Muscle Action VR

Muscle Action VR consists of four activity labs: Muscle Tracking Lab, Sandbox Lab, Guided Lesson Lab, and Game Lab. The Muscle Tracking Lab requires six body tracking points using VIVE headset, two VIVE controllers, and three VIVE trackers attached at the waist and ankles. A user can experience other labs without trackers.

Fig. 18.9   VIVE trackers setup

18.4.1.1 Muscle Tracking Lab

In this lab, a user with VIVE equipment including a VR headset, two controllers, three motion trackers, becomes a moving male écorché figure in VR (Fig. 18.10). The user is able to directly move their own body and see what muscles are contracting via a virtual mirror in the environment. Our system infers what muscles are being activated, and then highlights and displays these muscles so the user can get instant visual feedback. Users can learn about different muscles by using them directly and therefore gain an embodied understanding of muscle movements. The mirror system works by using a virtual camera that projects the figure onto a texture in front of the user. This provides more functionality than a regular mirror, giving the user the ability to switch to different angles. By switching angles, the user is able to see muscles from a different angle, which is crucial when viewing back or side muscles that are typically blocked from view.

The experience in this lab starts with a tutorial session. Going through the tutorial, the user learns how to rotate mirror views. The user can change the mirror views by selecting a different camera in the lab. Therefore, the user can view muscle details from other sides without turning their body. In addition, the user can change the muscle views by clicking the toggle button. The user can choose a mode to see either all muscles with highlighted muscles or specific muscles that are activated caused by a motion. After the tutorial, the user enacts motions that are demonstrated on the mirror and examines the visualization of the muscle activation via the mirror screen (Fig. 18.11). The last part of the session allows the user to freely move their body parts and learn how their movement affects muscle activation.

Fig. 18.10 A user can see themselves as a male muscle figure that moves based on their tracked movement

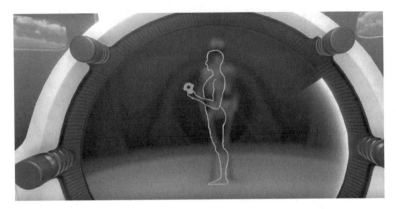

Fig. 18.11 The tutorial sessions show an animated visual reference to guide the user through the motion control system

18.4.1.2   Sandbox Lab

The Sandbox room allows the user to interactively generate muscles on different portions of the body to activate the muscles and to create specific actions. Each bone in the skeleton is a simulated physical object using a RigidBody component in Unity with joint connections to other bones. Users can see the mechanics behind how bones are connected and how muscles drive their motion and movement. The positioning of muscles relative to bones is a critical feature of what kind of motion will occur when the muscle is contracted or activated.

The Sandbox muscular system in the Sandbox Lab is comprised of three key components. The first component is the preparation step, where a developer marks all critical collision regions on the connecting bones objects. This is done to emphasize what sections of the skeleton the muscles should distort around. The deformation spheres used to mark these regions guide generated muscles around details such as joint connections or protruding bone features. The second key component is a convex hull algorithm. It takes a 3D line drawn by a user as input to generate a planar muscle that bends around the skeleton's deformation points. The start and endpoints of the user's line mark the insertion and origin points, respectively, of the generated muscle. Meanwhile, the line in between determines which side of the skeleton the algorithm will generate a muscle (Fig. 18.12). The third component is to activate each muscle to create movement. Each muscle can be individually contracted or relaxed, to produce flexion and extension of the skeletal joints. After a muscle is drawn, a slider is automatically created to allow the user to control the flexion and extension of the associated muscle. In addition, the entire skeleton's movement can be generated by activating multiple muscles simultaneously (Fig. 18.13).

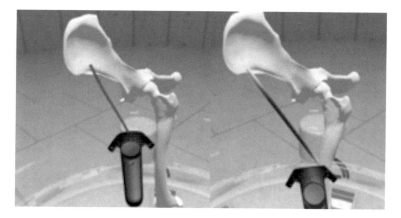

Fig. 18.12 Muscle drawing and generation

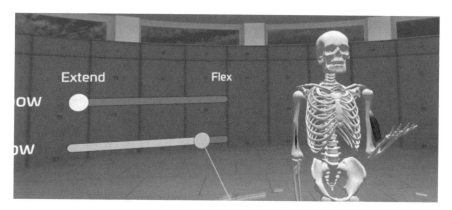

Fig. 18.13 Muscle activation interface

18.4.1.3   Guided Lesson Lab

Muscle Action VR also provides an interactive, but guided lesson lab about directional terms, and the basic biomechanics of muscles and their movements (Fig. 18.14).

18.4.1.4   Game Lab

The Game Lab allows the user to test their knowledge of human muscles and movements, covered in our application, through the form of a dodgeball game (Fig. 18.15). Users are challenged to move certain muscles by using a provided skeleton to deflect dodge balls. The user simply points at the provided muscles on the skeleton to contract and relax, so that the skeleton can move appropriately to play in the game. The user

Fig. 18.14 Guided lesson lab

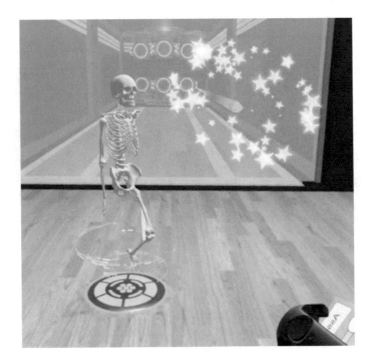

Fig. 18.15 Game lab

will continue to receive points for each dodge ball deflected, until three dodge balls have passed the player's perimeter, thus ending the game.

18.4.2 Development of Muscle Action VR

Similar to the *Anatomy Builder VR* project, *Muscle Action VR* was also developed in Unity3D through C# scripting and used the HTC VIVE as its VR platform. This development environment offered the same advantages from the previous project, such as being compatible with HTC VIVE through the OpenVR SDK and the SteamVR plugin, allowing easy integration and rapid prototyping. In addition to these advantages, the HTC VIVE was also chosen to utilize the VIVE tracker hardware, which is essential for the Body Tracking Lab to be possible.

The Muscle Tracking Lab utilizes six tracking points to drive the rig of the virtual body or "Muscle Man": a VIVE headset, two VIVE Controllers, and three VIVE trackers. Based on the positional and rotational values of these trackers, an IK solver from the FinalIK plugin was used to estimate and control different parts of the virtual body to mimic the user's movements in the real world. To create different muscles contracting and extending from certain actions, blend shapes were used to drive this effect, along with using the Shape Driver plugin to correctly create this effect based on the positional and rotational values of the virtual body's joints.

18.4.3 User Experiences in Muscle Action VR

We received very positive responses from preliminary studies done with several university students and anatomy experts. Most participants didn't have prior VR experience but they were able to navigate the application through and successfully learn key concepts and detailed visualizations. Students pointed out that this application would be greatly beneficial for learning anatomy and they would share their experience with peer students. In the open-ended interviews, participants described that the application was intuitive and engaging by providing innovative learning methods. Here is some feedback from participants:

> I wish we had this kind of learning aid when I was taking anatomy. This is fantastic.

> Even after I finished the VR experience, I still have an image in my mind so I can navigate the structure through.

> This looks so realistic. I feel like I am in the lab and dancing with the muscle man. I would like to show this to my roommate.

18.5 Conclusion

Our studies have focused on ways to utilize constructivist principles to design VR applications for learning anatomy. We created a learning platform that allows students the opportunity to not only visualize the individual components of the skeletal system but also interact with them in meaningful ways. Participants in our study enjoyed putting the skeleton together because the process was similar to that of completing a puzzle or building blocks. The virtual bones remained in place within the anti-gravity field, and joints were visually represented in a dynamic way. Another exciting aspect of the *Anatomy Builder VR* program is the inclusion of the "Sandbox Lab" for novices so that participants have an opportunity to place bones into an imaginative skeleton. Learners were encouraged to assemble the bones themselves, and they were free to make mistakes. This also provided a safe environment for active exploration.

For *Muscle Action VR*, we incorporated aspects from traditional art practices such as clay modeling and sculpting and dance to create an interactive and embodied learning environment. VR and traditional learning methods lack a way for learners to actually visualize and create movement. The musculoskeletal system is important because it is dynamic, yet cadaveric dissection and diagrams are static. Our VR application encourages learners to use their own body to visualize what is happening beneath the skin. We are extending the understanding of virtual reality design for anatomy education. In the future, *Anatomy Builder* and *Muscle Action VR* will include an even richer environment for self-evaluation and collaboration.

Acknowledgements We thank all the support from the Texas A&M TOP program, the Department of Visualization and the Department of Veterinary Integrative Biosciences. We appreciate the development team in the Department of Visualization at Texas A&M University.

References

Ajao KR, Oladosu OA, Popoola OT (2011) Using HOMER power optimization software for cost benefit analysis of hybrid-solar power generation relative to utility cost in Nigeria. Int J Res Rev Appl Sci 7(1):96–102

Anatomy in Clay Learning System (2019). https://www.anatomyinclay.com/. Retrieved 13 June 2019

Azer SA, Eizenberg N (2007) Do we need dissection in an integrated problem-based learning medical course? Perceptions of first-and second-year students. Surg Radiol Anat 29(2):173–180

Aziz M, Ashraf et al (2002) The human cadaver in the age of biomedical informatics. Anatom Rec Official Publ Am Assoc Anatom 269(1): 20–32

Blum T, Kleeberger V, Bichlmeier C, Navab N (2012) Mirracle: an augmented reality magic mirror system for anatomy education. In: 2012 IEEE Virtual Reality (VR). https://doi.org/10.1109/vr.2012.6180909

Cake MA (2006) Deep dissection: motivating students beyond rote learning in veterinary anatomy. J Veterin Med Educ 33(2):266–271

Canty DJ, Hayes JA, Story DA, Royse CF (2015) Ultrasound simulator-assisted teaching of cardiac anatomy to preclinical anatomy students: A pilot randomized trial of a three-hour learning exposure. Anat Sci Educ 2015(8):21–30

Drake RL, McBride JM, Lachman N, Pawlina W (2009) Medical education in the anatomical sciences: The winds of change continue to blow. Anatom Sci Educ 2(6):253–259

Heylings DJA (2002) Anatomy 1999–2000: the curriculum, who teaches it and how? Med Educ 36(8):702–710

Howell JN, Williams RL, Conatser RR, Burns JM, Eland DC (2005) The virtual haptic back (VHB): a virtual reality simulation of the human back for palpatory diagnostic training. SAE Techn Paper Series. https://doi.org/10.4271/2005-01-2679

Jang S et al (2016) Direct manipulation is better than passive viewing for learning anatomy in a three-dimensional virtual reality environment. Comput Educ

Jonassen D, Rohrer-Murphy L (1999) Activity theory as a framework for designing constructivist learning environments. ETR&D 47:61–79

Kerby J, Shukur ZN, Shalhoub J (2011) The relationships between learning outcomes and methods of teaching anatomy as perceived by medical students. Clinical Anat 24(4):489–497

Li J et al (2012) Maximizing modern distribution of complex anatomical spatial information: 3D reconstruction and rapid prototype production of anatomical corrosion casts of human specimens. Anatom Sci Educ 5:330–339

Malone E, Seo JH, Zahourek J, Pine M (2018) Effects of supplementing the deconstructive process of dissection with the constructive process of building muscle in clay. In: Presented as a Poster at the 130th Anniversary American Association of Anatomists (AAA) 2018 Annual Meeting at Experimental Biology, San Diego, California

Malone ER, Pine MD (2014) Poster presentation: creation of kinetic model for learning gross anatomy. In: National Conference on Undergraduate Research (NCUR). Lexington, Kentucky

Malone ER, Pine MD, Bingham G (2016a) Poster presentation: kinetic model for learning gross anatomy. american association of anatomists (AAA). Annual Meeting at Experimental Biology, San Diego, California

Malone ER, Seo JH, Pine MD, Smith BM (2016b) Poster presentation: an interactive simulation model to improve the students' ability to visualize movement. American Association of Anatomists (AAA) 2016 Regional Meeting, New York City, New York

Malone E et al (2017) Kinetic pelvic limb model to support students' understanding of spatial visualization in gross anatomy. In: Proceedings of the Eleventh International Conference on Tangible, Embedded, and Embodied Interactions, pp 361–365

McLachlan JC (2004) New path for teaching anatomy: living anatomy and medical imaging versus dissection. Anatom Rec Part B New Anatom Official Publ Am Assoc Anatom 281(1):4–5

Miller R (2000) Approaches to learning spatial relationships in gross anatomy perspective from wider principles of learning. Clinical Anat 13(6):439–443

Mione S, Valcke M, Cornelissen M (2016) Remote histology learning from static versus dynamic microscopic images. Anat Sci Educ 9:222–230

Mota M, Mata F, Aversi-Ferreira T (2010) Constructivist pedagogic method used in the teaching of human anatomy. Int J Morphol 28(2):369–374

Myers DL et al (2001) Pelvic anatomy for obstetrics and gynecology residents: an experimental study using clay models. Obstet Gynecol 97(2):321–324

Nicholson DT, Chalk C, Funnell WR, Daniel SJ (2006) Can virtual reality improve anatomy education? a randomised controlled study of a computergenerated three-dimensional anatomical ear model. Med Educ Med Educ 40(11):1081–1087. https://doi.org/10.1111/j.1365-2929.2006.02611.x

Noller C, Henninger W, Gronemeyer D, Budras K (2005) 3D reconstructions: new application fields in modern veterinary anatomy. Anat Histol Embryol J Veterin Med Ser C Anatom Histol Embryol 34(S1):30–38. https://doi.org/10.1111/j.1439-0264.2005.00669_86.x

NTP N (2011) Report on carcinogens. US department of health and human services, National Institutes of Health, National Institute of Environmental Health Sciences

Parikh M et al (2004) Three dimensional virtual reality model of the nor-mal female pelvic floor. Ann Biomed Eng 32:292–296

Pedersen K (2012) Supporting students with varied spatial reasoning abilities in the anatomy classroom. Teach Innovat Projects 2(2)

Petersson H, Sinkvist D, Wang C, Smedby Ö (2009) Web-based interactive 3D visualization as a tool for improved anatomy learning. Anatom Sci Educ Anat Sci Ed 2(2):61–68. https://doi.org/10.1002/ase.76

Preece D, Williams SB, Lam R, Weller R (2013) "Let's Get Physical": advantages of a physical model over 3D computer models and textbooks in learning imaging anatomy. Anatom Sci Educ Am Assoc Anatom 6(4):216–224. https://doi.org/10.1002/ase.1345

Preim Bernhard, Saalfeld Patrick (2018) A survey of virtual human anatomy education systems. Comput Graphics 71:132–153. https://doi.org/10.1016/j.cag.2018.01.005

Rizzolo LJ, Stewart WB, O'Brien M, Haims A, Rando W, Abrahams J, Aden M (2006) Design principles for developing an efficient clinical anatomy course. Med Teacher 28(2):142–151

Rose A et al (2015) Multi-material 3D models for temporal bone surgical simulation. Ann Otol Rhinol Laryngol 124(7):528–536

Ruoff CM (2011) Development of a computer program demonstrating the anatomy of the equine paranasal sinuses (Unpublished master's thesis). Texas A&M University

Skinder-Meredith Smith CF, Mathias HS (2010) What impact does anatomy education have on clinical practice? Clinical Anat 24(1):113–119

Sugand K, Abrahams P, Khurana A (2010) The anatomy of anatomy: a review for its modernization. Anatom Sci Educ 3(2):83–93

Tanaka K, Nishiyama K, Yaginuma H, Sasaki A, Maeda T, Kaneko SY, Tanaka M (2003) Formaldehyde exposure levels and exposure control measures during an anatomy dissecting course Kaibogaku zasshi. J Anat 78(2):43–51

Temkin B et al (2006) An interactive three-dimensional virtual body structures system for anatomical training over the internet. Clin Anat 19:267–274

Theodoropoulos G, Loumos V, Antonopoulos J (1994) A veterinary anatomy tutoring system. Comput Methods Progr Biomed 42(2):93–98. https://doi.org/10.1016/0169-2607(94)90045-0

Turney BW (2007) Anatomy in a modern medical curriculum. Annals Royal College Surg Engl 89(2):104–107

Viehdorfer M, Nemanic S, Mills S, Bailey M (2014) Virtual dog head. ACM SIGGRAPH 2014 Posters on SIGGRAPH'14. https://doi.org/10.1145/2614217.2614250

Virginia Tech students use VR technology to study anatomy (2019) Veted, dvm3660.com, pp 8–9

Waters JR et al (2011) Human clay models versus cat dissection: how the similarity between the classroom and the exam affects student performance. Adv Physiol Educ 35(2):227–236

Winkelmann A, Hendrix S, Kiessling C (2007) What do students actually do during a dissection course? First steps towards understanding a complex learning experience. Acad Med 82:989–995

Winterbottom M (2017) Active learning. Cambr Assessm Int Educ. www.cambridgeinternational.org/Images/271174-active-learning.pdf

Chapter 19
A Study of Mobile Augmented Reality for Motor Nerve Deficits in Anatomy Education

Margaret Cook, Jinsil Hwaryoung Seo, Michelle Pine, and Timothy Mclaughlin

Abstract Augmented reality applications for anatomy education have seen a large growth in their literature presence as an educational technology. However, the majority of these new anatomy applications limit their educational scope to the labelling of anatomical structures and layers, and simple identification interactions. There is a strong need for expansion of augmented reality applications, in order to give the user more dynamic control of the anatomy material within the application. To meet this need, the mobile augmented reality (AR) application, *InNervate AR*, was created. This application allows the user to scan a marker for two distinct learning modules; one for labelling and identification of anatomy structures, the other one for interacting with the radial nerve as it relates to the movement of the canine forelimb. A formal user study was run with this new application, which included the Crystal Slicing test for measuring visual-spatial ability, the TOLT test to measure critical thinking ability and both a pre- and post- anatomy knowledge assessment. Data analysis showed a positive qualitative user experience overall, and that the majority of the participants demonstrated an improvement in their anatomical knowledge after using *InNervate AR*. This implies that the application may prove to be educationally effective. In future, the scope of the application will be expanded, based on this study's analysis of user data and feedback, and educational modules for all of the motor nerves of the canine forelimb will be developed.

M. Cook
College of Veterinary Medicine & Biomedical Sciences, Texas A&M University, 4461 TAMU, College Station, TX 77843-4461, USA
e-mail: atmgirl@email.tamu.edu

J. H. Seo (✉) · T. Mclaughlin
Department of Visualization, Texas A&M University, 3137 Langford Building C, College Station, TX 77843-3137, USA
e-mail: hwaryoung@tamu.edu

T. Mclaughlin
e-mail: timm@viz.tamu.edu

M. Pine
Texas Veterinary Medical Center, Texas A&M University, TAMU 4458, College Station, TX 77843-4458, USA
e-mail: MPine@cvm.tamu.edu

© The Author(s), under exclusive license to Springer Nature Switzerland AG 2021
J.-F. Uhl et al. (eds.), *Digital Anatomy*, Human–Computer Interaction Series,
https://doi.org/10.1007/978-3-030-61905-3_19

367

19.1   Introduction

A 4th year veterinary student walks into a patient exam room on her very first day of veterinary clinic rotations. It's her last year of school before she becomes a licensed veterinarian, and now all of her knowledge will be put into practice. She sees a dog before her that is having difficulty bearing weight on its front leg. After examining the dog, she determines that no bones are broken, but that significant muscle loss has occurred in parts of the limb, meaning that the dog hasn't been able to use those muscles for a long time. Now she must reach far through the tunnel in her mind, back to her first year of veterinary education, and recall the intricate relationship between the muscles and nerves of the dog's limb.

A knowledge of anatomy is one of the foundational cornerstones of a veterinarian's or medical doctor's knowledge. This timeless subject is what guides their diagnostics and treatments for every patient that they take care of. It is no surprise then, that anatomy is a core component of the education that almost all healthcare professionals receive.

Traditionally, anatomy courses in veterinary medical education are primarily taught with the methods of didactic lectures and cadaver dissection. The anatomy classroom teaching materials are characterized by static and two-dimensional images. Laboratory involves dissection guides, cadavers and aids such as plastinated anatomical models (Peterson and Mlynarczyk 2016). However, decreased laboratory funding and laboratory time, and increased technology development, have led to limiting animal use to only teaching procedures which are considered essential (King 2004; Murgitroyd et al. 2015; Pujol et al. 2016). With the evolvement of learning theories in the classroom, as well as the growth of 3D interactive technology, there is a need for those who work in the anatomy higher education field to re-examine the learning tools that are used in anatomy courses (Azer and Azer 2016; Biassuto et al. 2006).

Augmented reality (AR) and virtual reality (VR) are the two recent 3D interactive technologies that are being researched for their merits as anatomy education tools. While both AR and VR technologies have their own advantages and disadvantages, it is unknown whether or not VR or AR is the better platform, particularly for 3D object manipulation. (Krichenbauer et al. 2018). For the purposes of this work, VR will refer to the traditional head-mounted-display (HMD) systems. VR has the advantage over AR of not having to wait for camera images, or perform rectification and correction of images (Krichenbauer et al. 2018). However, while VR has numerous benefits, it is unrealistic in large class sizes, due to its extensive set-up requirements. AR is defined as a technology that superimposes a computer-generated image on a user's view of the real world, thus providing a composite view. Especially, Mobile AR technology allows students to dynamically interact with digital content that is integrated with current print-based learning materials. The mobility of augmented reality on smartphones helps to eliminate constraints on time-of-use, size-of-location, or other demanding technical requirements (Fetaji et al. 2008). AR also does not elicit the same loss of depth perception that VR does, and it allows the user the advantage of being able to see their own body in their environment (Krichenbauer et al. 2018).

Overall, we chose to take the mobile AR approach to our project mainly due to its versatility and the realistic possibility of it being deployed in a large size class.

When exploring the literature, it is obvious that many research efforts have been made to capitalize on mobile AR's usefulness in anatomy education. However, the majority of these anatomy applications focus primarily on labelling of anatomical structures and layers, or simple identification interactions (Jamali et al. 2015; Kamphuis et al. 2014; Ma et al. 2016). While these are valuable interactions for learning, it is important that anatomy content in Mobile AR be expanded from simple identification questions, and labelled three-dimensional structures. Our team came together to address this need by building a mobile AR application (Cook et al. 2019) for smart mobile devices. Specifically how to visualize the deficits to canine muscle movement, in response to motor nerve damage. This mobile AR technology, *InNervate AR*, is innovative because rather than having another simple interaction and labelling interface, the user is able to take a more dynamic and interactive role in what information was being presented by their learning application. This paper presents a study of mobile AR for learning nerve deficits utilizing *InNervate AR*.

19.2   Background

19.2.1   Visual-Spatial Ability and Critical Thinking for Deeper Anatomy Knowledge

There are several cognitive processes that an anatomy student must use to successfully learn the anatomical relationships between the structures of the body. The first process involves engaging their visual-spatial ability. This skill is defined as the mental manipulation of objects in three-dimensional space. This knowledge is crucial for surgical skills, because anatomy education gives the baseline skillset for accurate diagnosis in organs and body systems (Azer and Azer 2016). Traditionally, this three-dimensional mental understanding has been taught with cadaver use. However, the amount of cadaver contact has been reduced in higher education, and so new three-dimensional models are being created to compensate. 3D modelling tools allow the user to add or remove structures and observe them from different angles in three-dimensional space, thus enhancing the teaching process of complicated anatomical areas (Pujol et al. 2016). Many studies have shown a positive relationship between the use of 3D technology and student performance, when related to visual-spatial ability. However, the literature has shown mixed results, indicating that more research needs to be done to explore this relationship (Hackett et al. 2016; Berney et al. 2015). With *InNervate AR*, we investigated how students' spatial visualization skills could impact learning anatomical content in this mobile AR environment.

A second mental process that a student must use to gain a deeper understanding of anatomy is practicing the use of their knowledge within critical thinking scenarios. One of the goals of *InNervate AR* is to take the anatomical material beyond pure identification, and into more complex and dynamic interaction, so that an element of critical thinking can be introduced. According to Abraham et al., "critical thinking is the process of actively and skillfully applying, relating, creating, or evaluating information that one has gathered". The ability to think critically is vital to science education, and is crucial for life-long learning (Abraham et al. 2004). Kumar and James support this argument by adding that critical thinking is a rational process, with personal reflection to reach a conclusion. This approach to learning has become a high focus in educational research (Kumar and James 2015). Critical reasoning is required to understand how motor innervation affects the relationship between nerve deficits and muscle movement. Therefore, we tried to create an application that supports students' critical reasoning skills, while learning about motor innervation of the canine limb. With this development in mind, we investigated how students' baseline critical thinking skills may impact their anatomical learning while using a mobile AR application.

19.2.2 Mobile Devices and Augmented Reality for Personalized Anatomy Education

Augmented reality (AR) usage and effectiveness are being increasingly studied in higher education. AR is a platform which combines the physical and virtual worlds, with user control over the interaction between the two. In order for this technology to be effectively implemented as an educational tool, specialists from both hard/software sectors and educational backgrounds must work together (Kesim and Ozarslan 2012). In a recent multi-university study, the mobile AR application, HuMAR was examined for its effectiveness as a tool for the higher education anatomy classroom. The intent of implementing HuMAR was to teach general human anatomy to students. Overall, they hoped to measure the user experience of the application in three different anatomy courses, across three different universities. They performed a pilot test, and after analyzing their pre- and post-surveys, determined that this mobile AR application could be effective in motivating and improving student learning (Jamali et al. 2015). Another research project was tested to see if mobile AR could be implemented in a Turkish medical school anatomy class as an educationally impactful tool. The researchers concluded that mobile AR decreases cognitive load, increases academic achievement, and can make the learning environment more flexible and satisfying (Küçük et al. 2016). The mobility of augmented reality on a smartphone helps to eliminate constraints on time-of-use, size-of-location, or other demanding technical requirements of educational technologies such as virtual reality (Fetaji et al. 2008). The ubiquitous nature of smartphones for students in higher education means that Mobile AR applications allow for a personalized education experience. The mobile AR application can be tailored to the student's personal needs, including specific requirements for their type of mobile device, and desired changes to the user interface of the application for ease of use while learning.

19.2.3 Existing User Interfaces in AR

In terms of the user interface of AR, most projects seem similar in nature. We conducted a literature review to explore what user interfaces were currently available. The *Miracle* system is an augmented reality mirror system, which is described as providing an identification of structures interaction and "a meaningful context compared with textbook description (Kamphuis et al. 2014)". The work done by Chien et al. includes a system that has "pop-up labelling" and an interactive 3D skull model that the users can rotate to view different angles of the model. They also found that the 3D display of AR helped students improve their spatial memory of the location of anatomical structures, as compared to a traditional 2D display (Chien et al. 2010). The *MagicMiror* project of Ma et al. is mapped to the user's own body, but it is still a simple point and click interface. The user is quizzed based on definitions and asked to identify structures (Ma et al. 2016). There is currently a lack of understanding as to how AR can support more complex learning in anatomy, and how to ensure that the AR system has strong usability in a classroom environment (Kamphuis et al. 2014; Cuendet et al. 2013). But in the review by Lee et al., this technology has demonstrated potential to serve in education, as it can make the environment more engaging, productive and enjoyable. Furthermore, it can provide a pathway for students to take control of their own learning and discovery process (Lee 2012).

Our team incorporated the positive elements of AR in a unique way. We designed the user interface of *InNervate AR* to be more dynamic and interactive in nature. The content displayed for the user is dependent on the input of the user, and rather than static selection or rotation of a still object, the element of actual anatomical movement is introduced.

19.3 Application: *InNervate AR*

19.3.1 Design

InNervate AR includes two learning sections, which are incorporated in user study handouts. Canine cadavers are accepted learning models in veterinary education, and participants in this study have used them in their coursework own coursework. Therefore, the anatomy of the canine was used in this anatomy mobile AR application. The application is divided into two learning sections.

Section 19.1 involves labelling and identification of the structures of the canine thoracic limb. This was purposefully developed to make sure that *InNervate AR* offered the same baseline tools for user experience and learning as the existing anatomy applications that are available. Figure 19.1 shows the view of the participant during both learning modules.

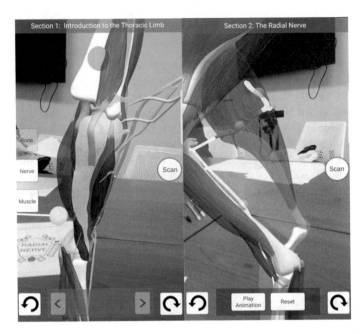

Fig. 19.1 The view of *InNervate AR* participant during the labelling module (left) and the radial nerve animation module (Right)

In Sect. 19.2, the user dynamically interacts with a musculoskeletal system of the canine thoracic limb, and plays animations of a healthy canine limb's range of movement. They can then visualize "damage" to different areas of the nerves of the limb, and be educated on what deficits exist. The "damage" is cuts to the nerve with the swipe of a finger on the device screen, and the resulting muscle action deficits are displayed with before and after animations of the muscles' ability (or inability) to move. Thus, the user can explore different combinations of effects upon the anatomy, and become more actively engaged in the educational process of the mobile AR application. Within an IRB Exemption for this user study, participants provided informed consent, took a pre-activities questionnaire, a literature cited Crystal Slicing Test, and Tobin & Capie Test of Logical Thinking (TOLT) (Ormand et al. 2017; Trifone 1987). Next, they interacted with the InNervate AR application on a mobile device, and finished the study with a post-activities' questionnaire. This initial push for expansion of anatomy content in mobile AR will hopefully encourage other researchers to add additional interactive content to their educational tools, and strengthen the presence of this technology in higher education anatomy curricula.

19.3.2   Key Elements of Design

To enhance the efficiency and learnability of *InNervate AR*, intentional steps were taken during the design process of both learning modules. The anatomy of any living being is so beautifully complex that artistic decisions have to be made when trying to recreate it in 3D. All of the assets were created to be as anatomically accurate as possible, with references from the TAMU VIBS 305 Anatomy course, which is based on Miller's Anatomy of the Dog, as well as feedback from anatomy experts in the project's team (Evans et al. 1996).

The cadavers that the students interact within a laboratory setting are very different colours from the 3D anatomical muscles that one usually sees in anatomy education tools. These artistic liberties are taken to make the structures life-like and aesthetically pleasing colour. Following this thread of thought, we chose a red-burgundy colour. The muscle striations were painted intentionally, to be anatomically accurate, and a normal map was applied to make the muscle feel more authentic and less like a perfectly modelled 3D object. The muscle striation references came from the TAMU VIBS 305 Anatomy dissection guide images.

In terms of efficiency, the screen size of the mobile device was of concern to us when we designed the user interface. With so many anatomical structures to choose between when using the select feature, a guide was needed to make sure that users wouldn't become frustrated by accidently selecting the wrong muscle, bone, or nerve. We solved this by adding blue selection spheres so that the user could touch the mobile device screen with the confidence that they would be selecting the structure that they intended. Furthermore, to prevent an overwhelming number of blue spheres from covering the thoracic limb in their environment, the muscles were divided into groups based on their location of attachment to the bones. Both of these user interface solutions are shown in Fig. 19.2.

To assist in the learnability of the radial nerve animation module, we put colour visual cues in place to improve user comprehension of the changes to the limb's movement based on their input. When the healthy range of motion animation plays, all of the muscles which are receiving innervation from a healthy radial nerve are highlighted with green. The user may then select a place along the length of the nerve to make a "cut". This selection is made using the same blue spheres that are implemented in the labelling learning module of the application. After the "cut" has created damage to the radial nerve, the muscles which are no longer functional are highlighted in red. These colour cues help to visualize why sections of the limb's movement have changed. This green and red visualizations are demonstrated in Fig. 19.3.

Fig. 19.2 Blue selection
spheres and muscle groups
solve user interface problems
with selecting anatomical
structures

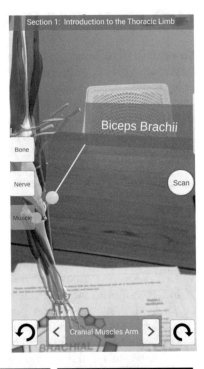

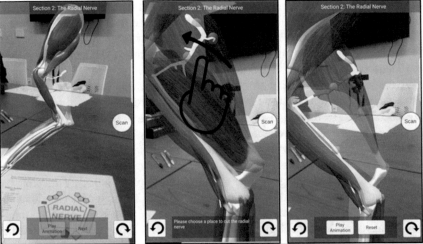

Fig. 19.3 The colour-based visual cues of the radial nerve animation module

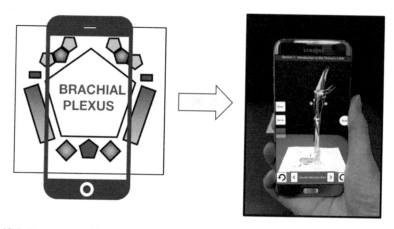

Fig. 19.4 The process of image recognition to load the *InNervate AR* application

19.3.3   Development

With the deployment of more robust devices, smartphones can be more easily used as platforms for augmented reality Therefore, *InNervate AR* was built on the platform of a smartphone. For the radial nerve animation module, a total of five animation sequences were created. The first animation sequence was the entire healthy range of motion of the canine thoracic limb. The other four scenarios involved changes in movement capabilities of the limb, based on the motor innervation provided by the radial nerve. These four radial nerve scenarios represented different possibilities of damage that could have occurred to the radial nerve. Due to the infinite number of possible damage scenarios to an organic animal's nerves, the number of nerve damage scenarios was narrowed down to a more finite set of four ranges. These four ranges would produce the most visually distinctive results between each of the scenarios. This was done so that the scenario possibilities would not overwhelm the user.

InNervate AR was designed as a marker-based system with Google ARCore software, utilizing image recognition developments from Viro Media. This means that the camera of the mobile device detects a shape on a piece of paper, known as the marker, and then the application loads the programmed learning module that corresponds to that marker (see Fig. 19.4).

19.3.4   Learning Objectives

Anatomy students struggle with combining several layers of their knowledge together to make logical conclusions about motor nerves and their relationship to the muscles which they innervate. An example of this difficulty is when the students are asked to answer an exam question about which muscle movement deficits would exist based on the information provided about an injury to a specific section of the thoracic limb. When answering that question, the student has to complete several mental steps. First, *they must correctly mentally visualize the muscles and nerves of the thoracic limb.* Next, *they must recall which motor nerves are located in the injured section of the thoracic limb.* Afterwards, *they must recall which muscles are specifically innervated by the motor nerves in that area of the thoracic limb.* By processing that information, they can recall what the actions of those muscles are, and then describe which muscle movements will be impaired. The final consideration that they must make is if the nerves which were damaged continued further down the limb, because if so, then further deficits might exist distally due to the linear relationship between nerve signals and the muscles that they communicate with.

InNervate AR was designed to give students a learning platform for seeing a 3D representation of these clinical reasoning scenarios. The AR technology allows the students to view all of the anatomical structures together, and then actually see how they work together when healthy, or become impaired when damaged.

19.4   User Study

19.4.1   Participants Recruitment

Students from a Texas A&M physiology course were recruited for a user study with *InNervate AR*. This course is the class that Texas A&M students are required to take in their degree plan after completion of their required anatomy course. The methods of recruitment included a class announcement, an email reminder of the user study announcement, and posted notices around the appropriate campus buildings.

19.4.2   Study Procedure

The appropriate Institutional Review Board (IRB) approval was acquired before the study. All participants were asked for informed consent before their participation in this user study. Each participant was allowed 90 min maximum to complete the activities of the user study. First, the participant was asked to complete a pre-activity questionnaire. The participant was then asked to complete the timed Crystal Slicing Test. They had 3 min to complete the test. The participant was next asked to complete the 38 min Tobin and Capie 1981 TOLT (Test of Logical Thinking) test.

After completion of the TOLT test, the participant was provided with a mobile device (SAMSUNG Galaxy) and a corresponding paper handout for how to interact with InNervate AR. This handout asked them to perform specific tasks, in a defined sequence, in order to ensure that the user had interacted with all parts of the application. The handout had a place for them to check-off when they had completed a task within the application. This handout also had image markers that Innervate AR could scan, to bring up the different learning modules that are built into the application.

The participant's duration of use of the application was recorded. The participant was free to ask the user study facilitator questions about navigation of the application. While the participant was using the application, another mobile application on the same device was recording the screen of the device. The participant's interaction with the mobile AR application was also recorded on video with a camera. After completing their interaction with *InNervate AR*, the participant was asked to complete a post-activity questionnaire.

19.4.3   Data Collection to Measure Application Effectiveness

For this user study's design, two peer-reviewed testing instruments were selected to test the participant's visual spatial ability and logical thinking ability. A pre- and post- questionnaire were also created.

19.4.3.1   Crystal Slicing Test

In order to assess mental visuo-spatial ability in this study, we elected to use the Crystal Slicing test. The participant was asked to choose which shape is produced as the result of an intersection between a plane and a crystal solid. This test was originally developed to provide spatial thinking practice to undergraduate geosciences students (Ormand et al. 2014). It has since been used to study the development of student's spatial thinking skills over time (Ormand et al. 2017). In addition, the Crystal Slicing Test has been positively reviewed as an instrument for quantifying the spatial ability of the test taker, as it relates to visualizing 3D objects (Gagnier et al. 2016).

19.4.3.2   Tobin and Capie 1981 TOLT (Test of Logical Thinking) Test

In order to assess formal reasoning and critical thinking abilities, we selected the Test of Logical Thinking (TOLT) test for this study. This test measures different categories of reasoning, including probabilistic, correlational and combinatorial reasoning. The statistical power of the results of the test is strengthened by the fact that the test taker must justify why they chose their answer (Trifone 1987).

19.4.3.3 Pre- and Post-Activity Questionnaires to Test Learning

Within the quasi-experimental design of this study, the non-equivalent groups design was followed. This means that no randomized control group exists, but a pre- and post-test is given to groups of people that are as similar as possible, in order to determine if the study intervention is effective or not. The pre-test was written to include anatomy knowledge questions, a free response question, demographics questions and Likert-Scale-based questions about their anatomy education experience. The post-test was written with five knowledge-based questions, three of which mirrored the anatomy knowledge questions of the pre-test, with the same concept being asked in a different way. The post-test also included Likert-Scale-based questions about their experience with the *InNervate AR* system, as well as a place to write additional feedback. The objective of these questionnaires was to obtain quantitative data based on the anatomy knowledge questions, and qualitative data based on the Likert and free response questions.

19.4.4 Data Analysis

The data from all of the user study participants was compiled and analyzed for patterns and trends. This involved grading the users' performance on the learning instruments used in this study, and identifying similarities and differences in the qualitative answers given during the pre- and post-questionnaires. We specifically wanted to identify how the critical thinking scores, and the visual spatial ability scores of the users affected their change in performance on the anatomical knowledge questions after using *InNervate AR*. Furthermore, the screen recordings and video recordings of the study were reviewed to analyze the overall user experience of the study participants.

19.5   Results & Discussion

19.5.1 Participant Demographics

There was a total of 22 participants in the user study for the Innervate AR application. All of the participants were Biomedical Sciences majors at Texas A&M University, and had taken the TAMU VIBS 305 Biomedical Anatomy course within the two previous academic years. Five of the participants were male, and 17 were female. When asked, 18% of these participants answered "Strongly Agree" and 59% of them answered "Agree" to the statement "I consider myself to have a high level of critical thinking ability". 11 of the participants obtained an "A" in the TAMU VIBS 305

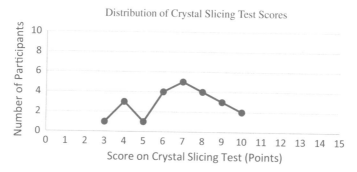

Fig. 19.5 Graph of distribution of Crystal Slicing Test Scores for the 22 participants

Anatomy course, 9 of the participants obtained a "B" and 2 participants obtained a "C" in the course.

19.5.2 Participant Crystal Slicing Test Results

The highest possible score that a participant could make on this 3 min test was 15 points. Only 9.09% of participants scored a 10 or better on this test. The majority of the user study pool (54.55%) made a score in the point range of 7–9. The next most common point range (22.73%) was a score of 5 or 6. The remainder of the participants (13.64%) scored less than 5 points. Figure 19.5 shows a distribution of these scores in graphical form. This data demonstrates that the participants in this user study had average or low visual-spatial ability in general.

19.5.3 Participant Test of Logical Thinking Results

With an allotted time of 38 min, the highest score that a participant could make on the TOLT was 10 points. A perfect score of 10 was made by 40.91% of the participants. A score of 9 was achieved by 31.82% of the participants. A score of 8 was made by 9.09% of the participants. Only one participant (9.09% scored a 7 on the test. The remaining participants (13.64%) scored a 6 or lower on the TOLT. Figure 19.6 shows a distribution of these scores in graphical form. This data showed that the participants trended toward having high critical thinking skills.

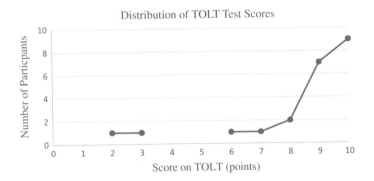

Fig. 19.6 Graph of distribution of TOLT Test Scores for the 22 participants

19.5.4 Participant Anatomical Knowledge Scores Results

In the pre-questionnaire, the participants had 3 anatomical knowledge test questions. In the post-questionnaire, the participant had 5 anatomical knowledge test questions, 3 of which were matched to the pre-questionnaire test questions. In other words, the same content was tested in those 3 questions, but asked in a different way. The scores of the participants were analyzed, and 77.27% of the participants' scores improved on the 3 matched questions, after using the *InNervate AR* application. 18.18% of the participants made the exact same score on the matched anatomy questions, and 4.55% of the participants had a lower score in the post-questionnaire on the 3 matched questions. This data is visualized in Fig. 19.7. This data shows that the majority of the user study participants showed an improvement in their performance on the matched anatomy knowledge questions in the post-questionnaire.

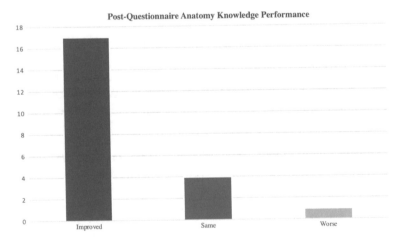

Fig. 19.7 Graphical distribution of matched anatomy knowledge question performance

19.5.5 User Experience with InNervate AR

In the post-study questionnaire, a series of Likert-Scale questions were asked about participants' perception of *InNervate AR*. The users' responses to these questions were all positive. Some of the categories they were asked to rate were as follows: the usefulness of the application as a visual aid for the spatial relationships between the anatomical structures, the flow and user interface of the application, and the usefulness of the application to practice critical reasoning scenarios.

The participants were also given free response questions. The first question asked "What did you like least about the *InNervate AR* application?" Common themes to how the participants answered this question included no "zoom" feature, problems with how to cut the nerves and problems with selecting the labelling spheres. The second question asked was "What did you like most about the *InNervate AR* application?" The most frequent responses to this question included getting to visualize the actions of the muscles with the animations, the graphic aesthetic of the application, how easy the application was to use and the accuracy of the anatomical content. The last free response question asked the participants if they had any further suggestions about the *InNervate AR* application. The responses included adding the ability to compare the healthy and damaged animation scenarios side-by-side, adding even more details about the muscles in the labelling module, and further customizing the visual user interface of the tool.

Finally, some of the verbal comments of the participants during their use of the *InNervate AR* application were as follows:

> "Oh man I wish I had had this when I was in anatomy lab...because it really connects it all together, especially with all of the bones articulating and everything being there. I remember having to draw so many layers". (User ID: 1001).

> "This is super helpful, I just can't get over it, it's one thing to see the words on paper, but to see a cut branch!" (User ID: 1001).

> "Nice way to look at the anatomy from different angles... Most apps don't have what would happen if something is wrong, they just have the structures" (User ID:1009).

19.5.6 User Interface Analysis

The participants in this study were video recorded while they used the *InNervate AR* application mobile device. In addition, the screen of the mobile device was recorded during their *InNervate AR* use. This data allowed for the user interface (UI) of the application to be analyzed for in the following usability categories: learnability and memorability, effectivity vs errors, and efficiency. These categories of analysis are important in a mobile learning application because they contribute to the user's ability to learn from the application, and influence the user's overall satisfaction (Fetaji et al. 2008).

19.5.6.1   Learnability and Memorability

These combined categories refer to how easily the user could master the use of the user interface (UI) without outside help, and how easily they could navigate the application if they made a mistake while using it (Fetaji et al. 2008).

One UI issue was immediately apparent, and will be addressed in the future for further study. Once the participant has "cut" the nerve in the radial nerve animation scenario, they can only view the damaged animations. In order to return to the healthy animation to compare and contrast, the user has to re-scan the marker and start the module over again. Furthermore, the rotation buttons for the UI might not have been intuitive enough. When a problem was experienced, the rotation function would either be asked about directly, or the study facilitator would inform the user that they could use the rotation buttons because they were leaning their entire body around the model with the phone in their hand. The UI button choice for rotation will be investigated for more intuitive options in the future.

19.5.6.2   Effectivity versus Errors

With the help of the screen and video footage, analysis was done to see how success-fully the tasks that the users were supposed to carry out were performed, and how many errors may have occurred. It is important to detect how many errors are made while the UI is being used, and analyze the importance of those errors (Fetaji et al. 2008). All of the users were able to successfully use both of the mobile AR anatomy modules. The most significant user error was that 36% of the users did not properly use the UI to switch between the layers of muscle groups available for learning in the labelling module of the application. This means that they missed information because they did not explore all of the possible muscle label settings.

19.5.6.3   Efficiency

The efficiency of the user interface refers to how quickly tasks can be completed using the interface of the application (Fetaji et al. 2008). There was no set time limit for the user study participants while they used the *InNervate AR* application. Overall, none of the video or screen recordings suggest that the users felt that completing tasks was inefficient or too slow. There seemed to be a correlation between increased *InNervate AR* usage time, and increased satisfaction or enjoyment while using the application. Shorter usage times might mean that the participant was in a hurry to complete the study, but none of the recorded responses suggest that any of the participants were dissatisfied with the efficiency of the *InNervate AR* application.

19.6 Conclusions and Summary

The goal of this project was to create a mobile anatomy AR application, *InNervate AR*, which provides more dynamic interactions than other mobile AR applications that have been previously created. This mobile AR technology is innovative because rather than having another simple viewing interaction and labelling interface, the user was able to take a more interactive role in what information was being presented by the application.

The results of this user study showed an extremely positive response from the participants, both in their qualitative feedback data, as well as their anatomical knowledge improvement. The majority of the participants tested for a high critical thinking ability, and there was one student with an average or low visual-spatial ability. Therefore, it was difficult to investigate how the base critical thinking ability could impact on learning anatomy using mobile AR. In terms of spatial visualization, there was no significant difference between high spatial visualization students and low spatial visualization students. The qualitative feedback from the participant's demonstrated areas where the *InNervate AR* application could use improvement, such as problems with how to cut the nerves, and problems with selecting the labelling spheres. However, the responses from participants were overwhelmingly positive in many categories. They enjoyed getting to visualize the actions of the muscles with the animations, the graphic aesthetic of the application, how easy the application was to use, and the accuracy of the anatomical content.

It is planned to use the data and feedback from this study as a guideline while further expanding *InNervate AR* to include all of the motor nerves of the limb as learning modules. Any future user studies will be completed in a classroom setting, so that a larger participant population can be guaranteed, and statistically significant results can be achieved. Furthermore, the limitations such as a low number of matched anatomy knowledge questions and gender bias will be addressed. Future user study and application design will also be more error tolerant, so that user errors with the technology, or differences in user background will not have huge consequences when analyzing results (Rouse 1990).

This study was a wonderful learning opportunity because it showed the great potential that *InNervate AR* has for anatomy higher education, and brought to light what weaknesses in the technology and research study design should be worked on in the future. It is our hope that this initial push for expansion of anatomy content in mobile AR will encourage other researchers to add additional interactive content to their educational tools, and strengthen the presence of this technology in higher education anatomy curricula.

Acknowledgements We would like to thank all of the contributors who have worked on the project. We also appreciate the support from the Department of Visualization at Texas A&M University.

References

Abraham RR, Upadhya S, Torke S, Ramnarayan K (2004) Clinically oriented physiology teaching: strategy for developing critical-thinking skills in undergraduate medical students. Adv Physiol Educ 28(3):102–104

Azer SA, Azer S (2016) 3D anatomy models and impact on learning: a review of the quality of the literature. Health Profess Educ 2(2):80–98

Berney S, Bétrancourt M, Molinari G, Hoyek N (2015) How spatial abilities and dynamic visualizations interplay when learning functional anatomy with 3D anatomical models. Anatom Sci Educ 8(5):452–462. https://doi.org/10.1002/ase.1524

Biassuto SN, Caussa LI, Criado del Río LE (2006) Teaching anatomy: Cadavers versus computers? Annals Anat 188(2):187–190

Chien C-H, Chen C-H, Jeng T-S (2010) An interactive augmented reality system for learning anatomy structure. In: The Proceedings of the International Multi-Conference of Engineers and Computer Scientists, pp 1–6

Cook M, Payne A, Seo JH, Pine M, McLaughlin T (2019) InNervate AR: dynamic interaction system for motor nerve anatomy education in augmented reality. In: Stephanidis C (eds) HCI International 2019–Posters. HCII 2019. Communications in Computer and Information Science, vol 1033. Springer, Cham. https://doi.org/10.1007/978-3-030-23528-4_49

Cuendet S, Bonnard Q, Do-Lenh S, Dillenbourg P (2013) Designing augmented reality for the classroom. Comput Educ 68:56–557

Evans HE, DeLahunta A, Miller ME (1996) Millers guide to the dissection of the dog. Saunders, Philadelphia

Fetaji M, Dika Z, Fetaji B (2008) Usability Testing and Evaluation of a Mobile Software Solution: A Case Study. In: The Proceedings of the ITI 2008 30th International Conferences on Information Technology Interfaces, June 23–26, Cavtat, Croatia, pp 501–506

Gagnier KM, Shipley TF, Tikoff B, Garnier BC, Ormand C, Resnick I (2016) Chapter 2: Training Spatial Skills in Geosciences: A Review of Tests and Tools. In 3-D structural interpretation: Earth, mind, and machine: Tulsa, OK: AAPG Memoir, vol 111, pp 7–23)

Hackett M, Proctor M, Proctor M (2016) Three-dimensional display technologies for anatomical education: a literature review. J Sci Educ Technol 25(4):641–654. https://doi.org/10.1007/s10956-016-9619-3

Jamali SS, Shiratuddin MF, Wong KW, Oskam CL (2015) Utilising mobile- augmented reality for learning human anatomy. Proced Soc Behav Sci 197:659–668

Kamphuis C, Barsom E, Schijven M, Christoph N (2014) Augmented reality in medical education? Perspect Med Educ 3(4):300–311

Kesim M, Ozarslan Y (2012) Augmented reality in education: current technologies and the potential for education. Proced Soc Behav Sci 47(222):297–302. https://doi.org/10.1016/j.sbspro.2012.06.654

King, L. A. (2004). Ethics and welfare of animals used in education: An overview. *Animal Welfare*, 13(SUPPL.), 221–227

Küçük S, Kapakin S, Göktaş Y (2016) Learning anatomy via mobile augmented reality: effects on achievement and cognitive load. Anatom Sci Educ 9(5):411–421

Kumar R, James R (2015) Evaluation of critical thinking in higher education in oman. Int J Higher Educ 4(3):33–43

Krichenbauer M, Yamamoto G, Taketom T, Sandor C, Kato H (2018) Augmented reality versus virtual reality for 3d object manipulation. IEEE Trans Vis and Comput Graph 24(2):1038–1048

Lee K (2012) Augmented reality in education and training. Tech Trends 56(2):2–13

Ma M, Fallavollita P, Seelbach I, Von Der Heide AM, Euler E, Waschke J, Navab N (2016) Personalized augmented reality for anatomy education. Clin Anat 29(4):446–453

Murgitroyd E, Madurska M, Gonzalez J, Watson A (2015) 3D digital anatomy modelling - Practical or pretty? Surgeon 13(3):177–180

Ormand CJ, Manduca C, Shipley TF, Tikoff B, Cara L, Atit K, Boone AP (2014) Evaluating geoscience students' spatial thinking skills in a multi-institutional classroom study. J Geosci Educ 62(1):146–154

Ormand CJ, Shipley TF, Tikoff B, Dutrow B, Goodwin LB, Hickson T, Resnick I (2017) The spatial thinking workbook: a research-validated spatial skills curriculum for geology majors. J Geosci Educ 65(4):423–434

Peterson DC, Mlynarczyk GSA (2016) Analysis of traditional versus three-dimensional augmented curriculum on anatomical learning outcome measures. Anatom Sci Educ 9(6):529–536

Pujol S, Baldwin M, Nassiri J, Kikinis R, Shaffer K (2016) Using 3D modeling techniques to enhance teaching of difficult anatomical concepts. Acad Radiol 23(4):193–201. https://doi.org/10.1016/j.molmed.2014.11.008.Mitochondria

Rouse WB (1990) Designing for human error: concepts for error tolerant systems. In: Booher HR (ed) Manprint. Springer, Dordrecht

Trifone JD (1987) The test of logical thinking: applications for teaching and placing science students. Am Biol Teacher 49(8):411–416

Printed in the United States
by Baker & Taylor Publisher Services